STUDY GUIDE AND SOLUTIONS MANUAL

RUSSELL

GENETICS

FOURTH EDITION

CHET S. FORNARI, Ph.D.
DEPT. BIOLOGICAL SCIENCES
DEPAUW UNIVERSITY
GREENCASTLE, IN 46135

Prepared with assistance of

Bruce A. Chase
UNIVERSITY OF NEBRASKA AT OMAHA

HarperCollinsCollegePublishers

HarperCollins® and ® are registered trademarks of HarperCollins Publishers Inc.

Study Guide and Solutions Manual to accompany Russell, *Genetics,* Fourth Edition.

Copyright © 1996 by HarperCollins College Publishers

All rights reserved. Printed in the United States of America. No part of this book may be reproduced in any manner whatsoever without written permission with the following exception: classroom material may be copied for classroom use. For information, address HarperCollins College Publishers, 10 East 53rd Street, New York, NY 10022.

ISBN 0-673-52361-6

96 97 98 99 9 8 7 6 5 4 3 2

Table of Contents

Preface		iv
Chapter 1:	Genetics: An Introduction	1
Chapter 2:	Mendelian Genetics	5
Chapter 3:	Chromosomal Basis of Inheritance, Sex Determination, and Sex Linkage	31
Chapter 4:	Extensions of Mendelian Genetic Analysis	59
Chapter 5:	Linkage, Crossing-Over, and Gene Mapping in Eukaryotes	75
Chapter 6:	Advanced Genetic Mapping in Eukaryotes	111
Chapter 7:	Chromosomal Mutations	129
Chapter 8:	Genetic Mapping in Bacteria and Bacteriophages	147
Chapter 9:	The Beginnings of Molecular Genetics: Gene Function	169
Chapter 10:	The Structure of Genetic Material	187
Chapter 11:	The Organization of DNA in Chromosomes	199
Chapter 12:	DNA Replication and Recombination	213
Chapter 13:	Transcription, RNA Molecules, and RNA Processing	231
Chapter 14:	The Genetic Code and the Translation of the Genetic Message	245
Chapter 15:	Recombinant DNA Technology and the Manipulation of DNA	263
Chapter 16:	Regulation of Gene Expression in Bacteria and Bacteriophages	285
Chapter 17:	Regulation of Gene Expression and Development in Eukaryotes	301
Chapter 18:	Gene Mutation	317
Chapter 19:	Transposable Elements, Tumor Viruses, and Oncogenes	337
Chapter 20:	Extranuclear Genetics	345
Chapter 21:	Population Genetics	359
Chapter 22:	Quantitative Genetics	383

PREFACE

Genetics is a fascinating subject. For many students, an understanding of genetics leads to entirely new ways of thinking about life. Still, like many things that are really worthwhile, understanding genetics well is no mean feat. While it can be exhilarating finally to understand a genetic issue and discover a solution to a tough problem, the path leading to understanding can be tortuous. Since there is often more than one path that can be productive, this study guide is not written to show the "correct path." Rather, it aims to help the reader see a clearer approach, and learn how to map out a smoother, more productive route.

Each chapter of the guide contains a set of features designed with this aim in mind. First, an outline of the text material is presented to provide an organizational overview. Second, key terms are grouped contextually and the reader is challenged to distinguish between related terms. Third, suggestions for analytical approaches and strategies for problem solving are presented. In some chapters, these are substantial, and can provide a foundation for analytical thinking that extends far beyond the presented material. Fourth, a set of multiple choice questions is offered. Some of these are straightforward and are designed to stimulate recall of recently read material, while others probe for a deeper understanding and are designed to stimulate clearer thinking. Fifth, a set of thought questions is posed. Some of these ask the student to garner evidence and argue for a viewpoint. Other questions focus on material that is not extensively revisited in the problems. Still other questions ask the student to speculate, to imagine possibilities, or to identify the kind of evidence that would be required to substantiate a hypothesis. Sixth, complete, expansive solutions to all of the problems are provided. I have tried to anticipate students' questions in composing the solutions. Readers arriving at the correct answer, but without clear explanations for their answer, are encouraged to consult the solutions. It is not just the answer, but also the logic leading to the answer that must be understood. Multiple approaches are sometimes outlined to encourage flexibility. When all of the features of the study guide are utilized, a well-rounded, thorough understanding of the material should be obtained.

Two points cannot be made strongly enough. First, independent thought and reading are essential. Mimicking an instructor's views or just satisfying an instructor's standard will not lead to an understanding of genetics. Understanding comes from thoughtful, careful and individual work. This starts with thoughtful reading of the text, which takes time and cannot be rushed. Second, solving the problems, and working through the exercises in this guide will facilitate a thorough understanding of the material as well as the development of substantial analytical skills. The reader should work through the problems in order, systematically and routinely. The challenges that are met in problem solving will build confidence and a solid conceptual understanding. For most of us, analytical skills are learned. There should be no doubt in the value of taking the requisite time to develop these skills.

I am indebted to Peter Russell for providing exceptionally clear and thoughtful material for me to work with, to Gail Patt for the use of her previous edition of this guide, and to J/B Woolsey and Associates for the use of their art. I am also grateful to Anita Bennett for the preparation of this manuscript and Donna Campion, Supplements Editor for HarperCollins College Publishers. Their help, patience and support have been essential to the completion of this guide. Finally, I would welcome comments or corrections from the users of this manual.

Bruce Chase
June, 1995
Omaha, Nebraska
bachase@cwis.unomaha.edu

CHET S. FORNARI, Ph.D.
DEPT. BIOLOGICAL SCIENCES
DEPAUW UNIVERSITY
GREENCASTLE, IN 46135

CHAPTER 1

GENETICS: AN INTRODUCTION

I. CHAPTER OUTLINE

THE BRANCHES OF GENETICS
RELATIONSHIP OF GENETICS TO OTHER AREAS OF BIOLOGY
GENETICISTS AND GENETICS RESEARCH
What Are Geneticists?
What Is Basic and Applied Research?
What Organisms Are Suitable for Genetic Experimentation?
THE DEVELOPMENT OF THE FIELD OF GENETICS
What Is the Nature of the Genetic Material?
How Is the Genetic Material Transmitted from Generation to Generation?
What Are Genetic Maps?
What Are Genes at the Molecular Level?
How Is Gene Expression Controlled?
What Experimental Approaches are Commonly Used In Genetics Research Today?

II. REVIEW OF KEY TERMS AND CONCEPTS

Once you have an idea of the scope of genetics, you will be able to relate studies in different areas of genetics. To better understand the scope of genetics, try the following exercise after you have completed reading Chapter 1 in the text.

Without consulting the text and in your own words, write a brief definition of each term in the groups below. Then, either using a short phrase or a simple diagram, identify the relationship(s) between specific pairs of terms within a set. Finally, consult the text (and perhaps a friend who has also done the exercise) to check your answers.

1	2	3	4
genetics	genetics	eukaryotes	chromosomes
genes	hypothetico-deductive investigation	nucleus	centrioles
transmission genetics	basic research	prokaryotes	endoplasmic reticulum
molecular genetics	applied research	chromosomes	ribosomes
population genetics	model organisms	bacteria	Golgi apparatus
	recombinant DNA		mitochondria
	DNA typing		chloroplasts

5	6	7	8
mutation	enzyme	nucleotide	messenger RNA
recombination	polypeptide	transcription	transfer RNA
selection	protein	RNA	ribosomal RNA
phenotype	one gene-one enzyme hypothesis	DNA	small nuclear RNA
genetic map	one gene–one polypeptide hypothesis	RNA polymerase	ribosomes
genotype	gene	restriction enzymes	genetic code
		cloning vectors	translation

III. QUESTIONS FOR PRACTICE

A. Multiple Choice Questions

1. Transmission genetics is primarily concerned with
   a. the distribution and behavior of genes in populations.
   b. the passing of genes from generation to generation and their recombination.
   c. the structure and function of genes at the molecular level.
   d. the means by which mutations are retained in nature.

2. Which of the following is not a criterion to choose an organism for genetic experimentation?
   a. The organism should be able to be used in applied research.
   b. The organism should be easy to handle.
   c. The organism should exhibit genetic variation.
   d. The organism should have a relatively short life cycle.

3. Which one of the following eukaryotic cell structures does not contain DNA?
   a. a nucleus
   b. a mitochondrion
   c. the endoplasmic reticulum
   d. a chloroplast

4. Which of the following is not an accurate description of a chromosome?
   a. It is a colored body localized in the nucleus.
   b. It is a protein and nucleic acid complex.
   c. It is the cellular structure that contains the genetic material.
   d. In eukaryotes, it is composed of many DNA molecules attached end to end.

5. Which of the following is not true?
   a. genes encode proteins
   b. genes are proteins
   c. genes are transcribed into RNA molecules
   d. mRNA molecules are translated into proteins on ribosomes

Answers: 1b; 2a; 3c; 4d; 5b

B. Thought Questions

1. Why is it difficult to draw a sharp boundary between molecular genetics, transmission genetics and population genetics?
2. Geneticists typically use the hypothetico-deductive method of investigation. How does this rational method of investigation still allow for research projects to go in exciting, unpredictable directions?
3. Consider how basic and applied research might be interrelated in the following research areas:
   a. studies of mutations that affect the production of alcohol (ethanol) in yeast
   b. recombinant DNA technology
   c. studies of mutations that result in pesticide resistance in *Drosophila melanogaster*
   d. studies on jumping genes in corn
4. What features make an organism well-suited for genetic experimentation?
5. What is the relationship between a genotype and a phenotype?
6. What is a gene?

B. Thought Questions

1. What is the difference [illegible] boundary between a molecular [illegible] genetics and population genetics?
2. Geneticists typically use the hypothetico-deductive method of investigation. How does this rational method of investigation still allow for serendipity to play a role in answering [illegible] questions?

   a. studies of mutations that alter the production of alcohol dehydrogenase in yeast

CHAPTER 2

MENDELIAN GENETICS

I. CHAPTER OUTLINE

GENOTYPE AND PHENOTYPE
MENDEL'S EXPERIMENTAL DESIGN
MONOHYBRID CROSSES AND MENDEL'S PRINCIPLE OF SEGREGATION
The Principle of Segregation
Representing Crosses with a Branch Diagram
Confirming the Principle of Segregation: The Use of Testcrosses
DIHYBRID CROSSES AND THE MENDELIAN PRINCIPLE OF INDEPENDENT ASSORTMENT
The Principle of Independent Assortment
Branch Diagram of Dihybrid Crosses
Trihybrid Crosses
"REDISCOVERY" OF MENDEL'S PRINCIPLES
STATISTICAL ANALYSIS OF GENETIC DATA: THE CHI-SQUARE TEST
MENDELIAN GENETICS IN HUMANS
Pedigree Analysis
Examples of Human Genetic Traits

II. REVIEW OF KEY TERMS, SYMBOLS AND CONCEPTS

Geneticists speak their own language. They communicate with each other very precisely by using specialized terms and symbols that have very fine meanings. To help you learn the "language" of genetics, try the following exercise after you have completed reading Chapter 2 in the text.

Without consulting the text and in your own words, write a brief definition of each term in the groups below. Then, either using a short phrase or a simple diagram, specify the relationship(s) between pairs of terms within each set. Finally, consult the text (and perhaps a friend who has also done the exercise) to check your answers.

1	2	3	4
hereditary trait	genotype	self-fertilization	P, F_1, F_2, F_3
gene	phenotype	F_1 cross	reciprocal cross
locus	dominant	cross-fertilization	gamete
allele	recessive	true-breeding	zygote
allelomorph	heterozygote	P generation	allele

5	6	7	8
probability product rule sum rule independent assortment equal segregation	branch diagram probability product rule	heterozygous dominant hybrid testcross	pedigree analysis genetic counseling 1:1, 3:1, all:none wild-type allele

9	10
hypothesis expected values *P* significance	recessive pedigree rare homozygote *AA A–, Aa aa*

III. THINKING ANALYTICALLY

To learn genetics is much more than just learning a set of facts. Geneticists use analytical reasoning together with previously learned information to interpret new findings. To understand a new situation, they relate past knowledge to the new situation. In this process, they expand their understanding of old concepts.

Solving problems is therefore essential to learn genetics. It reinforces what you have read, and allows you to understand the inter-relationships between terms, ideas and facts.

Unfortunately, learning to reason analytically is not always straightforward. It also takes time. The time will be well spent however, as you will grow tremendously in your reasoning ability.

Here are some pointers that will be helpful to keep in mind as you work on problems.

1. Read the problem straight through without pausing. Get the sense of what it is about.
2. Then read it again, slowly and critically. This time,
   a. Jot down pertinent information.
   b. Assign descriptive gene symbols. Use a dash (–) when unsure of a genotype.
   c. Scan the problem from start to finish, and then from the end to the beginning, and carefully analyze the *terms* used in the problem. Often, clues to the answer can be found in the way the problem is stated. (Geneticists use very precise language!)
3. If you are unsure how to continue, ask yourself, "What are the options here?"
   a. Write down all the options you perceive.
   b. Carefully analyze whether a particular option will provide a solution.

4. When you think you have a solution, read through the problem once more, and make sure that your analysis is consistent with *all* the data in the problem. Try to see if there is a more general principle behind the solution, or for that matter, a more elegant solution.

5. If you do get stuck (and everyone does!), DON'T GO IMMEDIATELY TO THE ANSWER KEY. This will only give you the *illusion* of having solved the problem. On an exam, or in real life, there won't be an answer key available. Learn to control the temptation "to just know the answer" and go on to another problem. Come back to this one another time, perhaps the next day. Your mind may be able to use the time to sort through loose ends.

6. If you can't solve the problem after coming back to it a second time, read through the answer key. Then, close the answer key and try the problem once more. Then, the next day, try it a third time without the answer key.

IV. QUESTIONS FOR PRACTICE

A. Multiple Choice Questions

1. A dominant gene is one that
   a. suppresses the expression of genes at all loci.
   b. masks the expression of neighboring gene loci.
   c. masks the expression of a recessive allele.
   d. masks the expression of all the foregoing.

2. A dihybrid is an individual that
   a. is heterozygous throughout its genotype.
   b. is heterozygous for two genes under study.
   c. is the result of a testcross.
   d. is used for a testcross.

3. The cross of an uncertain genotype with a homozygous recessive genotype at the same locus is a
   a. pure-breeding cross.
   b. monohybrid cross.
   c. testcross.
   d. dihybrid cross.

4. The genotypic ratio of the progeny of a monohybrid cross is typically
   a. 1:2:1
   b. 9:3:3:1
   c. 27:9:9:9:3:3:3:1
   d. 3:1

5. A typical phenotypic ratio of a dihybrid cross with dominant and recessive alleles is
   a. 9:1
   b. 1:2:1
   c. 3:1
   d. 9:3:3:1

6. Pedigrees showing rare recessive traits
   a. have about half of the progeny affected when one parent is affected.
   b. have heterozygotes that are phenotypically affected.
   c. have about 3/4 of the progeny affected when both parents are affected.
   d. often skip a generation.

7. In a chi-square test, a P value equal to 0.04 tells one that
   a. there is a 4 percent chance the hypothesis is correct.
   b. there is a 4 percent chance the hypothesis is incorrect.
   c. if the experiment were repeated, chance deviations from the expected values as large as those observed would be seen only 4 percent of the time.
   d. if the experiment were repeated, chance deviations from the expected values as large as those observed would be seen at least 96 percent of the time.

8. In a chi-square test, a P value equal to 0.04 indicates that
   a. the hypothesis is unlikely to be true.
   b. the hypothesis is false and must be rejected.
   c. the hypothesis is true.
   d. the hypothesis is likely to be true.

Answers: 1c; 2b; 3c; 4a; 5d; 6d; 7c; 8a

B. Thought Questions

1. When are Mendel's principles realized during meiosis?
2. How might have the discovery of chromosomes towards the end of the nineteenth century led to the re-discovery of Mendel's work?
3. Over a period of many years, numerous attempts were made to try and understand how physical traits are passed from one generation to the next. Gregor Mendel was first to make a breakthrough. How do you account for his success, in light of years of failure before him?
4. As presented in Chapter 1, genes specify products, and often these products are proteins that act as enzymes which catalyze biochemical reactions. How might a particular allele of a gene that encodes an enzyme be dominant, while another is recessive?
5. a. Clearly distinguish between an allele, a gene and a locus.
   b. Can you see an analogy between a gene and an allele and
      - a digit and an index finger?
      - a canine and a German Shepherd?
      - a dime and a mint, 1954 dime?
   Where do these analogies break down?
6. Why might the term "degrees of freedom" be so-named, and why (for the problems in this chapter) is it equal to $(n - 1)$?

V. SOLUTIONS TO TEXT PROBLEMS

2.1 In tomatoes, red fruit color is dominant to yellow. Suppose a tomato plant homozygous for red is crossed with one homozygous for yellow. Determine the appearance of (a) the F_1; (b) the F_2; (c) the offspring of a cross of the F_1 back to the red parent; (d) the offspring of a cross of the F_1 back to the yellow parent.

Answer: Consider two ways to solve this problem.

i. First, notice that this is a situation akin to Mendel's crosses. In such crosses, dominant traits mask the appearance of recessive traits, and when a true breeding dominant plant is crossed to a true breeding recessive plant, all the F_1 progeny show the dominant trait. Here the dominant trait is red, the recessive trait is yellow and the parents are true breeding, since they are homozygous.
(a) Just as in Mendel's crosses, the F_1 plants will be heterozygous and all show the dominant, red trait.
(b) The F_2 plants that result from the cross of two F_1 heterozygotes will also show the same ratios seen by Mendel, that is 3 dominant:1 recessive, or 3 red:1 yellow.
(c) When an F_1 plant is crossed back to the red parent, a heterozygous plant is crossed to a true-breeding dominant plant. Here, each offspring receives a

dominant allele from the red parent and so all the progeny must show the dominant, red phenotype.

(d) When an F_1 plant is crossed back to the yellow parent, a heterozygous plant is crossed back to a true-breeding recessive plant. This is akin to one of Mendel's test crosses, and so the progeny should show a 1:1 ratio of red:yellow plants.

ii. Assign *R* as the allele symbol for dominant red color, and *r* as the allele symbol for recessive yellow color.

(a) Then the initial cross between two homozygous plants can be depicted as *RR* x *rr*, and the F_1 progeny obtain an *R* allele from the red parent, and a *r* allele from the yellow parent. Therefore, they are all heterozygotes and are *Rr*. As the *R* (red) allele masks the appearance of the *r* (yellow) allele, the *Rr* progeny are red.

(b) The F_2 are obtained from *Rr* x *Rr*, and so will be composed of 1 *RR*:2 *Rr*:1 *rr* types of progeny. There will be 3 red (*RR* or *Rr*):1 yellow (*rr*).

(c) The F_1 crossed back to the red parent can be depicted as *Rr* x *RR*. All the progeny will obtain an *R* allele from the red parent, the progeny can be written as *R*– (either *RR* or *Rr*) and all will be red.

(d) The F_1 crossed back to the yellow parent can be depicted as *Rr* x *rr*. There will be two equally frequent types of progeny, *Rr* and *rr*. Thus, half the progeny will be red and half yellow.

2.2 Refer to the preceding question, and assume again that red fruit color is dominant to yellow. A red-fruited tomato plant, when crossed with a yellow-fruited one, produces progeny about half of which are red-fruited and half of which are yellow-fruited. What are the genotypes of the parents?

Answer: Notice that in the preceding question, and for that matter, in all Mendelian crosses that involve a single pair of traits, there are only three types of progeny ratios, depending on the types of parents involved.

i. When one parent is homozygous dominant (*RR* x *R*–, *RR* x *rr*), the progeny ratio is all dominant to no recessive (all *R*–).

ii. When both parents are heterozygous (*Rr* x *Rr*), as in a cross between two F_1 progeny, the progeny ratio is 3 dominant to 1 recessive (3 *R*–:1 *rr*).

iii. When one parent is heterozygous and the other homozygous recessive (*Rr* x *rr*), as in a test-cross, the progeny ratio is 1 dominant to 1 recessive (1 *Rr*:1 *rr*).

Here a dominant red plant is crossed to a recessive yellow plant, and so the cross can be depicted as *R*– x *rr*. Given the phenotype of the parents, this cross is either type (i) *RR* x *rr*, or type (iii) *Rr* x *rr*. We can tell which by looking at the progeny ratios. As the progeny show a 1:1 red:yellow ratio, this must be a type (iii) cross, and so the parent genotypes are *Rr* x *rr*.

2.3 In maize, a dominant gene *A* is necessary for seed color as opposed to colorless (*a*). A recessive gene *wx* results in waxy starch as opposed to normal starch (*Wx*). The two genes segregate independently. Give phenotypes and relative frequencies for offspring resulting when a plant of genetic constitution *Aa WxWx* is testcrossed.

Answer: A plant that is genotypically *Aa WxWx* has two types of gametes: *A Wx* and *a Wx* (notice that *Wx* is a symbol for one allele, not two!). In a test cross, the other parent will be homozygous for the recessive alleles at the color gene as well as the waxy gene, *aa wxwx* and have gametes that are all *a wx*. This cross can be illustrated in the following Punnett square:

		Gametes of *Aa WxWx*	
		A Wx	*a Wx*
Gametes of *aa wxwx*	*a wx*	*Aa Wxwx*	*aa Wxwx*

Progeny will be of two equally frequent genotypes: *Aa Wxwx* and *aa Wxwx*. Thus, half will be colored (*Aa*), half will be colorless (*aa*) and all will have normal starch (*Wxwx*).

2.4 F_2 plants segregate 3/4 colored:1/4 colorless. If a colored plant is picked at random and selfed, what is the probability that more than one type will segregate among a large number of its progeny?

Answer: In the F_2, a 3:1 colored:colorless phenotypic ratio reflects a 1 *CC*:2 *Cc*:1 *cc* genotypic ratio, where colored (*C*) is dominant to colorless (*c*). Thus, there are two types of colored plants that are present in a 1 *CC*:2 *Cc* ratio. If one is picked at random, there is a 1/3 chance of picking a homozygous *CC* plant, and a 2/3 chance of picking a heterozygous *Cc* plant. When this plant is selfed and a large number of progeny sampled, the *CC* plant will only produce *CC* (colored) plants, while the *Cc* plant will produce both colored and colorless plants (in a 3:1 ratio). Thus, to get more than one type of plant segregating, a *Cc* plant must be chosen initially. The chance of doing this is 2/3. [Notice that the question asks for the chance that a particular colored plant will have two types of progeny, and *not* for the ratio of progeny types when more than one is seen].

2.5 In guinea pigs rough coat (*R*) is dominant over smooth coat (*r*). A rough-coated guinea pig is bred to a smooth one, giving eight rough and seven smooth progeny in the F_1.

a. What are the genotypes of the parents and their offspring?
b. If one of the rough F_1 animals is mated to its rough parent, what progeny would you expect?

Answer: a. Since rough (*R*) is dominant over smooth (*r*), one can depict a cross between a rough and smooth guinea pig as *R*– x *rr*, where *R*– represents either *RR* or *Rr*. If the cross is *RR* x *rr*, then all the progeny will be *Rr* and be rough. If the cross is *Rr* x *rr*, then half the progeny will be *Rr* (rough) and half will be *rr* (smooth). The 8 rough:7 smooth progeny ratio approximates a 1:1 *Rr*:*rr* ratio, and so the initial cross must have been *Rr* x *rr* and the rough progeny must be *Rr* while the smooth progeny must be *rr*.

b. If one of the rough F_1 animals is mated back to its rough parent, one would have *Rr* x *Rr*. This cross would produce both rough (*R*–) and smooth (*rr*) progeny in a 3:1 ratio.

2.6 In cattle the polled (hornless) condition (*P*) is dominant over the horned (*p*) phenotype. A particular polled bull is bred to three cows. Cow A, which is horned, produces a horned calf; a polled cow B produces a horned calf; and cow C produces a polled calf. What are the genotypes of the bull and the three cows, and what phenotypic ratios do you expect in the offspring of these three matings?

Answer: In this problem it helps to first depict the phenotypes as genotypes. This can be done by realizing that a polled animal exhibiting the dominant, hornless condition can be depicted as *P*–, and any horned animal exhibits the recessive condition and must be *pp*. Therefore, the three crosses and their progeny can be depicted as

cow phenotype	cow genotype	x	polled bull genotype	progeny phenotype	progeny genotype
A: horned	*pp*	x	*P*–	horned	*pp*
B: polled	*P*–	x	*P*–	horned	*pp*
C: horned	*pp*	x	*P*–	polled	*P*–

One can now follow how each set of parents contributed alleles to their progeny. In the crosses with cows A and B, a horned, *pp* offspring can only be obtained if each parent contributes a recessive *p* allele to their progeny. Therefore both the polled bull and cow B must be heterozygous and are *Pp*. With this information, both the crosses with cows A and C appear to be test-crosses (*Pp* x *pp*) and will produce 1:1 phenotypic ratios of polled (*Pp*) and horned (*pp*) progeny. The cross with cow B is a cross between two heterozygotes (*Pp* x *Pp*), and will produce a 3:1 phenotypic ratio of polled to horned progeny.

2.7 In the Jimsonweed purple flowers are dominant to white. When a particular purple-flowered Jimsonweed is self-fertilized, there are 28 purple-flowered and 10 white-flowered progeny. What proportion of the purple-flowered progeny will breed true?

Answer: The ratio of 28 purple to 10 white plants is close to 3:1. A 3:1 ratio means that the purple plant that was initially selfed must have been heterozygous, or *Pp*. A cross of *Pp* x *Pp* would yield progeny with a genotypic ratio of 1 *PP*:2 *Pp*:1 *pp*. Thus, among the purple progeny, there would be 1/3 *PP* homozygotes and 2/3 *Pp* heterozygotes. Since the homozygous *PP* plants are the only plants that will breed true, only 1/3 of the purple-flowered progeny will breed true.

2.8 Two black female mice are crossed with the same brown male. In a number of litters female X produced 9 blacks and 7 browns and female Y produced 14 blacks. What is the mechanism of inheritance of black and brown coat color in mice? What are the genotypes of the parents?

Answer: Notice that the two crosses give very different results. Use both sets of results to answer the question. In the cross of black female X with the brown male, the 9 black and 7 brown progeny approximate a 1:1 phenotypic ratio. This suggests that this cross is similar to a test cross, and might be depicted *Bb* x *bb*. However, it is not clear from this cross alone which color trait is dominant. This question can be answered by considering the cross of black female Y with the brown male, where only black progeny are seen. This is like a cross between two true-breeding individuals where black is dominant to brown,

$BB \times bb$. Thus, the brown male is homozygous recessive (*bb*), female X is heterozygous (*Bb*), and female Y is homozygous dominant (*BB*).

2.9 Bean plants may differ in their symptoms when infected with a virus. Some show local lesions that do not seriously harm the plant. Other plants show general systemic infection. The following genetic analysis was made:

P	local lesions x systemic lesions
F_1	all local lesions
F_2	785 local lesions:269 systemic lesions

What is probably the genetic basis of this difference in beans? Assign gene symbols to all the genotypes occurring in the above experiment. Design a testcross to verify your assumptions.

Answer: In this question, note that the trait that is being examined is the *response* of plants to a virus. One envisions that crosses are performed between plants, and then, instead of looking at flower color or seed shape to score a trait, plants are individually tested for their response to viral infection. A cross of a plant showing only local lesions to one showing systemic lesions gives only plants that show local lesions, indicating that a local lesion response might be considered to be dominant to a systemic lesion response. If the allele for the local lesion response is depicted as *L*, and the allele for the systemic lesion response as *l*, the parental generation can be written as $LL \times ll$. Then, the F_1 would be all *Ll* and the F_2 would be 1 *LL*:2 *Ll*:1 *ll*. This fits with the observed numbers of 785 local and 269 systemic (3 *L*–:1 *ll*) lesion plants.

To test the hypothesis that the local lesion response is dominant to the systemic lesion response, testcross the F_2 plants to true breeding plants that show a systemic lesion response (i.e., are *ll*). The progeny of this testcross can then be assayed for their response to viral infection so that the genotype of the F_2 can be determined. One would expect 2/3 of those F_2 plants showing a local lesion response to be heterozygotes, i.e., *Ll*. When they are testcrossed, half of their progeny should show local responses and half systemic responses. The remaining 1/3 will be *LL*, and give only progeny with local responses.

2.10 A normal *Drosophila* has both brown and scarlet pigment granules in the eyes, which appear red as a result. Brown (*bw*) is a recessive gene on chromosome 2 which, in homozygous condition, results in the absence of scarlet granules (so that the eyes are brown). Scarlet (*st*) is a recessive gene on chromosome 3 which, when homozygous, results in scarlet eyes due to the absence of brown pigment. Any fly homozygous for recessive brown and recessive scarlet alleles produces no eye pigment and has white eyes. The following results are obtained from crosses:

P	brown-eyed fly x scarlet-eyed fly
F_1	red eyes (both brown and scarlet pigment present)
F_2	9/16 red:3/16 scarlet:3/16 brown:1/16 white

a. Assign genotypes to the P and F_1 generations.
b. Design a testcross to verify the F_1 genotype, and predict the results.

Answer: Here, two independently assorting genes affecting eye pigment are being followed. Use the symbols suggested in the problem to depict alleles at each locus. For the locus controlling scarlet granules, use *bw* to represent the recessive allele and *Bw* to represent the dominant allele. For the locus controlling brown granules, use *st* to represent the recessive allele and *St* to represent the dominant allele. When writing the genotypes of the flies being crossed, be certain to write the alleles present at both genes.

a. Our first task is to assign genotypes to the P and F_1 generations. Brown-eyed flies lack scarlet granules, so they must be *bwbw* (no scarlet) and *St–*; these flies do make brown granules. Note that they can either be *bwbw StSt* or *bwbw Stst*, and that we cannot tell their exact genotype yet. Using similar reasoning, scarlet-eyed flies must be *Bw– stst*. We can determine the exact genotype of these P-generation flies by considering what progeny they produce. Since each parent can be either of two genotypes, four different crosses are possible. These are illustrated in the following branch diagrams:

(i) brown eyes x scarlet eyes
bwbw StSt x *BwBw stst*

bwbw x *BwBw*   *StSt* x *stst*

Bwbw —— *Stst* ——→ *Bwbw Stst*
(scarlet granules) (brown granules) (all red eyes)

(ii) brown eyes x scarlet eyes
bwbw Stst x *BwBw stst*

bwbw x *BwBw*   *Stst* x *stst*

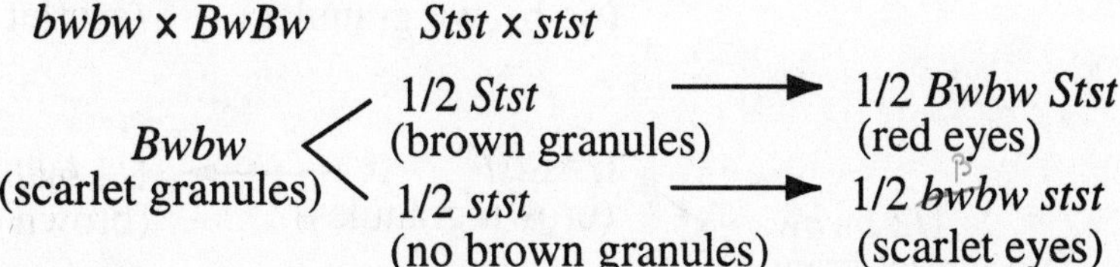

(iii) brown eyes x scarlet eyes
bwbw StSt x *Bwbw stst*

bwbw x *Bwbw*   *StSt* x *stst*

1/2 *bwBw* —— *Stst* ——→ 1/2 *bwBw Stst*
(scarlet granules) (brown granules) (red eyes)

1/2 *bwbw* —— *Stst* ——→ 1/2 *bwbw Stst*
(no scarlet granules) (brown granules) (brown eyes)

(iv) brown eyes x scarlet eyes

bwbw x *Bwbw* *Stst* x *stst*

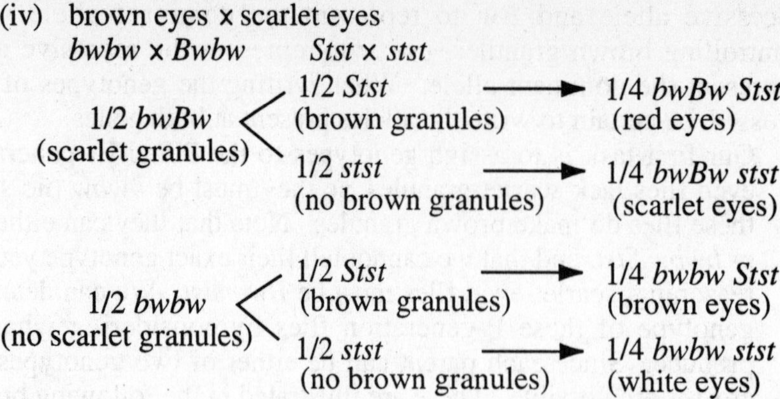

Notice that even though the parents in each of the above crosses look the same, their progeny are not. The only way all red-eyed progeny can be obtained is if the cross is that shown in (i). To see if the F_2 that is observed is consistent with this view, use a branch diagram as follows:

red eyes x red eyes

Bwbw Stst x *Bwbw Stst*

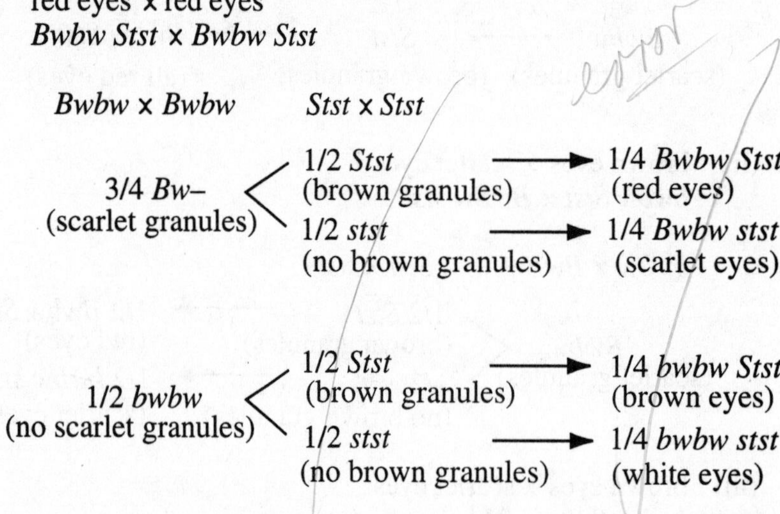

b. To verify the F_1 genotype, cross the red-eyed F_1 to doubly homozygous recessive white-eyed flies of the genotype *bwbw stst*. One expects to see a 1:1:1:1 ratio of red:brown:scarlet:white flies, as shown in the following branch diagram.

red eyes x white eyes
Bwbw Stst x *bwbw stst*

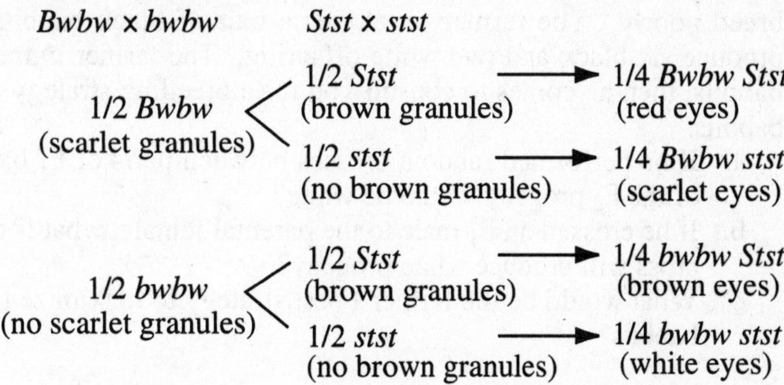

2.11 Grey seed color (*G*) in garden peas is dominant to white seed color (*g*). In the following crosses, the indicated parents with known phenotypes but unknown genotypes produced the listed progeny. Give the possible genotype(s) of each female parent based on the segregation data.

PARENTS	PROGENY		FEMALE PARENT
FEMALE X MALE	GREY	WHITE	GENOTYPE
grey x white	81	82	?
grey x grey	118	39	?
grey x white	74	0	?
grey x grey	90	0	?

Answer: First notice that white seeds are always *gg* while grey seeds are either *GG* or *Gg*. Thus, one can assign partial genotypes (either *G–* or *gg*) based on the phenotypes of the parents. The problem is to determine whether a grey parent is *GG* or *Gg*. In monohybrid crosses, there are three kinds of phenotypic ratios that can be generated. The cross *Gg* x *Gg* gives a 3 grey (*G–*):1 white (*gg*) ratio, the cross *Gg* x *gg* gives a 1 grey (*Gg*):1 white (*gg*) ratio, and the crosses *GG* x *gg* and *GG* x *G–* give all grey (*G–*) progeny. As the first cross of grey x white gives a 1:1 ratio, it must be *Gg* x *gg*. As the second cross of grey x grey gives a 3:1 ratio, it must be *Gg* x *Gg*. As the third cross of grey x white gives all grey, it must be *GG* x *gg*. In the fourth cross, only grey progeny are produced. One of the parents must be *GG*, but the other's genotype can only be specified as *G–*. One has:

PARENTS	PROGENY		FEMALE PARENT
FEMALE X MALE	GREY	WHITE	GENOTYPE
grey x white	81	82	*Gg*
grey x grey	118	39	*Gg*
grey x white	74	0	*GG*
grey x grey	90	0	*GG* or *G–*

2.12 Fur color in babbits, a furry little animal and popular pet, is determined by a pair of alleles, *B* and *b*. *BB* and *Bb* babbits are black, and *bb* babbits are white.

A farmer wants to breed babbits for sale. True-breeding white (*bb*) female babbits breed poorly. The farmer purchases a pair of black babbits, and these mate and produce six black and two white offspring. The farmer immediately sells his white babbits, then he comes to consult you for a breeding strategy to produce more white babbits.

a. If he performed random crosses between pairs of F_1 babbits, what proportion of the F_2 progeny would be white?

b. If he crossed an F_1 male to the parental female, what is the probability that this cross will produce white progeny?

c. What would be the farmer's best strategy to maximize the production of white babbits?

Answer: As the farmer starts with a pair of black babbits, he must be starting with animals that are either *BB* or *Bb*. Since his pair is not true breeding, and indeed, gives a 3:1 ratio of black to white progeny, he must have two heterozygotes, or *Bb* babbits. Therefore, one expects to find a 1 *BB*:2 *Bb* genotypic ratio among the black F_1 progeny that were not sold.

a. If one picks randomly among these black F_1 progeny, there will be a 1/3 chance of picking a *BB* individual, and a 2/3 chance of picking a *Bb* individual. There are four possible types of crosses. By using the product rule, one can determine that the probability of each is:

(i) *BB* x *BB* $p = 1/3 \times 1/3 = 1/9$
(ii) *Bb* x *BB* $p = 2/3 \times 1/3 = 2/9$
(iii) *BB* x *Bb* $p = 1/3 \times 2/3 = 2/9$
(iv) *Bb* x *Bb* $p = 2/3 \times 2/3 = 4/9$
(Notice that, since each pick is independent, ii $\neq$ iii)

Only cross (iv) will give any white progeny, and then only 1/4 of the time. Therefore, the chance of obtaining white progeny in the F_2 is 4/9 x 1/4 = 1/9.

b. If he crosses an F_1 male to the parental female, there are two types of crosses, with probabilities:

(i) *Bb* (F_1 male) x *Bb* (parental female) $p = 2/3 \times 1 = 2/3$
(ii) *BB* (F_1 male) x *Bb* (parental female) $p = 1/3 \times 1 = 1/3$

Here, only the first cross will produce white progeny, and then only 1/4 of the time. Using the product rule, the chance that this strategy will yield white progeny is 2/3 x 1/4 = 1/6.

c. As neither of the above strategies work very well, the farmer should re-mate his initial two black babbits, and obtain white progeny 25 percent of the time. When he obtains a white male, he should retain this male (as male fertility is not affected) for breeding with his initial two black babbits. (This cross would be *Bb* x *bb*). Half of the offspring of this cross would be white, and the other half would be heterozygous. If the farmer kept some of the heterozygous females as well as a few of the white, homozygous male offspring, he should be able to repeat this cross and sustain a steady supply of white babbits.

2.13 In Jimsonweed, purple flower (*P*) is dominant to white (*p*), and spiny pods (*S*) are dominant to smooth (*s*). In a cross between a Jimsonweed homozygous for white flowers and spiny pods and one homozygous for purple flowers and smooth pods, determine the phenotype of (a) the F_1; (b) the F_2; (c) the progeny of a cross of the F_1

back to the white, spiny parent; (d) the progeny of a cross of the F_1 back to the purple, smooth parent.

Answer: Since you are told that each parent is homozygous, the parental genotypes can be determined from their phenotypes. One has a white (*pp*) spiny (*SS*) plant crossed to a purple (*PP*) smooth (*ss*) plant. This initial information can be used with branch diagrams to solve each part of the problem as follows:

P: white, spiny x purple, smooth
pp SS x *PP ss*
Gametes: *p S* *P s*
F_1: *Pp Ss* (purple, spiny)
F_1 x F_1: *Pp Ss* x *Pp Ss*

Pp x *Pp* *Ss* x *Ss*

3/4 *P–* < 3/4 *S–* → 9/16 *P– S–* (purple, spiny)
3/4 *P–* < 1/4 *ss* → 3/16 *P– ss* (purple, smooth)
1/4 *pp* < 3/4 *S–* → 3/16 *pp S–* (white, spiny)
1/4 *pp* < 1/4 *ss* → 1/16 *pp ss* (white, smooth)

F_1 x white, spiny: *Pp Ss* x *pp SS*
Pp x *pp* *Ss* x *SS*

1/2 *Pp* —— *S–* → 1/2 *Pp Ss* (purple, spiny)
1/2 *pp* —— *S–* → 1/2 *pp ss* (white, spiny)

F_1 x purple, smooth: *Pp Ss* x *PP ss*
Pp x *PP* *Ss* x *ss*

P– < 1/2 *Ss* → 1/2 *P– Ss* (purple, spiny)
P– < 1/2 *ss* → 1/2 *P– ss* (purple, smooth)

2.14 What progeny would you expect from the following Jimsonweed crosses? You are encouraged to use the branch diagram approach.

a. *PP ss* x *pp SS*
b. *Pp SS* x *pp ss*
c. *Pp Ss* x *Pp SS*
d. *Pp Ss* x *Pp Ss*
e. *Pp Ss* x *Pp ss*
f. *Pp Ss* x *pp ss*

Answer: Using branch diagrams, one has:

a. *PP ss* x *pp SS*

PP x *pp* *ss* x *SS*

Pp —— *Ss* ——▶ all *Pp Ss* (purple, spiny)

b. *Pp SS* x *pp ss*

Pp x *pp* *Ss* x *ss*

1/2 *Pp* —— *Ss* ——▶ 1/2 *Pp Ss* (purple, spiny)

1/2 *pp* —— *Ss* ——▶ 1/2 *pp Ss* (white, spiny)

c. *Pp Ss* x *Pp SS*

Pp x *Pp* *Ss* x *SS*

3/4 *P–* —— *S–* ——▶ 3/4 *P– S–* (purple, spiny)

1/4 *pp* —— *S–* ——▶ 1/4 *pp S–* (white, spiny)

d. *Pp Ss* x *Pp Ss*

Pp x *Pp* *Ss* x *Ss*

3/4 *P–* —— *S–* ——▶ 3/4 *P– S–* (purple, spiny)

1/4 *pp* —— *S–* ——▶ 1/4 *pp S–* (white, spiny)

e. *Pp Ss* x *Pp ss*

Pp x *Pp* *Ss* x *ss*

3/4 *P–* < 1/2 *Ss* ——▶ 3/8 *P–* Ss (purple, spiny)

3/4 *P–* < 1/2 *ss* ——▶ 3/8 *P– ss* (purple, smooth)

1/4 *pp* < 1/2 *Ss* ——▶ 1/8 *pp* Ss (white, spiny)

1/4 *pp* < 1/2 *ss* ——▶ 1/8 *pp ss* (white, smooth)

f. *Pp Ss* x *pp ss*

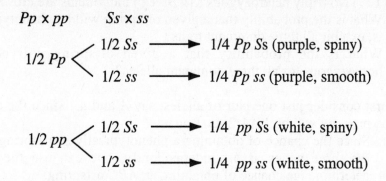

Pp x *pp*   *Ss* x *ss*

1/2 *Pp* < 1/2 *Ss* → 1/4 *Pp Ss* (purple, spiny)
1/2 *ss* → 1/4 *Pp ss* (purple, smooth)

1/2 *pp* < 1/2 *Ss* → 1/4 *pp Ss* (white, spiny)
1/2 *ss* → 1/4 *pp ss* (white, smooth)

2.15 In summer squash white fruit (*W*) is dominant over yellow (*w*), and disk-shaped fruit (*D*) is dominant over sphere-shaped fruit (*d*). In the following problems the appearances of the parents and their progeny are given. Determine the genotypes of the parents in each case.

a. White, disk x yellow, sphere gives 1/2 white, disk and 1/2 white, sphere.
b. White, sphere x white, sphere gives 3/4 white, sphere and 1/4 yellow, sphere.
c. Yellow, disk x white, sphere gives all white, disk progeny.
d. White, disk x yellow, sphere gives 1/4 white, disk; 1/4 white, sphere; 1/4 yellow, disk; and 1/4 yellow, sphere.
e. White, disk x white, sphere gives 3/8 white, disk; 3/8 white, sphere; 1/8 yellow, disk; and 1/8 yellow, sphere.

Answer: In this problem, it is very helpful to define symbols for the alleles specifying each trait and use information about which allele is dominant or recessive to make initial assignments of possible genotypes. For example, if a plant is white, it must have a dominant *W* allele, but as it may be either *WW* or *Ww*, it would be initially noted as *W–*. If a plant is yellow, it has to be *ww*. Then, by considering just one pair of allelic traits at a time, and recalling the Mendelian progeny ratios you have seen [a 1:1 ratio follows from a testcross (*Aa* x *aa*), an all to none ratio follows if at least one parent is homozygous dominant (*AA* x *A–* or *aa*) and a 3:1 ratio follows from a monohybrid cross (*Aa* x *Aa*)], you can ascertain whether a parental *W–* plant can be assigned to be *Ww* or *WW*. In this way, you can determine that

Cross	Parents	Progeny
a.	white, disk x yellow, sphere *Ww Dd* *ww dd*	1/2 white, disk; 1/2 white, sphere
b.	white, sphere x white, sphere *Ww dd* *Ww dd*	3/4 white, sphere; 1/4 yellow, sphere
c.	yellow, disk x white, sphere *ww DD* *WW dd*	all white, disk
d.	white, disk x yellow, sphere *Ww Dd* *ww dd*	1/4 white, disk; 1/4 white, sphere; 1/4 yellow disk; 1/4 yellow sphere
e.	white, disk x white, sphere *Ww Dd* *Ww dd*	3/8 white, disk; 3/8 white, sphere 1/8 yellow, disk; 1/8 yellow, sphere

2.16 Genes *a*, *b* and *c* assort independently and are recessive to their respective alleles *A*, *B*, and *C*. Two triply heterozygous (*Aa Bb Cc*) individuals are crossed.

a. What is the probability that a given offspring will be phenotypically *ABC*, that is, exhibit all three dominant traits?

b. What is the probability that a given offspring will be genotypically homozygous for all three dominant alleles?

Answer: First consider just one pair of alleles, say *A* and *a*. Since the cross is *Aa* x *Aa*, the progeny will be 1 *AA*:2 *Aa*:1 *aa*.

a. Since the chance of obtaining a phenotypically *A* offspring will be 3/4, and the three genes assort independently, one can use the product rule to determine the chance of obtaining an *ABC* offspring.

p = (chance of *A*–)(chance of *B*–)(chance of *C*–)
p = (3/4)(3/4)(3/4)
p = (27/64)

b. Since the chance of obtaining an *AA* offspring will be 1/4, and the three genes assort independently, use the product rule to determine the chance of obtaining an *AA BB CC* offspring.

p = (chance of *AA*)(chance of *BB*)(chance of *CC*)
p = (1/4)(1/4)(1/4)
p = (1/64)

You can also solve this problem by setting up a branch diagram, or much more laboriously, by setting up a Punnett square.

2.17 In garden peas tall stem (*T*) is dominant over short stem (*t*), green pods (*G*) are dominant over yellow pods (*g*), and smooth seeds (*S*) are dominant over wrinkled seeds (*s*). Suppose a homozygous short, green, wrinkled pea plant is crossed to a homozygous tall, yellow, smooth one.

a. What will be the appearance of the F_1?

b. What will be the appearance of the F_2?

c. What will be the appearance of the offspring of a cross of the F_1 back to its short, green, wrinkled parent?

d. What will be the appearance of the offspring of a cross of the F_1 back to its tall, yellow, smooth parent?

Answer: First note that the problem asks only for the *appearance* of the offspring of the various crosses, and not the genotypes. Then write out the genotype of the parental cross: *tt GG ss* x *TT gg SS*.

a. For each pair of traits, a homozygous dominant plant is crossed to a homozygous recessive plant. Thus, all the F_1 progeny will show the dominant trait, and be tall, green and smooth (*Tt Gg Ss*).

b. The F_2 results from selfing the F_1, i.e., *Tt Gg Ss* x *Tt Gg Ss*. The appearance of the F_2 is most readily determined by employing a branch diagram.

Tt x *Tt*   *Gg* x *Gg*   *Ss* x *Ss*

- 3/4 *T–*
  - 3/4 *G–*
    - 3/4 *S–* → 27/64 *T– G– S–* (tall, green, smooth)
    - 1/4 *ss* → 9/64 *T– G– ss* (tall, green , wrinkled)
  - 1/4 *gg*
    - 3/4 *S–* → 9/64 *T– gg S–* (tall, yellow, smooth)
    - 1/4 *ss* → 3/64 *T– gg ss* (tall, yellow, wrinkled)

- 1/4 *tt*
  - 3/4 *G–*
    - 3/4 *S–* → 9/64 *tt G– S–* (short, green, smooth)
    - 1/4 *ss* → 3/64 *tt G– ss* (short, green , wrinkled)
  - 1/4 *gg*
    - 3/4 *S–* → 3/64 *tt gg S–* (short, yellow, smooth)
    - 1/4 *ss* → 1/64 *tt gg ss* (short, yellow, wrinkled)

c. The branch diagram of the cross *Tt Gg Ss* x *tt GG ss* is:

Tt x *tt*   *Gg* x *GG*   *Ss* x *ss*

- 1/2 *Tt* — all *G–*
  - 1/2 *Ss* → 1/4 *Tt G– Ss* (tall, green, smooth)
  - 1/2 *ss* → 1/4 *Tt G– ss* (tall, green , wrinkled)
- 1/2 *tt* — all *G–*
  - 1/2 *Ss* → 1/4 *tt G– Ss* (short, green, smooth)
  - 1/2 *ss* → 1/4 *tt G– ss* (short, green , wrinkled)

d. The branch diagram of the cross *Tt Gg Ss* x *TT gg SS* is:

Tt x *TT*   *Gg* x *gg*   *Ss* x *SS*

- all *T–*
  - 1/2 *Gg* — all *S–* → 1/2 *T– Gg S–* (tall, green, smooth)
  - 1/2 *gg* — all *S–* → 1/2 *T– gg S–* (tall, yellow, wrinkled)

2.18 *C/c, O/o,* and *I/i* are three independently segregating gene pairs of alleles in chickens. *C* and *O* are dominant genes, both of which are necessary for pigmentation. *I* is a dominant inhibitor of pigmentation. Individuals of genotype *cc*, or *oo*, or *Ii*, or *II* are white, regardless of what other genes they possess.

Assume that White Leghorns are *CC OO II*, White Wyandottes are *cc OO ii*, and White Silkies are *CC oo ii*, what types of offspring (white or pigmented) are possible, and what is the probability of each, from the following crosses:

a. White Silkie x White Wyandotte?
b. White Leghorn x White Wyandotte?
c. (Wyandotte–Silkie F_1) x White Silkie?

Answer: In order for an individual to be pigmented, they must have one of each of the dominant *C* and *O* alleles and not have any of the dominant *I* alleles. That is, they must be *C– O– ii*.

a. White Silkie (*CC oo ii*) x White Wyandotte (*cc OO ii*). These are two true-breeding strains at each of three genes, and differ from each other at two loci, the *C/c* and *O/o* loci. The progeny will be dihybrids, are all *Cc Oo ii* and are all pigmented.
b. White Leghorn (*CC OO II*) x White Wyandotte (*cc OO ii*). These two are also true-breeding at each of three genes, and differ from each other at two loci, the *C/c* and *I/i* loci. Their progeny will be dihybrids, are all *Cc OO Ii* and are all white (because of *Ii*).
c. Wyandotte-Silkie F_1 (*Cc Oo ii*) x White Silkie (*CC oo ii*). As both parents are *ii*, all the progeny will be as well, and so, if they are *C– O–*, they will be pigmented. The proportions of white and pigmented progeny can be determined by using a branch diagram, or considering each locus separately as follows: At the *C/c* locus, the cross is *Cc* x *CC*, so all the progeny will be *C–*. At the *O* locus, the cross is *Oo* x *oo*, a testcross, so half the progeny will be *Oo* and half will be *oo*. Thus, half of the progeny will be *C– Oo ii* and be pigmented, and half will be *C– oo ii* and be white.

2.19 Two homozygous strains of corn are hybridized. They are distinguished by six different pairs of genes, all of which assort independently and produce an independent phenotypic effect. The F_1 hybrid is selfed to give an F_2.
a. What is the number of possible genotypes in the F_2?
b. How many of these genotypes will be homozygous at all six gene loci?
c. If all gene pairs act in a dominant-recessive fashion, what proportion of the F_2 will be homozygous for all dominants?
d. What proportion of the F_2 will show all dominant phenotypes?

Answer: While this problem can be solved using a branch diagram, one can also approach it in a different manner. A general relationship exists between the number of possible genotypes and phenotypes. At any one locus having alleles that act in a dominant-recessive fashion, there are two possible phenotypes (*A–* and *aa*) and three possible genotypes (*AA*, *Aa* and *aa*). When several independently assorting loci are considered, the number of possibilities grows according to the number of combinations of alleles possible. For two loci, independent assortment of the alleles at each locus allows for 2 x 2 = 4 phenotypes (two possible phenotypes at each locus assorted with either of two possible phenotypes at a second locus) and 3 x 3 genotypes (three possible genotypes at one locus assorted with any of three possible genotypes at a second locus). For three loci, independent assortment of the alleles at each locus allows for 2 x 2 x 2 = 8 phenotypes and 3 x 3 x 3 = 27 genotypes. For *n* loci, the relationship becomes 2^n possible phenotypes and 3^n possible genotypes.

To determine the frequency of a particular phenotypic or genotypic class, consider that at one locus, an F_1 hybrid cross gives 3/4 phenotypically dominant and 1/4 phenotypically recessive progeny and 1/4 homozygous dominant, 1/2 heterozygous and 1/4 homozygous recessive progeny. The fraction of progeny of a particular genotypic or phenotypic class can be determined by considering the combinations that are possible. For example, for two independently assorting loci *A/a* and *B/b*, 3/4 of the progeny of a cross of *Aa Bb* x *Aa Bb* will show the *A–* phenotype and 3/4 of the progeny will show the *B–* phenotype. When independent assortment of these two loci is considered, 3/4 x 3/4 = 9/16 of the progeny will show both the *A–* and the *B–* phenotypes. (Multiply the individual probabilities using the product rule.) Thus, when *n* loci are involved,

$(3/4)^n$ of the progeny will show all dominant phenotypes, $(1/4)^n$ of the progeny will show all recessive phenotypes, $(1/4)^n$ of the progeny will be homozygous dominant or homozygous recessive at all loci and $(1/2)^n$ of the progeny will be heterozygous at all the loci.

a. For six loci, $3^6 = 729$ possible genotypes will be seen.

b. Notice that the question asks for how many of the *genotypes* will be homozygous. For just one locus, two (*AA*, *aa*) of three possible (*AA*, *Aa*, *aa*) genotypes are homozygous. Thus, for six loci, $2^6 = 64$ genotypes will be homozygous for all loci. Note that included in these genotypes are those that are homozygous dominant at each locus, those that are homozygous recessive at each locus and those that show combinations in between.

c. In a monohybrid cross, 1/4 of the progeny will be homozygous dominant. For the F_2 in this cross, $(1/4)^6 = 1/4{,}096$ of the progeny will be homozygous dominant. Notice that this fraction of the progeny derives from a single genotype.

d. In a monohybrid cross, 3/4 of the progeny show a dominant phenotype. For the F_2 in this cross, $(3/4)^6 = 729/4{,}096 = 17\%$ of the progeny will show all six dominant phenotypes.

2.20 The coat color of mice is controlled by several genes. The agouti pattern, characterized by a yellow band of pigment near the tip of the hairs, is produced by the dominant allele *A*; homozygous *aa* mice do not have the band and are non-agouti. The dominant allele *B* determines black hairs, and the recessive allele *b* determines brown. Homozygous c^hc^h individuals allow pigment to be deposited only at the extremities (e.g., feet, nose, and ears) in a pattern called Himalayan. The genotype *C*– allows pigment to be distributed over the entire body.

a. If a true-breeding black mouse is crossed with a true-breeding brown, agouti, Himalayan mouse, what will be the phenotypes of the F_1 and F_2?

b. What proportion of the black agouti F_2 will be of genotype $Aa\ BB\ Cc^h$?

c. What proportion of the Himalayan mice in the F_2 are expected to show brown pigment?

d. What proportion of all agoutis in the F_2 are expected to show black pigment?

Answer: a. The cross is $aa\ BB\ CC \times AA\ bb\ c^hc^h$. Since the parents are true breeding at each of three different loci, the F_1 must be a trihybrid, is $Aa\ Bb\ Cc^h$. It will be agouti with black hairs and pigmented extremities. In the F_2, nine possible phenotypes can be obtained. One can either write out all the possibilities or, to be more thorough, use a branch diagram.

b. Among the F_2 *A*– progeny, 2/3 will be *Aa*. Among the F_2 *B*– progeny, 1/3 will be *BB*. Among the F_2 *C*– progeny, 2/3 will be Cc^h. To determine the probability of all these criteria being met, multiply the individual probabilities. Therefore, among the black agouti F_2 progeny, 2/3 x 1/3 x 2/3 = 4/9 will be $Aa\ BB\ Cc^h$.

c. This can be determined in two different ways. First, one can consider the branch diagram: 3/64 + 1/64 = 4/64 = 1/16 of the Himalayan F_2 show brown pigment. Alternatively, one can calculate the percentage of the F_2 that are $bb\ c^hc^h$. There is a 1/4 chance that a mouse is *bb* and a 1/4 chance that a mouse is c^hc^h, giving a 1/16 chance that a mouse is $bb\ c^hc^h$.

d. An agouti, black mouse must be *A– B–*. One can either consult the branch diagram, or calculate the probability of obtaining F_2 mice that are *A– B–*. The chance of obtaining an F_2 mouse that is *A–* and *B–* is 3/4 x 3/4 = 9/16.

2.21 In cocker spaniels, solid coat color is dominant over spotted coat. Suppose a true-breeding, solid-colored dog is crossed with a spotted dog, and the F_1 dogs are interbred.

a. What is the probability that the first puppy born will have a spotted coat?
b. What is the probability that if four puppies are born, all of them will have a solid coat?

Answer: First, assign symbols to the alleles, and write down the cross. Let solid color be *S* and spotted be *s*. Then, if a true-breeding, solid-colored dog is bred to a spotted dog, the cross is *SS* x *ss*, the F_1 all *Ss* and interbreeding the F_1 will give 3/4 *S–* (solid) and 1/4 *ss* (spotted) progeny.

a. The chance that the first puppy born is spotted is just the chance of getting an *ss* offspring, or 1/4.
b. The chance of getting four puppies all having solid coats is the chance of getting an *S–* offspring the first time, *and* the second time, *and* the third time *and* the fourth time. Apply the product rule to get p = 3/4 x 3/4 x 3/4 x 3/4 = 81/256.

2.22 In the F_2 of his cross of red-flowered x white-flowered *Pisum*, Mendel obtained 705 plants with red flowers and 224 with white.

a. Is this result consistent with his hypothesis of factor segregation, from which a 3:1 ratio would be predicted?
b. In how many similar experiments would a deviation as great as or greater than this one be expected? (Calculate χ^2 and obtain the approximate value of P from the table.)

Answer: a. It is important to *think through* this problem, and not just plug numbers into a chi-square formula. Here, we want to test the hypothesis of factor segregation. The hypothesis states that each allele in a monohybrid will segregate into gametes independently, so that zygotes have a half-chance of obtaining either allele from a heterozygous parent. Drawing a Punnett square shows that a 3:1 ratio of progeny phenotypes would be expected from a F_1 x F_1 cross. Mendel observed 705 red and 224 white plants in the F_2, a ratio of 3.14:1. Thus, this ratio seems consistent with his hypothesis.

b. To test *how* significant this result is, use the χ^2 test to determine how frequently these types of numbers would be obtained in similar experiments.

Class	Observed	Expected	d	d^2	d^2/e
red	705	697	8	64	0.09
white	224	232	-8	64	0.27
Total	929	929	0	–	χ^2 =0.36

$\chi^2 = 0.36$; df = 1; $P = 0.60$

According to the χ^2 test, then, Mendel's result is consistent with this hypothesis. More specifically, in approximately 60 percent of similar experiments, one would expect a deviation (i.e., a value of χ^2) as great as or greater than this one. One therefore does not reject the hypothesis.

2.23 In tomatoes, cut leaf and potato leaf are alternative characters with cut (*C*) dominant to potato (*c*). Purple stem and green stem are another pair of alternative characters with purple (*P*) dominant to green (*p*). A true-breeding cut, green tomato plant is crossed with a true-breeding potato, purple plant and the F_1 plants are allowed to interbreed. The 320 F_2 plants were phenotypically 189 cut, purple; 67 cut, green; 50 potato, purple; and 14 potato, green. Propose an hypothesis to explain the data, and use the χ^2 test to test the hypothesis.

Answer: First, use the symbols defined in the problem to write the genotypes of the parents and expected offspring if the two loci assort independently. One has:

P: $CC\,pp \times cc\,PP$

$F_1 \times F_1$: $Cc\,Pp \times Cc\,Pp$

F_2: $9\ C\text{–}\ P\text{–}:3\ C\text{–}\ pp:3\ cc\ P\text{–}:1\ cc\ pp$
(9 cut, purple:3 cut, green:3 potato, purple:1 potato, green)

If the two loci assort independently then, one expects to see a 9:3:3:1 phenotypic ratio in the F_2. Therefore, test the hypothesis of independent assortment using the χ^2 test.

Class	Observed	Expected	d	d^2	d^2/e
cut, purple	189	180	9	81	0.45
cut, green	67	60	7	49	0.81
potato, purple	50	60	-10	100	1.66
potato, green	14	20	-6	36	1.80
Total	320	320	0	–	$\chi^2 = 4.72$

$\chi^2 = 4.72$; df = 3; $0.10 < P < 0.20$

Therefore, in experiments similar to this one, a deviation at least as great as that observed here would be seen about 20 percent of the time. Thus, the hypothesis of two independently assorting genes is accepted as being possible.

2.24 The simple case of just two mating types (male and female) is by no means the only sexual system known. The ciliated protozoan *Paramecium bursaria* has a system of four mating types, controlled by two genes (*A* and *B*). Each gene has a dominant and a recessive allele.

The four mating types are expressed under the following scheme:

GENOTYPE	MATING TYPE
AA BB	*A*
Aa BB	*A*
AA Bb	*A*
Aa Bb	*A*
AA bb	*D*
Aa bb	*D*
aa BB	*B*
aa Bb	*B*
aa bb	*C*

It is clear, therefore, that some of the mating types result from more than one possible genotype. We have four strains of known mating type —"*A*," "*B*," "*C*," and "*D*"— but unknown genotype. The following crosses were made with the indicated results:

	MATING TYPE OF PROGENY			
Cross	*A*	*B*	*C*	*D*
"*A*" x "*B*"	24	21	14	18
"*A*" x "*C*"	56	76	55	41
"*A*" x "*D*"	44	11	19	33
"*B*" x "*C*"	0	40	38	0
"*B*" x "*D*"	6	8	14	10
"*C*" x "*D*"	0	0	45	45

Assign genotypes to "*A*," "*B*," "*C*," and "*D*."

Answer: First write down what you know about the mating types.

$A = A\text{–}\ B\text{–}$
$B = aa\ B\text{–}$
$C = aa\ bb$
$D = A\text{–}\ bb.$

Now, rewrite the table in terms of what is known.

CROSS	MATING TYPE OF PROGENY A *A– B–*	B *aa B–*	C *aa bb*	D *A– bb*
"*A*" x "*B*" *A– B–* x *aa B–*	24	21	14	18
"*A*" x "*C*" *A– B–* x *aa bb*	56	76	55	41
"*A*" x "*D*" *A– B–* x *A– bb*	44	11	19	33
"*B*" x "*C*" *aa B–* x *aa bb*	0	40	38	0
"*B*" x "*D*" *aa B–* x *A– bb*	6	8	14	10
"*C*" x "*D*" *aa bb* x *A– bb*	0	0	45	45

Notice that a cross involving mating type *C* is similar to a test cross, as *C* is *aa bb*. The progeny ratios in crosses involving "*C*" will therefore reflect the gametes from the non-"*C*" parent. In the cross of "*B*" x "C", both the parents are *aa*, and the 1 B:1 C ratio in the progeny reflects the fact that "*B*" is *Bb*. Therefore, "B" is *aa Bb*. In the cross of "*C*" and "*D*," both parents are *bb*, and the 1 C:1 D progeny ratio indicates that "*D*" is *Aa*. "*D*" is therefore *Aa bb*.

To determine the genotype of "*A*," consider that the cross of "*A*" x "*C*" produces progeny of all four mating types. This would be expected only if "*A*" were *Aa Bb*. However, the ratios that are observed do not quite appear to be the 1:1:1:1 ratios that would be expected if the cross were *Aa Bb* x *aa bb*. To test whether these ratios are significantly different from those expected by chance, use the χ^2 test. The hypothesis here is that strain "*A* " is *Aa Bb*. The expectation is that there will be four equally frequent classes of progeny when strain "*A*" is crossed with strain "*C*."

Class	Observed	Expected	d	d^2	d^2/e
"*A*"	56	57	-1	1	0.02
"*B*"	76	57	19	361	6.33
"*C*"	55	57	-2	4	0.07
"*D*"	41	57	-16	256	4.49
Total	228	228	0	–	$\chi^2 = 10.91$

$\chi^2 = 10.91$; df = 3; $0.05 < P < 0.01$.

This value of *P* indicates that from 1 to 5 times out of 100 (1 to 5 percent of the time) one would expect these values by chance if the hypothesis were true. This is not very frequent, one would therefore consider the hypothesis unlikely to be true. However, what other explanation might there for the data? In a cross of *A– B–* x *aa bb*, one can only obtain both *A (A– B–)* and *C (aa bb)* progeny if the *A– B–* parent is *Aa Bb*. The deviation seen here may therefore reflect chance, or some other factor that affects the frequency of recovery of one or more of the mating types. It also illustrates that χ^2 tests do not *prove* or disprove a hypothesis.

One means by which to further check whether the assigned genotypes "fit" is to try to fit the genotype assignments to the remaining crosses. In cross "*A*" x "*B*", one would have *Aa Bb* x *aa Bb* and expect 3/8 *Aa B–* ("*A*"), 3/8 *aa B–* ("*B*"), 1/8 *Aa bb* ("*D*") and 1/8 *aa bb* ("*C*"). In the cross "*A*" x "*D*", one would have *Aa Bb* x *Aa bb* and expect 3/8 *A– Bb* ("*A*"), 3/8 *A– bb* ("*D*"), 1/8 *aa Bb* ("*B*") and 1/8 *aa bb* ("*C*"). In both cases, the numbers do not fit exactly, but they do show the trends expected. (Try the χ^2 test on these data to check how well they fit with expectations.)

2.25 In bees, males (drones) develop from unfertilized eggs and are haploid. Females (workers and queens) are diploid and come from fertilized eggs. *W* (black eyes) is dominant over *w* (white eyes). Workers of genotype *RR* or *Rr* use wax to seal crevices in the hive; *rr* workers use resin instead. A *Ww Rr* queen founds a colony after being fertilized by a black-eyed drone bearing the *r* allele.

a. What will be the appearance and behavior of workers in the new hive, with their relative frequencies?
b. Give the genotypes of male offspring, with relative frequencies.
c. Fertilization normally takes place in the air during a "nuptial flight," and any bee unable to fly would effectively be rendered sterile. Suppose a recessive mutation, *c*, occurs spontaneously in a sperm that fertilizes a normal egg, and that the effect of the mutant gene is to cripple the wings of any adult not bearing the normal allele *C*. The fertilized egg develops into a normal queen named Madonna. What is the probability that wingless males will be *found in a hive* founded two generations later by one of Madonna's granddaughters?
d. By one of Madonna's great-great-granddaughters?

Answer: The initial cross is *Ww Rr* x *W r*. Note that since males are haploid, they have only one set of chromosomes. As a consequence, their gametes are only of one kind (e.g., the *W r* male will have only *W r* gametes) and their phenotype can be used to directly infer their genotype.

a. The progeny females will be 1/2 *W– Rr* (black-eyed, wax sealers) and 1/2 *W– rr* (black-eyed, resin sealers).
b. As males arise from unfertilized eggs, they receive chromosomes only from their mother. The progeny males will be 1/4 *W R* (black-eyed, wax sealers), 1/4 *W r* (black-eyed, resin sealers), 1/4 *w r* (white-eyed, resin sealers) and 1/4 *w R* (white-eyed, wax sealers).
c. The egg fertilized by the mutation-bearing sperm results in a *Cc* female. Since fertilization occurs in flight, only males that are *C* contribute genes to the next generation. Hence the first generation arises from the cross *Cc* x *C*. There is a 1/2 chance of obtaining daughters that are *Cc*. Such a daughter can only be fertilized by a *C* male, so that the chance of her having a *Cc* daughter is also 1/2. The chance of having a *Cc* granddaughter is thus 1/2 x 1/2 = 1/4. Such granddaughters will have wingless males if they are prolific, so the probability that wingless males will be found in a hive founded two generations later by one of Madonna's granddaughters is 1/4.
d. $(1/2)^4 = 1/16$.

2.26 For pedigrees A and B below, indicate whether the trait involved in each case could be (a) recessive or (b) dominant. Explain your answer.

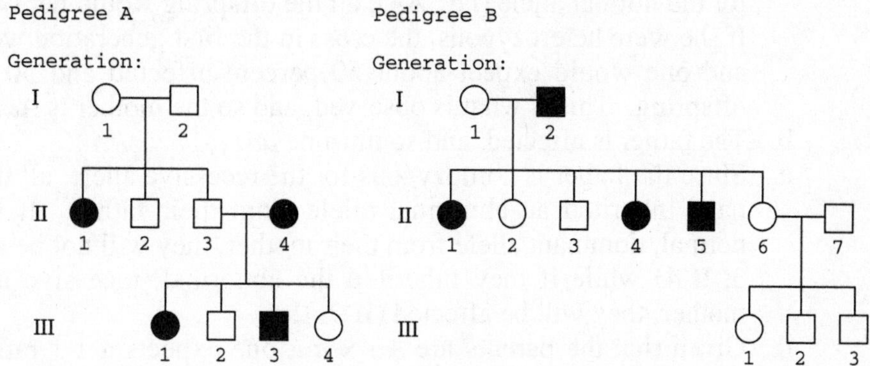

Answer: For pedigree A, parents not showing the trait in generation I have a daughter that exhibits the trait. This is typical of a cross of two heterozygotes giving a homozygous recessive, affected individual: *Aa* x *Aa* (two normal parents) give 3 *A–* (normal):1 *aa* (affected). If this were the case, individuals II-2 and II-3 could be carriers (i.e., be *Aa*). If individual II-3 is a carrier and his spouse is homozygous recessive (*aa*), generation III would arise from the cross *Aa* x *aa*, and the 1:1 ratio of affected to unaffected individuals seen would be expected. Thus, pedigree A is consistent with the trait being recessive.

In pedigree B, the trait is present in half of the individuals in generations I and II, but in none of those of generation III. However, the parents of generation III do not themselves exhibit the trait. This is consistent with the trait being dominant. With this view, individual I-1 would be *aa* and individual I-2 would be *Aa*. Half of their progeny would be expected to show the trait, as is observed. Furthermore, unaffected parents should have unaffected offspring, as is seen in generation III. Notice however, that the pedigree is also consistent with a recessive trait that is not rare. In this case, individual I-1 would be heterozygous, and the cross in generation I would be *Aa* x *aa*, giving the 50 percent affected progeny seen in generation II. Although individual II-6 would have to be *Aa*, if her spouse was *AA*, all the children would be normal.

2.27 Consider the pedigree below. The allele responsible for the trait (*a*) is recessive to the normal allele (*A*).

a. What is the genotype of the mother?
b. What is the genotype of the father?
c. What are the genotypes of the children?
d. Given the mechanism of inheritance involved, does the ratio of children with the trait to children without the trait match what would be expected?

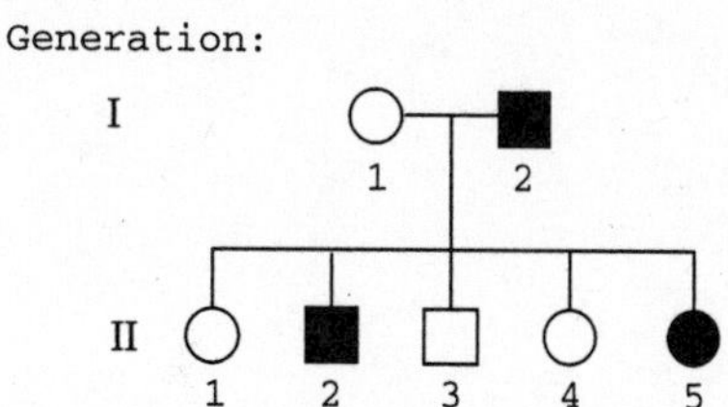

Answer:

a. If one knows that the allele responsible for the trait is recessive to the normal allele, then to have affected offspring in generation II, the mother (individual I-1) must be heterozygous (i.e., be *Aa*). If she were homozygous for the normal allele (i.e., *AA*), all the offspring would be *A*–, and be normal. If she were heterozygous, the cross in the first generation would be *Aa* x *aa*, and one would expect about 50 percent affected and 50 percent normal offspring. This is what is observed, and so the mother is *Aa*.

b. The father is affected, and so must be *aa*.

c. Since the father is homozygous for the recessive allele, all the children must have inherited an abnormal allele from their father. If they inherited a normal, dominant allele from their mother, they will not be affected (II-1, II-3, II-4) while if they inherited the abnormal, recessive allele from their mother, they will be affected (II-2, II-5).

d. Given that the parents are *Aa* x *aa*, one expects a 1:1 ratio of affected to unaffected children. For the five offspring shown, one sees a ratio of 2:3, which, given the number of offspring, approximates a 1:1 ratio. Remember that each birth is independent of the others, and while, if many progeny are seen, one should see about a 1:1 ratio, for small numbers of progeny, the ratios may be significantly off from what is expected. Indeed, even if the couple were to have more unaffected children, the pedigree would still be consistent with the inheritance of a recessive trait. (Can you show that if the pedigree is taken on its own, *without any additional information about the trait from the question*, the pedigree is also consistent with the inheritance of a dominant trait?)

CHAPTER 3

CHROMOSOMAL BASIS OF INHERITANCE, SEX DETERMINATION, AND SEX LINKAGE

I. CHAPTER OUTLINE

MITOSIS AND MEIOSIS
- Chromosome Complement of Eukaryotes
- Asexual and Sexual Reproduction
- Mitosis
- Genetic Significance of Mitosis
- Meiosis
- Genetic Significance of Meiosis
- Locations of Meiosis in the Life Cycle

CHROMOSOME THEORY OF INHERITANCE
- Sex Chromosomes
- Sex Linkage
- Nondisjunction of X Chromosomes

SEX DETERMINATION
- Genotypic Sex Determination Systems
- Environmental Sex Determination Systems

ANALYSIS OF SEX-LINKED TRAITS IN HUMANS
- X-Linked Recessive Inheritance
- X-Linked Dominant Inheritance
- Y-Linked Inheritance

II. REVIEW OF KEY TERMS, SYMBOLS AND CONCEPTS

Without consulting the text and in your own words, write a brief definition of each term in the groups below. Then, either using a short phrase or a simple diagram, identify the relationship(s) between specific pairs of terms within a set. Finally, consult the text (and perhaps a friend who has also done the exercise) to check your answers.

1	2	3	4
mitosis meiosis M, S, G_1, G_2 karyokinesis cytokinesis	diploid haploid gamete zygote genome	homologous chromosome non-homologous chromosome sex-chromosome autosome diploid haploid	asexual reproduction vegetative reproduction sexual reproduction somatic cells germ cells

5	6	7	8
prophase metaphase anaphase telophase cytokinesis karyokinesis	centromere metacentric chromosome submetacentric chromosome acrocentric chromosome telocentric chromosome satellite	tubulin aster kinetochore microtubule spindle metaphase plate scaffold	disjunction nondisjunction anaphase segregation independent assortment

9	10	11	12
meiosis I prophase I anaphase I metaphase I telophase I pairing disjunction	synapsis crossing-over pairing synaptonemal complex recombinant chromosome bivalent chiasma tetrad	leptonema zygonema pachynema diplonema diakinesis pairing scaffold synapsis crossing-over terminalization	homolog chromatid bivalent dyad segregation

13	14	15	16
meiosis II prophase II anaphase II metaphase II telophase II mitosis	sperm spermatogenesis primary spermatocyte secondary spermatocyte spermatid mitosis meiosis	oogenesis primary oogonia secondary oogonia mitosis meiosis primary oocyte secondary oocyte first polar body second polar body ovum	gametophyte sporophyte megasporogenesis microsporogenesis alternation of generations meiosis mitosis haploid diploid

17	18	19	20
pistil ovary ovule megaspore synergid egg cell polar nuclei central cell antipodal cell embryo sac	stamen microspore pollen generative cell tube cell pollen tube micropyle double fertilization	chromosome theory sex chromosome X chromosome Y chromosome homogametic sex heterogametic sex sex-linkage X-linkage	wild type mutant allele Mendelian symbolism *Drosophila* symbolism

21	22	23	24
hemizygous homozygous criss-cross inheritance reciprocal cross X-linked recessive trait X-linked dominant trait Y-linked trait holandric trait	primary nondisjunction secondary nondisjunction aneuploidy karyotype Turner syndrome Klinefelter syndrome XYY syndrome	genotypic sex determination environmental sex determination Y chromosome sex determination X chromosome-autosome balance system testis-determining factor (*TDF*) gene *Tdy* gene Sex-determining region Y (*SRY*) transgene transgenic organism	hermaphroditic monoecious dioecious mating type sex-determination

25
dosage compensation Barr Body lyonization X-inactivation center *Xist*

III. THINKING ANALYTICALLY

The problems in this chapter are more complicated than the ones you encountered in Chapter 2, and so it is even more important to approach your solutions analytically and systematically. Always, after you have arrived at a solution, go back and test it against the data presented in the problem.

A solid understanding of chromosome behavior during meiosis and mitosis forms a foundation on which to understand many aspects of transmission genetics. It is especially important to understand how chromosome behavior during these cell division processes relates to the segregation and independent assortment of traits in genetic crosses.

IV. QUESTIONS FOR PRACTICE

A. Multiple Choice Questions

1. All of the following occur during prophase I of meiosis *except*
   a. chromosome condensation.
   b. pairing of homologs.
   c. chiasma formation.
   d. terminalization.
   e. segregation.

2. The chromosome theory of inheritance holds that
   a. chromosomes are inherited.
   b. the chromosomes contain the hereditary material.
   c. the genes are inherited.
   d. the chromosomes are DNA.

3. Proof that genes lie on chromosomes was obtained by
   a. correlating aneuploids resulting from non-disjunction with inheritance patterns.
   b. showing that some genes appear to be sex-linked.
   c. showing that males have an X and a Y, while females have two X's.
   d. showing that in *Drosophila*, males are the heterogametic sex.

4. An individual that produces both ova and sperm
   a. has the XY genotype.
   b. reproduces asexually.
   c. is said to be hermaphroditic.
   d. both b and c, above.

5. Which of the following is *not true*?
   a. Crossing-over begins in leptonema.
   b. Chiasma are visible manifestations of crossing-over.
   c. Chiasma are visible in pachynema.
   d. Chiasma are visible before terminalization.

6. The somatic cells of an individual having the karyotype 48, XXXY would have
   a. four Barr bodies.
   b. three Barr bodies.
   c. two Barr bodies.
   d. no Barr bodies.

7. Independent assortment of two genes occurs if
   a. they are located on the sex chromosomes.
   b. they consist of different alleles.
   c. they reside in homologous chromosomes.
   d. they are located on different chromosome pairs.

8. Which of the following statements is true about mitosis?
   a. The nucleolus disappears during metaphase.
   b. Nuclear membranes reform at the end of telophase.
   c. Centromeres are aligned on an equatorial plane at prophase.
   d. The nucleolus reforms at the end of anaphase.

9. Normally alleles are segregated into separate cells
   a. at the time of fertilization.
   b. during meiosis.
   c. by mitosis.
   d. by crossing-over.

10. Sister chromatids are
   a. synonymous with homologous chromosomes.
   b. present only in meiosis and not in mitosis.
   c. identical products of chromosome duplication held together by a replicated, but unseparated centromere.
   d. visible in interphase just after S phase.

Answers: 1e; 2b; 3a; 4c; 5c; 6c; 7d; 8b; 9b; 10c.

B. Thought Questions

1. When during meiosis does the chromosomal behavior that underlies each of Mendel's laws occur?

2. Explain why meiosis I is considered to be a reductional division, while meiosis II is considered to be an equational division.

3. What is meant by criss-cross inheritance, and what clinical significance does it have?

4. Develop a flowchart illustrating the decision-making path you would employ to ascertain whether a pedigree showed autosomal dominant, autosomal recessive, X-linked dominant, X-linked recessive or Y-linked inheritance.

5. Define the term lyonization, give an example of it and discuss its functional significance.

6. Why do you think that most flowering plants are monoecious, whereas in most animals, the sexes are separate?

7. If all mammals have evolved from a common ancestor, how do you account for the variety of chromosome numbers that they possess? By what mechanism or mechanisms do you think changes in chromosome number could have occurred? (Consider aneuploidy resulting from nondisjunction, and perhaps other accidental changes.)

8. Except for the loci involved with sex-determination in mammalian males, there is little clear evidence for genetic loci on the Y chromosome. Why might this be the case and what might this tell you about the Y chromosome in mammals?

9. What are some of the different mechanisms by which sex-type can be determined? Do you think it is significant that so many different mechanisms can be employed?

10. Even though X-inactivation occurs in XXY individuals, they do not have the same phenotype as XY males. Why might this be the case?

V. SOLUTIONS TO TEXT PROBLEMS

In 3.1 through 3.3, select the correct answer.

3.1 Interphase is a period corresponding to the cell cycle phases of
   a. mitosis.
   b. S.
   c. $G_1 + S + G_2$.
   d. $G_1 + S + G_2 + M$.

Answer: c.

3.2 Chromatids joined together by a centromere are called
   a. sister chromatids.
   b. homologs.
   c. alleles.
   d. bivalents (tetrads).

Answer: a.

3.3 Mitosis and meiosis always differ in regard to the presence of
   a. chromatids.
   b. homologs.
   c. bivalents.
   d. centromeres.
   e. spindles.

Answer: c.

3.4 State whether each of the following statements is true or false. Explain your choice.
   a. The chromosomes in a somatic cell of any organism are all morphologically alike.
   b. During mitosis the chromosomes divide and the resulting sister chromatids separate at anaphase, ending up in two nuclei, each of which has the same number of chromosomes as the parental cell.
   c. At zygonema, any chromosome may synapse with any other chromosome in the same cell.

Answer: a. False. Chromosomes within a cell can be different sizes, and can have their centromeres positioned differently. Hence, for many organisms, this statement will be false. However, within any diploid cell, a pair of homologous chromosomes will be morphologically alike.
   b. True.
   c. False. In meiosis I, only homologous chromosomes will synapse together.

3.5 For each mitotic event described below, write the name of the event in the blank provided in front of the description. Then put the events in the correct order (sequence); start by placing a 1 next to the description of interphase, and continue through 6 which should correspond to the last event in the sequence.

NAME OF EVENT		ORDER OF EVENT
	The cytoplasm divides and the cell contents are separated into two separate cells.	
	Chromosomes become aligned along the equatorial plane of the cell.	
	Chromosome replication occurs.	
	The migration of the daughter chromosomes to the two poles is complete.	
	Replicated chromosomes begin to condense and become visible under the microscope.	
	Sister chromatids begin to separate and migrate toward opposite poles of the cell.	

Answer: Try to visualize the events of mitosis dynamically as you do this problem.

NAME OF EVENT		ORDER OF EVENT
cytokinesis	The cytoplasm divides and the cell contents are separated into two separate cells	6
metaphase	Chromosomes become aligned along the equatorial plane of the cell	3
interphase	Chromosome replication occurs	1
telophase	The migration of the daughter chromosomes to the two poles is complete	5
prophase	Replicated chromosomes begin to condense and become visible under the microscope	2
anaphase	Sister chromatids begin to separate and migrate toward opposite poles of the cell	4

3.6 Decide whether the answer to these statements is *yes* or *no*. Then explain the reasons for your decision.
  a. Can meiosis occur in a haploid species?
  b. Can meiosis occur in a haploid individual?

Answer: a. Yes. Providing that a sexual mating system exists in the species, two haploid cells can fuse to produce a diploid cell that can then go through meiosis to produce haploid progeny. The fungi *Neurospora crassa* and *Saccharomyces cerevisiae* exemplify this type of life cycle.
  b. No. Meiosis only can occur in a diploid cell. A haploid individual cannot form a diploid cell, so meiosis cannot occur.

3.7 The general life cycle of a eukaryotic organism has the sequence
  a. 1N→meiosis→2N→fertilization→1N.
  b. 2N→meiosis→1N→fertilization→2N.
  c. 1N→mitosis→2N→fertilization→1N.
  d. 2N→mitosis→1N→fertilization→2N.

Answer: b. Only diploid (2N) cells can undergo meiosis, and haploid cells (1N) fuse at fertilization, regenerating the diploid state.

3.8 Which statement is true?
  a. Gametes are 2N; zygotes are 1N.
  b. Gametes and zygotes are 2N.
  c. The number of chromosomes can be the same in gamete cells and in somatic cells.
  d. The zygotic and the somatic chromosome numbers cannot be the same.
  e. Haploid organisms have haploid zygotes.

Answer: c. A haploid organism provides an example of how (c) is true.

3.9 All of the following happen in prophase I of meiosis *except*
  a. chromosome condensation.
  b. pairing of homologs.
  c. chiasma formation.
  d. terminalization.
  e. segregation.

Answer: e.

3.10 Give the name of the stages of mitosis or meiosis at which the following events occur:
  a. Chromosomes are located in a plane at the center of the spindle.
  b. The chromosomes move away from the spindle equator to the poles.

Answer: a. Metaphase in mitosis, metaphase I and metaphase II in meiosis.
  b. Anaphase in mitosis, anaphase I and anaphase II in meiosis.

3.11 Given the diploid, meiotic mother cell shown below, diagram the chromosomes as they would appear

a. in late pachynema;
b. in a nucleus at prophase of the second meiotic division;
c. in the first polar body resulting from oogenesis in an animal.

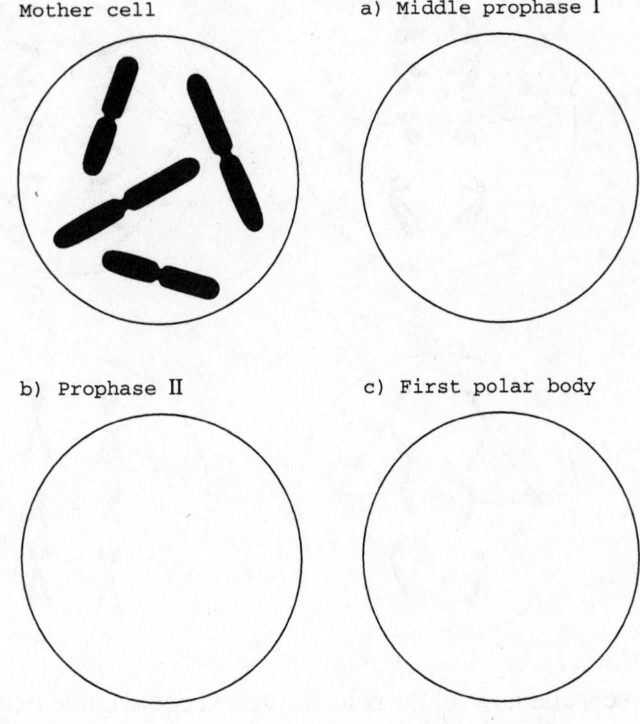

Answer:

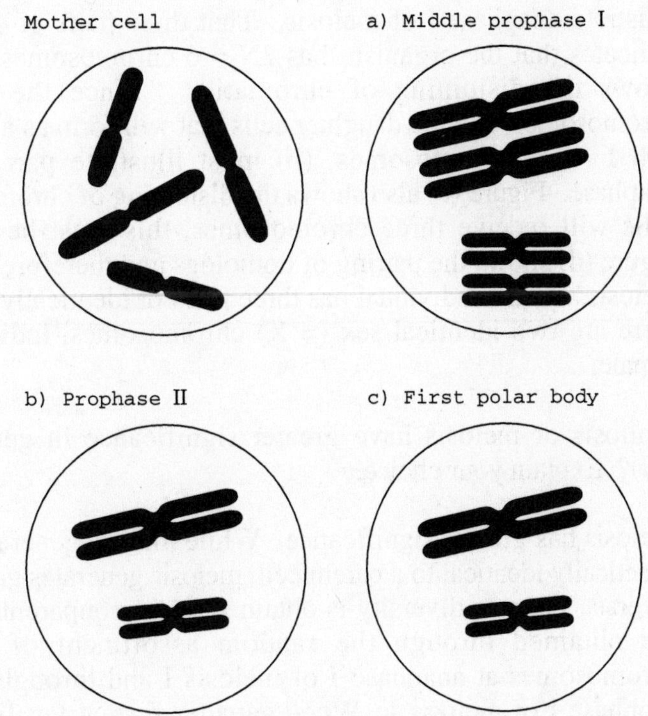

3.12 The cells in the figure below were all taken from the same individual (a mammal). Identify the cell division events happening in each cell, and explain your reasoning. What is the sex of the individual? What is the diploid chromosome number?

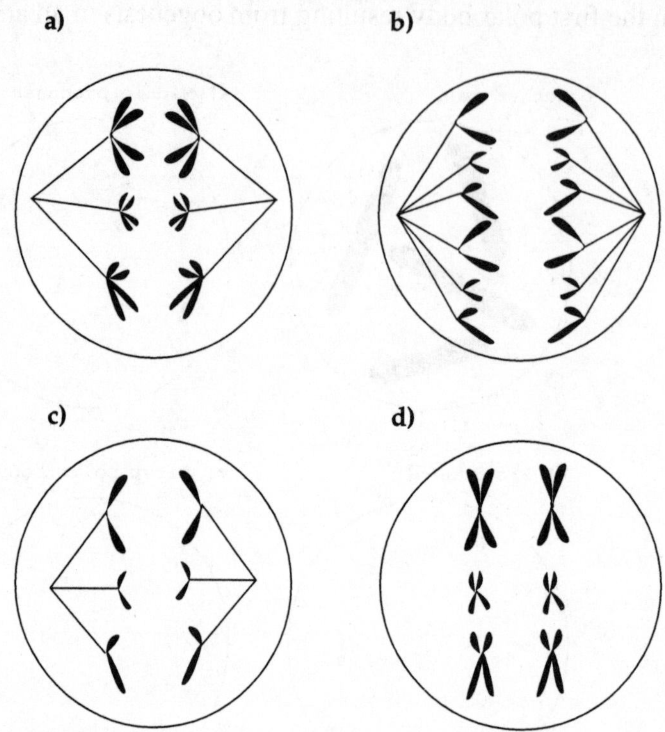

Answer: To reconcile how all the cells illustrated could come from one individual, consider that the cells shown could come from either somatic or germline cells. Cell (a) shows 3 pairs of previously synapsed homologs disjoining and must therefore illustrate anaphase I of meiosis. That three pairs of chromosomes are present indicates that the organism has 2N = 6 chromosomes, so that N = 3. Cell (b) shows the disjoining of chromatids. Since the organism has 2N = 6 chromosomes, and the daughter cells that will form as a result of this cell division will have 6 chromosomes, (b) must illustrate part of mitosis, specifically anaphase. Figure (c) also shows the disjoining of chromatids. Since the daughter cells will receive three chromosomes, this must be anaphase II of meiosis. Figure (d) shows the pairing of homologs and therefore illustrates metaphase I of meiosis. As the individual has three pairs of identically appearing chromosomes, there are two identical sex (= X) chromosomes, indicating that the animal is female.

3.13 Does mitosis or meiosis have greater significance in genetics (i.e., the study of heredity)? Explain your choice.

Answer: Meiosis has greater significance. While mitosis generates progeny cells that are genetically identical to a parent cell, meiosis generates gametes that are genetically diverse. Gamete diversity is obtained when nonparental combinations of genes are obtained through the random assortment of maternal and paternal chromosomes at anaphase I of meiosis I and through crossing-over events in Prophase I of meiosis I. When gametes fuse at fertilization, more diversity is obtained. Hence, meiosis provides a means for genetic variation.

3.14 Consider a diploid organism that has three pairs of chromosomes. Assume that the organism receives chromosomes A, B, and C from the female parent and A', B' and C' from the male parent. To answer the following questions, assume that no crossing-over occurs.

a. What proportion of gametes of this organism would be expected to contain all the chromosomes of maternal origin?
b. What proportion of the gametes would be expected to contain some chromosomes from both maternal and paternal origin?

Answer: a. The chance a gamete would have a particular maternal chromosome is 1/2. Therefore, the chance of obtaining a gamete with all three maternal chromosomes is $(1/2)^3 = 1/8$.
b. As shown in (a), the chance of getting a gamete with only maternal chromosomes is 1/8. This is also the chance of getting a gamete with only paternal chromosomes. All other gametes will have both maternal and paternal chromosomes. Hence, the chance of obtaining a gamete with both maternal and paternal chromosomes is $1 - (1/8 + 1/8) = 3/4$.

3.15 Normal diploid cells of a theoretical mammal are examined cytologically at the mitotic metaphase stage for their chromosome complement. One short chromosome, two medium-length chromosomes, and three long chromosomes are present. Explain how the cells might have such a set of chromosomes.

Answer: Since the cells are normal and the cell is diploid, the chromosomes should exist in pairs. The cell could have come from a male mammal with one medium-length pair, one long-length pair and a heteromorphic sex-chromosome (X, Y) pair.

3.16 Is the following statement true or false? Explain your decision. "Meiotic chromosomes may be seen after appropriate staining in nuclei from rapidly dividing skin cells."

Answer: False. Skin cells consist entirely of mitotic, not meiotic, cells.

3.17 Is the following statement true or false? Explain your decision. "All of the sperm from one human male are genetically identical."

Answer: False. Genetic diversity in the male's sperm is achieved during meiosis. There is crossing-over between non-sister chromatids as well as independent assortment of the male's maternal and paternal chromosomes during spermatogenesis.

3.18 The horse has a diploid set of 64 chromosomes, and a donkey has a diploid set of 62 chromosomes. Mules (viable but usually sterile progeny) are produced when a male donkey is mated to a female horse. How many chromosomes will a mule cell contain?

Answer: The horse will contribute 32 chromosomes and the donkey will contribute 31 chromosomes to the mule, giving it a total of 63 chromosomes.

3.19 The red fox has 17 pairs of large, long chromosomes. The arctic fox has 26 pairs of shorter, smaller chromosomes.

a. What do you expect to be the chromosome number in somatic tissues of a hybrid between these two foxes?
b. The first meiotic division in the hybrid fox shows a mixture of paired and single chromosomes. Why do you suppose this occurs? Can you suggest a possible relationship between this fact and the observed sterility of the hybrid?

Answer: a. The red fox will have gametes that have 17 chromosomes, and the arctic fox will have gametes that have 26 chromosomes. The hybrid will be formed with both gametes, and have 17 + 26 = 43 chromosomes. Once the fertilized egg is formed, it will divide by mitosis to produce daughter cells with 43 chromosomes.

b. Some of the chromosomes are able to pair, while others are not. This could occur because of the evolutionary similarity of some of the chromosomes, which would allow them to pair. Those that are unable to pair may lack similarity. It is possible that the sperm or eggs of the hybrid do not form properly, as they lack a proper number of chromosomes and are missing genes that are necessary for sperm or oocyte development. If gametes are formed, sterility might arise because those gametes that are formed typically lack a complete set of chromosomes from either parent, making it improbable that a fertilized embryo would have a complete set of chromosomes of either the hybrid, the red fox or the arctic fox. A lack of one or more chromosomes results in missing genes that are necessary for viability.

3.20 At the time of synapsis, preceding the reduction division in meiosis, the homologous chromosomes align in pairs and one member of each pair passes to each of the daughter nuclei. In an animal with five pairs of chromosomes, assume that chromosomes 1, 2, 3, 4, and 5 have come from the father, and 1', 2', 3', 4', and 5' have come from the mother. In what proportion of the germ cells of this animal will all the paternal chromosomes be present together?

Answer: The chance of getting a particular paternal chromosome is 1/2. The chance of getting all five paternal chromosomes is $(1/2)^5 = 1/32$.

3.21 In a male *Homo sapiens*, from which grandparent could each sex-chromosome have been derived? (Indicate yes or no for each option.)

	MOTHER'S		FATHER'S	
	Mother	Father	Mother	Father
X chromosome	____	____	____	____
Y chromosome	____	____	____	____

Answer: Try diagramming the inheritance of the sex-chromosomes after considering that a male normally receives his mother's X and his father's Y, while a female normally receives an X from each parent.

	MOTHER'S		FATHER'S	
	Mother	Father	Mother	Father
X chromosome	yes	yes	no	no
Y chromosome	no	no	no	yes

3.22 In *Drosophila*, white eyes are a sex-linked character. The mutant allele for white eyes (w) is recessive to the wild-type allele for brick-red eye color (w^+).

a. A white-eyed female is crossed with a red-eyed male. An F_1 female from this cross is mated with her father, and an F_1 male is mated with his mother. What will be the eye color of the offspring of these last two crosses?

b. A white-eyed female is crossed with a red-eyed male, and the F_2 from this cross is interbred. What will be the eye color of the F_3?

Answer: a. The initial cross is $ww \times w^+Y$, so that the F_1 females are ww^+ and the F_1 males are wY. The second set of crosses are therefore $w^+w \times w^+Y$ and $wY \times ww$. The former will give all brick-red females ($w^+–$) and half white (wY) and half brick-red (w^+Y) males. The latter will give only white eyed males and females (wY and ww).

b. The initial cross is $ww \times w^+Y$, so that the F_1 progeny are white-eyed males and brick-red-eyed females (wY and w^+w). If the F_1 is interbred, one obtains an F_2 that consists of 1/2 white-eyed (wY) and 1/2 brick-red-eyed (w^+Y) males, and 1/2 white-eyed (ww) and 1/2 brick-red-eyed (w^+w) females. If these are allowed to interbreed at random to give an F_3, one will obtain 5/16 brick-red-eyed females (1/16 will be w^+w while 1/4 will be w^+w^+), 3/16 white-eyed females (ww) and 3/8 white-eyed males (wY) and 1/8 brick-red-eyed males (w^+Y). Note that the ratio of males to females remains 1:1 in each generation.

3.23 One form of color blindness (c) in humans is caused by a sex-linked recessive mutant gene. A woman with normal color vision (c^+) and whose father was color-blind marries a man of normal vision whose father was also color-blind. What proportion of their offspring will be color-blind? (Give your answer separately for males and females.)

Answer: Since fathers always give their X-chromosome to their daughters, the woman must be heterozygous for the color-blindness trait, and be genotypically c^+c. The man she marries receives his X-chromosome from his mother, not his father. As he has normal color vision, he is c^+Y. The cross is therefore $c^+c \times c^+Y$. All the daughters of this couple will receive the paternal X bearing the c^+ allele and have normal color vision. As the sons will receive the maternal X, half will be colorblind and half will have normal color vision, as half will be cY and half will be c^+Y.

3.24 In humans, red-green color blindness is due to an X-linked recessive gene. A color-blind daughter is born to a woman with normal color vision and a father who is color-blind. What is the mother's genotype with respect to the alleles concerned?

Answer: Let c be the color-blind allele and c^+ be the normal allele. The color-blind daughter must be cc, and have received an X bearing the c allele from each parent.

Since her mother has normal color-vision, she must have one normal c^+ allele for this recessive trait. Her mother is therefore a carrier, and is a cc^+ heterozygote.

3.25 In humans, red-green color blindness is recessive and X-linked, while albinism is recessive and autosomal. What types of children can be produced as the result of marriages between two homozygous parents, a normal-visioned albino woman and a color-blind, normally pigmented man?

Answer: Let c and c^+ be the color-blind and normal vision alleles, respectively, and let a and a^+ be the albino and normal pigmentation alleles, respectively. Then the cross can be represented as $c^+c^+ \ aa \times c\text{Y} \ a^+a^+$. As all the offspring will be a^+a, all will have normal pigmentation. The offspring will either be c^+c or $c^+\text{Y}$, and have normal color vision. The daughters will however, be carriers for the color-blindness trait.

3.26 In *Drosophila*, vestigial (partially formed) wings (vg) are recessive to normal long wings (vg^+), and the gene for this trait is autosomal. The gene for the white eye trait is on the X chromosome. Suppose a homozygous white-eyed, long-winged female fly is crossed with a homozygous red-eyed, vestigial-winged male.

a. What will be the appearance of the F_1?
b. What will be the appearance of the F_2?
c. What will be the appearance of the offspring of a cross of the F_1 back to each parent?

Answer: a. The initial cross is $ww \ vg^+vg^+ \times w^+\text{Y} \ vgvg$. The F_1 consists of $w\text{Y} \ vg^+vg$ (white, normal-winged) males and $ww^+ \ vg^+vg$ (red, normal winged) females.

b. The F_2 would be produced by crossing $w\text{Y} \ vg^+vg$ males and $w^+w \ vg^+vg$ females. In both the male and the female progeny, 1/8 will be white and vestigial, 1/8 will be red and vestigial, 3/8 will be white and normal-winged, and 3/8 will be red and normal-winged.

c. If the F_1 males are crossed back to the female parent, one would have $w\text{Y} \ vg^+vg \times ww \ vg^+vg^+$. All the progeny would be white and normal-winged. If the F_1 females are crossed back to the male parent, one would have $ww^+ \ vg^+vg \times w^+\text{Y} \ vgvg$. Among the male progeny, there would be 1/4 white, vestigial; 1/4 red, vestigial; 1/4 white, normal-winged and 1/4 red, normal-winged. Among the female progeny, half would be red and normal winged and half would be red and vestigial.

3.27 In *Drosophila*, two red-eyed, long-winged flies are bred together and produce the offspring given in the following table:

	FEMALES	MALES
red-eyed, long-winged	3/4	3/8
red-eyed, vestigial-winged	1/4	1/8
white-eyed, long-winged	–	3/8
white-eyed, vestigial winged	–	1/8

What are the genotypes of the parents?

Answer: If you have done problem 3.24, you already know that *w* is X-linked while *vg* is autosomal. Even if you haven't done problem 3.24, you can determine this just by examining the frequencies of progeny phenotypes associated with each trait. Notice that the ratio of long-winged to vestigial-winged progeny is 3:1 (3/4 to 1/4) in both sexes, while the ratio of red-eyed to white-eyed progeny is all to none in females and 1:1 in males. This is consistent with *vg* being autosomal and *w* being X-linked. The 3:1 ratio of long-winged to vestigial-winged progeny indicates that each parent was heterozygous at the *vg* locus. Since both parents had red-eyes, both had (at least) one w^+ allele. Since half of the sons are white-eyed, the mother must have been heterozygous. Therefore, the parents were $w^+w\ vg^+vg$ x w^+Y vg^+vg .

3.28 In poultry a dominant sex-lined gene (*B*) produces barred feathers, and the recessive allele (*b*), when homozygous, produces nonbarred feathers. Suppose a nonbarred cock is crossed with a barred hen.

a. What will be the appearance of the F_1 birds?
b. If an F_1 female is mated with her father, what will be the appearance of the offspring?
c. If an F_1 male is mated with his mother, what will be the appearance of the offspring?

Answer:
a. In poultry, sex-type is determined by a ZZ (male) and ZW (female) system. The cross can be depicted as *bb* (nonbarred cock) x *B*W (barred hen). The F_1 progeny will be *b*W (nonbarred) hens and *Bb* (barred) cocks.
b. The cross can be represented as *b*W x *bb*. All the progeny will be nonbarred.
c. The cross can be represented as *Bb* x *B*W. The progeny will be 1/2 barred cocks (1/4 *BB*, 1/4 *Bb*), 1/4 barred hens (*B*W) and 1/4 nonbarred hens (*b*W).

3.29 In chickens, barred plumage (*B*) is dominant over non-barred (solid-color) (*b*); the locus for this plumage phenotype is located on the sex chromosomes. (In birds, the female is the heterogametic sex.) The phenotypes can be distinguished in newly-hatched chicks. Commercial chicken breeders in England have used this difference to separate male and female chicks, otherwise a difficult task. In order to accomplish this, what must be the genotype of: (a) the female parent; and (b) the male parent?

Answer: Consider the two options:

(1) male barred x female non-barred (2) male non-barred x female barred

Z^BZ^B x Z^bW — Z^bZ^b x Z^BW

Z^BW (barred females) — Z^bW (non-barred females)

Z^BZ^b (barred males) — Z^BZ^b (barred males)

There will be criss-cross inheritance if the second option is taken. The dominant barred trait must be in the female parent and the recessive non-barred trait must be in the male parent. The dominant barred trait will appear in the male progeny, and the recessive non-barred trait will appear in the female progeny.

3.30 A man (A) suffering from defective tooth enamel, which results in brown-colored teeth, marries a normal woman. All their daughters have brown teeth, but the sons are normal. The sons of man A marry normal women, and all their children are normal. The daughters of man A marry normal men, and 50 percent of their children have brown teeth. Explain these facts.

Answer: Notice that the trait is transmitted from the father to his daughters, indicating criss-cross inheritance. This is typical of an X-linked trait. Since man A marries a normal woman and all of their daughters have the trait, the trait must be dominant. Man A's X chromosome bearing the defective tooth enamel allele was inherited by all of his daughters, and none of his sons. All of his daughters would therefore have defective tooth enamel, and be heterozygous for the dominant, defective enamel gene. These daughters would transmit the dominant, defective enamel gene 50 percent of the time, giving rise to 50 percent of their children being affected.

3.31 In humans, differences in the ability to taste phenylthiourea are due to a pair of autosomal alleles. Inability to taste is recessive to ability to taste. A child who is a nontaster is born to a couple who can both taste the substance. What is the probability that their next child will be a taster?

Answer: Since the inability to taste the substance is recessive, the nontaster child must be homozygous for the recessive allele and each of his parents must have given the child a recessive allele. Since both parents can taste, they must also bear a dominant allele. Let T represent the dominant (taster) allele, and t represent the recessive (non-taster) allele (Note: Mendelian notation is used here for convenience, but also because there is no value in assigning a normal (+) and abnormal allele). Then the cross can be written as Tt x Tt. The chance that their next child will be a taster is the chance that the child will be TT or Tt, or 3/4.

3.32 Cystic fibrosis is inherited as an autosomal recessive. Two noncystic fibrosis parents have two children with cystic fibrosis and three children who do not have cystic fibrosis. They come to you for genetic counseling.

a. What is the numerical probability that their next child will have cystic fibrosis?
b. Their non-affected children are concerned about being heterozygous. What is the numerical probability that a given non-affected child in the family is heterozygous?

Answer: a. Since the disease is autosomal recessive, and unaffected parents have affected offspring, both parents must be heterozygous. Let c^+ represent the normal allele and c represent the disease allele. Then the parental cross can be represented as $c^+c \times c^+c$, and there is a 1/4 chance that any conception will produce a cc (affected) child.

b. If a child is not affected, the child is either c^+c^+ (p = 1/3) or c^+c (p = 2/3). Thus, there is a 2/3 chance that a non-affected child is heterozygous.

3.33 Huntington's disease is a human disease inherited as a Mendelian autosomal dominant. The disease results in choreic (uncontrolled) movements, progressive mental deterioration, and eventually death. The disease affects the carriers of the trait anytime between 15 and 65 years of age. The American folk singer Woody Guthrie died of Huntington's disease, as did one of his parents. Marjorie Mazia, Woody's wife, had no history of this disease in her family. The Guthries have three children. What is the probability that a particular Guthrie child will die of Huntington's disease?

Answer: Let H represent the disease allele, and let h represent the normal allele. Since we are told that Woody's father died of the disease and are not told anything about his mother, we may assume that his mother did not have the disease allele. Thus, Woody must have been Hh. His children are the progeny of a cross that can be represented as $Hh \times hh$. Each will have a 50 percent chance of receiving the H allele.

3.34 Suppose gene A is on the X chromosome, and genes B, C, and D are on three different autosomes. Thus A– signifies the dominant phenotype in the male or female. An equivalent situation holds for B–, C–, and D–. The cross $AA\ BB\ CC\ DD$ females x aY $bb\ cc\ dd$ males is made.

a. What is the probability of obtaining an A– individual in the F_1?
b. What is the probability of obtaining an a male in the F_1?
c. What is the probability of obtaining an A– B– C– D– female in the F_1?
d. How many different F_1 genotypes will there be?
e. What proportion of F_2s will be heterozygous for the four genes?
f. Determine the probabilities of obtaining each of the following types in the F_2: (1) A– $bb\ CC\ dd$ (female); (2) aY $BB\ Cc\ Dd$ (male); (3) AY $bb\ CC\ dd$ (male); (4) $aa\ bb\ Cc\ Dd$ (female).

Answer: a. Since one is only concerned with a single trait, one can consider just part of the cross: $AA \times a$Y. The progeny will either be AY or Aa, and all be A–. Thus, the chance of obtaining an A– individual in the F_1 is 1.

b. As shown in (a), there is no chance (p = 0) of obtaining an aY individual in the F_1.

c. The F_1 progeny will be A– $Bb\ Cc\ Dd$. Half will be female, so p = 1/2.

d. Two, $Aa\ Bb\ Cc\ Dd$ (females) and AY $Bb\ Cc\ Dd$ males.

e. For the X chromosome trait, the F_1 cross is AY x Aa, and half of the female offspring (1/4 or the total) will be heterozygous Aa individuals. For each of the autosomal traits, 1/2 of the offspring will be heterozygous (e.g., $Bb \times Bb$ gives 1/2 Bb individuals). Thus, the chance that an F_2 individual will be heterozygous at all four traits is 1/4 x 1/2 x 1/2 x 1/2 = 1/32.

f. Before determining the probabilities, consider that at any autosomal gene, one will have a 1/4 chance of obtaining either type of homozygote (e.g., BB, bb) and a 1/2 chance of obtaining a heterozygote. At the A gene, the cross is AY x

Aa, and so there will be a 1/4 chance of obtaining an *A*Y male, a 1/4 chance of obtaining an *a*Y male, a 1/4 chance of obtaining an *Aa* female and a 1/4 chance of obtaining an *AA* female (there will be a 1/2 chance of obtaining an *A*– female).

Then the chance of obtaining (1) a *A– bb CC dd* (female) is *p* = (1/2 x 1/4 x 1/4 x 1/4) = 1/128; (2) a *a*Y *BB Cc Dd* (male) is *p* = (1/4 x 1/4 x 1/2 x 1/2) = 1/64; (3) a *A*Y *bb CC dd* (male) is *p* = (1/4 x 1/4 x 1/4 x 1/4) = 1/256; (4) an *aa bb Cc Dd* (female) is *p* = (0 x 1/4 x 1/2 x 1/2) = 0.

3.35 As a famous mad scientist, you have cleverly devised a method to isolate *Drosophila* ova that have undergone primary nondisjunction of the sex chromosomes. In one experiment you used females homozygous for the sex-linked recessive mutation causing white eyes (*w*) as your source of nondisjunction ova. The ova were collected and fertilized with sperm from red-eyed males. The progeny of this "engineered" cross were then backcrossed separately to the two parental strains (this is called "backcrossing"). What classes of progeny (genotype and phenotype) would you expect to result from these backcrosses? (The genotype of the original parents may be denoted as *ww* for the females and w^+Y for the males.)

Answer: First consider what kinds of animals will be produced in the "engineered" cross. Primary nondisjunction of sex chromosomes in the meiosis of a female will result in eggs that either have no X chromosomes or two X chromosomes. Thus, the eggs will either lack the *w* allele or be *ww*. Sperm from a red-eyed male will either bear a Y chromosome or bear the w^+ allele. Upon fertilization then, four classes of embryos will be obtained: Y-only, w^+O, *ww*Y and w^+ww. Animals with either no X chromosome or three X chromosomes are inviable, and so only w^+ (red, XO males) and *ww*Y (white, XXY females) animals will be obtained.

Now consider the backcrosses. The "engineered" red males will be sterile, as they lack the Y chromosome that is necessary for male fertility. Hence, the only backcross that will give progeny is the mating between the *ww*Y females and the w^+Y males. The sex chromosome constitution of the female's gametes is dependent on the pairing and subsequent disjoining of the two X chromosomes and the Y chromosome during meiosis. If there is normal pairing and disjunction of the two X chromosomes and the Y chromosome assorts independently of them, XY (*w*Y) and X (*w*) eggs will be produced. If, as described in the text, less frequent secondary nondisjunction occurs, XX-bearing (*ww*) and Y-bearing eggs will be produced. To visualize the results of these eggs being fertilized by sperm from a w^+Y male, examine the following Punnett square:

		Gametes of *ww*Y Female			
		Normal Disjunction		Secondary Nondisjunction	
		w (XY)	*w* (X)	*ww* (XX)	Y
Gametes of w^+Y Male	w^+ (X)	w^+wY (XXY) red female	w^+w (XX) red female	w^+ww (XXX) inviable	w^+Y (XY) red male
	Y	*w*YY (XYY) white male	*w*Y (XY) white male	*ww*Y (XXY) white female	(YY) inviable

3.36 In *Drosophila* the bobbed gene (bb^+) is located on the X chromosome: *bb* mutants have shorter, thicker bristles than wild-type flies. Unlike most X-linked genes, however, the Y chromosome also carries a bobbed gene. The mutant allele *bb* is recessive to bb^+. If a wild-type F_1 female that resulted from primary nondisjunction in oogenesis in a cross of a bobbed female with a wild-type male is mated to a bobbed male, what will be the phenotypes and their frequencies in the offspring? List males and females separately in your answer. (Hint: Refer to the chapter for information about the frequency of nondisjunction in *Drosophila*.)

Answer: The cross that gave rise to the wild-type F_1 female can be denoted as $X^{bb}X^{bb}$ x $X^{bb^+}Y^{bb^+}$. Nondisjunction in meiosis of the $X^{bb}X^{bb}$ female would produce an $X^{bb}X^{bb}$ egg. Fertilization by a Y^{bb^+}-bearing sperm would produce a bb^+ ($X^{bb}X^{bb}Y^{bb^+}$) female. We are concerned with the progeny that result when this female is mated to a *bb* (X^{bb} Y^{bb}) male. Normal disjunction (which occurs about 96 percent of the time) in this female will produce $X^{bb}Y^{bb^+}$ and X^{bb} bearing eggs. If these eggs are fertilized by sperm from a *bb* male (i.e., either X^{bb} or Y^{bb}), half bb^+ ($X^{bb}X^{bb}Y^{bb^+}$ and $X^{bb}Y^{bb^+}Y^{bb}$) males and females and half *bb* ($X^{bb}X^{bb}$ and $X^{bb}Y^{bb}$) males and females will be produced. If secondary nondisjunction occurs (about 4 percent of the time), then $X^{bb}X^{bb}$ and Y^{bb^+} eggs will be produced. When these eggs are fertilized by sperm from a *bb* male, the only viable progeny will be *bb* females ($X^{bb}X^{bb}Y^{bb}$) and bb^+ males ($X^{bb}Y^{bb^+}$). This is diagrammed below:

		Gametes of $X^{bb}X^{bb}Y^{bb^+}$ Female			
		Normal Disjunction–92%		Secondary Nondisjunction–8%	
		46% $X^{bb}Y^{bb+}$	46% X^{bb}	4% $X^{bb}X^{bb}$	4% Y^{bb^+}
Gametes of X^{bb} Y^{bb} Male	50% X^{bb}	$X^{bb}X^{bb}Y^{bb^+}$ *bb*+ female 23%	$X^{bb}X^{bb}$ *bb* female 23%	$X^{bb}X^{bb}X^{bb}$ inviable 2%	$X^{bb}Y^{bb^+}$ *bb*+ male 2%
	50% Y^{bb}	$X^{bb}Y^{bb^+}Y^{bb}$ *bb*+ male 23%	$X^{bb}Y^{bb}$ *bb* male 23%	$X^{bb}X^{bb}Y^{bb}$ *bb* female 2%	YY inviable 2%

There will be 23% normal females, 25% bobbed females, 25% normal males and 23% bobbed males.

3.37 A Turner syndrome individual would be expected to have the following number of Barr bodies in the majority of cells:

a. 0
b. 1
c. 2
d. 3

Answer: a.

3.38 An XXY Klinefelter syndrome individual would be expected to have the following number of Barr bodies in the majority of cells:

a. 0
b. 1
c. 2
d. 3

Answer: b.

3.39 In human genetics the pedigree is used for analysis of inheritance patterns. The female is represented by a circle and the male by a square. The following figure presents three, two generation family pedigrees for a trait in humans. Normal individuals are represented by unshaded symbols and people with the trait by shaded symbols. For each pedigree (A, B, and C), state, by answering yes or no in the appropriate blank space, whether transmission of the trait can be accounted for on the basis of each of the listed simple modes of inheritance:

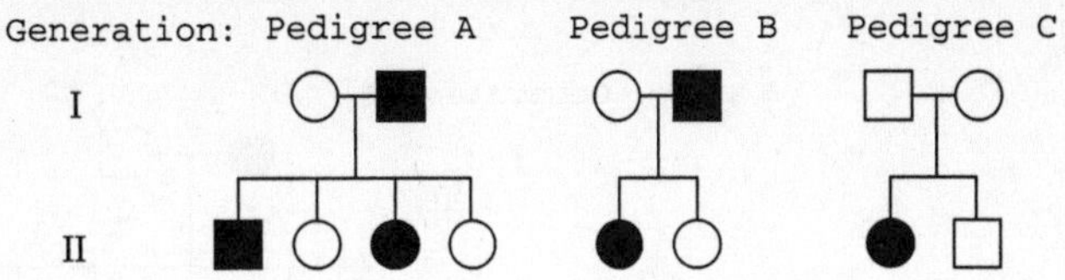

Generation:	Pedigree A	Pedigree B	Pedigree C
I			
II			

	Pedigree A	Pedigree B	Pedigree C
Autosomal recessive			
Autosomal dominant			
X-linked recessive			
X-linked dominant			

Answer: This problem elegantly brings up the issue that the precise mode of inheritance of a trait cannot often be determined when there are small numbers of individuals in a pedigree, and if little information is given concerning the frequency of the trait in the population being examined. For example, Pedigree A could easily fit a autosomal dominant trait (*AA* and *Aa* individuals are affected): the affected father would be heterozygous for the trait (*Aa*), the mother unaffected (*aa*) and one would expect half of their offspring to be affected. However, if the trait were autosomal recessive (i.e., *aa* were the affected individuals), the mother were heterozygous (*Aa*) and the father homozygous (*aa*), one would similarly observe half affected offspring. One could also fit the pedigree to an X-linked recessive trait: the mother would be heterozygous (X^AX^a), the father hemizygous (X^aY) and half of the progeny affected (either X^aX^a or X^aY). An X-linked dominant trait would not fit the pedigree, as it would require all the daughters of the affected father to be affected (as they all receive their father's X), and not allow the son to be affected (as he does not receive his father's X). Pedigrees B and C can be solved by similar analytical reasoning.

	Pedigree A	Pedigree B	Pedigree C
Autosomal recessive	Yes	Yes	Yes
Autosomal dominant	Yes	Yes	No
X-linked recessive	Yes	Yes	No
X-linked dominant	No	No	No

3.40 Looking at the following pedigree, in which shaded symbols represent a "trait," which of the progeny (as designated by numbers) eliminate X-linked recessiveness as a mode of inheritance for the trait?

a. I.1 and I.2
b. II.4
c. II.5
d. II.2 and II.4

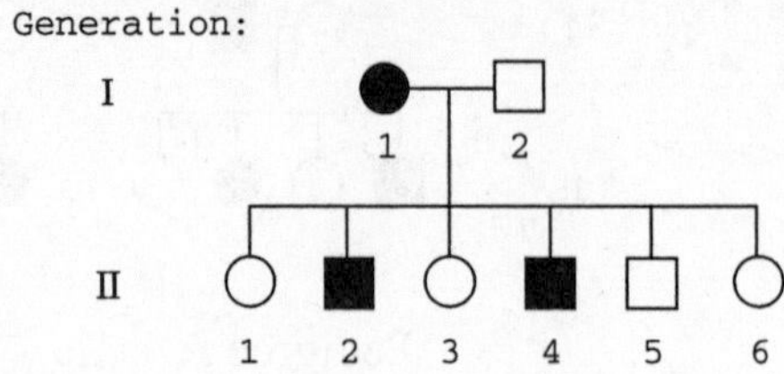

Answer: If the trait were X-linked recessive, then II.5 should be affected, as he would receive his mother's X chromosome. Therefore the correct answer is (c), II.5.

3.41 When constructing human pedigrees, geneticists often refer to particular persons by a number. The generations are labeled by roman numerals and the individuals in each generation by Arabic numerals. For example, in the pedigree in the figure below, the female with the asterisk would be I.2. Use this means to designate specific individuals in the pedigree. Determine the probable inheritance mode for the trait shown in the affected individuals (the shaded symbols) by answering the following questions. Assume the condition is caused by a single gene.

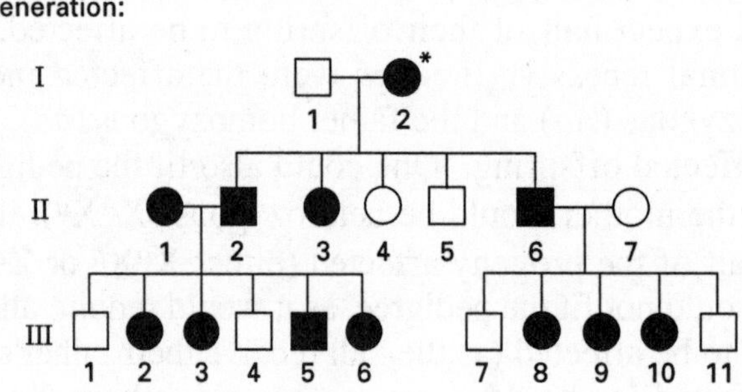

a. Y-linked inheritance can excluded at a glance. What two other mechanisms of inheritance can be definitely excluded? Why can these be excluded?
b. Of the remaining mechanisms of inheritance, which is the most likely? Why?

Answer:
a. Y-linked inheritance can be excluded because females are affected. X-linked recessive inheritance can also be excluded, as an affected mother (I.2) has a normal son (II.5). Autosomal recessive inheritance can also be excluded, as in such a case, two affected parents such as II.1 and II.2 would be expected to have all affected offspring, which they do not.
b. The remaining two mechanisms of inheritance are X-linked dominant and autosomal dominant. Genotypes can be written to satisfy both mechanisms of inheritance. Of these two, X-linked dominance may be more likely, as II.6 and II.7 have only affected daughters, indicating criss-cross inheritance. If the trait were to show autosomal dominance, one would expect half of the daughters and half of the sons to be affected.

3.42 A three-generation pedigree for a particular human trait is shown in the following figure.

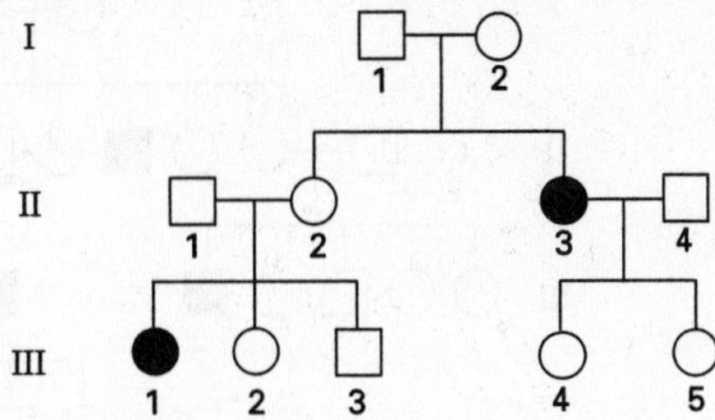

a. What is the mechanism of inheritance for the trait?
b. Which persons in the pedigree are known to be heterozygous for the trait?
c. What is the probability that III.2 is a carrier (heterozygous)?
d. If III.3 and III.4 marry, what is the probability that their first child will have the trait?

Answer: a. The trait must be recessive, as unaffected parents (I.1 and I.2, II.1 and II.2) have affected offspring. It also must be autosomal, as affected female daughters are born to unaffected fathers.

b. Given the answer to (a), individuals I.1 and I.2, as well as individuals II.1 and II.2 must be heterozygous. Also, individuals III.4 and III.5 must be heterozygous as they have an affected, homozygous mother.

c. One can denote the parents of III.2 as *Aa* x *Aa*. We know individual III.2 is *A*–, so she is either *AA* (p = 1/3) or *Aa* (p = 2/3).

d. By the same reasoning as described in (c), individual III.3 has a 2/3 chance of being *Aa*. The chance that he will contribute an *a* allele to his offspring is thus 2/3 x 1/2 = 1/3. The chance that individual III.4 will contribute an *a* allele to her offspring is 1/2, as she is heterozygous. Thus, the chance that progeny of III.3 and III.4 will be affected is 1/3 x 1/2 = 1/6.

3.43 For the more complex pedigrees shown in the following figure, indicate the probable mechanism of inheritance: autosomal recessive, autosomal dominant, X-linked recessive, X-linked dominant, Y-linked.

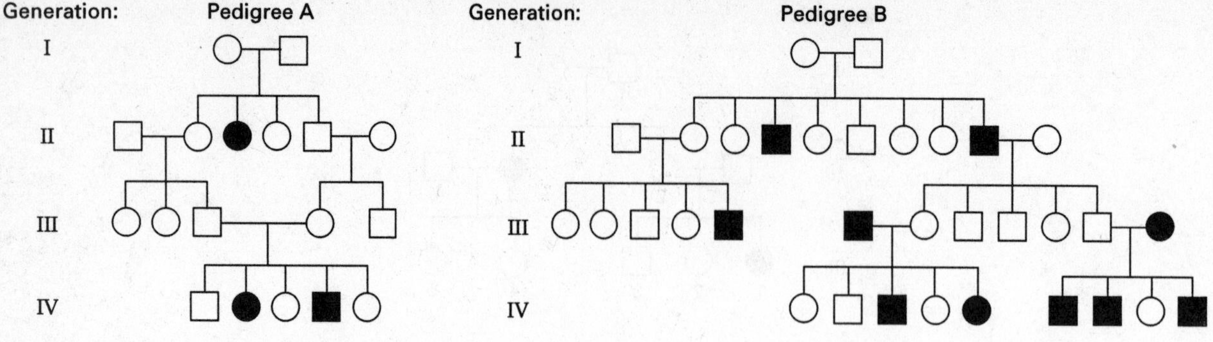

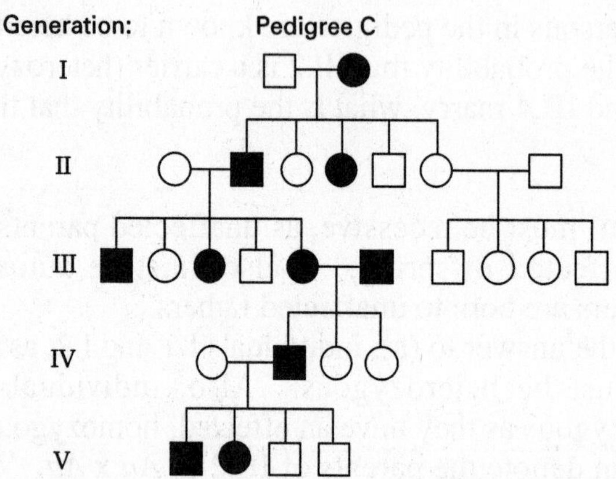

Answer: In complex pedigrees, an appropriate strategy is to initially ask whether the trait is Y-linked, then if it is dominant or recessive, and finally whether it is autosomal or X-linked. Proceeding logically through these choices helps limit the options. Before concluding is important to verify that the phenotype of every member of the pedigree fits the inferred inheritance pattern. Often, there may be only a single individual that can be used to distinguish one type of inheritance pattern from another.

Pedigree A: Clearly, the pedigree is not Y-linked, as males are not the sole transmitters of the trait. The most striking aspect of this pedigree is that the trait skips generations, i.e., unaffected parents have affected offspring. This indicates that the pedigree must be recessive. Now, one is left to determine if it is autosomal or X-linked. It does not fit an X-linked pedigree, as two affected females have normal (albeit heterozygous) fathers. For recessive pedigrees, both parents must contribute an abnormal allele, so that affected daughters must have affected fathers. The pedigree is therefore autosomal recessive. (Check by assigning appropriate genotypes to each pedigree member.)

Pedigree B: This pedigree is also not Y-linked, and must be recessive, as twice there are unaffected parents who have affected offspring. The preponderance of affected males suggests that it might be X-linked. The only way to determine this for certain is to try to assign genotypes and see if such an inheritance mode fits. It does. Could the pedigree be autosomal recessive? This would be a possibility if the trait were relatively common, as one would have to invoke that the individuals who married into the family in generation II were heterozygotes. This may be the case, as several homozygotes do marry into the family in generation III. What is the best guess then? X-linked recessive is the most likely, partly because of the frequency of affected males, but mostly because of the three affected male offspring of the homozygous female in generation III.

Pedigree C: The trait is expressed in every generation, making it a good candidate for a dominant trait. It is expressed equally by males and females, making it a good candidate for an autosomal trait. Indeed, it must be autosomal, for two reasons: First, individuals III.5 and III.6 have unaffected daughters, which would not be possible if the trait were X-linked and dominant, and second, individual IV.2, a male, has an affected offspring that is male, which would not be possible if the trait were X-linked. Thus, the trait is autosomal dominant.

3.44 If a rare genetic disease is inherited on the basis of an X-linked dominant gene, one would expect to find the following:

a. Affected fathers have 100 percent affected sons.
b. Affected mothers have 100 percent affected daughters.
c. Affected fathers have 100 percent affected daughters.
d. Affected mothers have 100 percent affected sons.

Answer: Since daughters always receive their father's X chromosome, if a father carries a dominant trait, it will always appear in his daughters. The correct answer is (c).

3.45 If a genetic disease is inherited on the basis of an autosomal dominant gene, one would expect to find the following:

a. Affected fathers have only affected children.
b. Affected mothers never have affected sons.
c. If both parents are affected, all of their offspring have the disease.
d. If a child has the disease, one of his or her grandparents also had the disease.

Answer: Answer (a) is untrue because the affected father could be heterozygous: only half of his offspring are expected to be affected. Answer (b) is untrue because, if the mother is heterozygous, half of her offspring, regardless of sex-type, would be expected to be affected. Answer (c) is untrue as, if the parents are each heterozygous, one would expect 1/4 of their offspring to carry only the recessive, normal trait. Answer (d) is the most likely expectation. Note however, that if the mutation was new in either the child or his/her parents, his/her grandparent could have been unaffected.

3.46 If a genetic disease is inherited as an autosomal recessive, one would expect to find the following:

a. Two affected individuals never have an affected child.
b. Two affected individuals have affected male offspring but no affected female children.
c. If a child has the disease, one of his or her grandparents will have had it.
d. In a marriage between an affected individual and an unaffected one, all the children are unaffected.

Answer: Answer (a) is untrue because two affected individuals will always have affected children (*aa* x *aa* can give only *aa* offspring). Answer (b) is untrue as an autosomal trait is inherited independent of sex-type. Answer (c) need not be true, as the trait could be masked by normal dominant alleles through many generations before two normal heterozygotes marry and produce affected, homozygous offspring. Answer (d) is the most likely situation. If the trait is rare, then it is likely that an unaffected individual marrying into the pedigree is homozygous for a normal allele. Since the trait is recessive, and the children receive the dominant, normal allele from the unaffected parent, the children will be normal.

3.47 Which of the following statements is *not* true for a disease that is inherited as a rare X-linked dominant?

a. All daughters of an affected male will inherit the disease.
b. Sons will inherit the disease only if their mothers have the disease.
c. Both affected males and affected females will pass the trait to half the children.
d. Daughters will inherit the disease only if their father has the disease.

Answer: The only untrue statement is (d). Since daughters receive an X chromosome from each of their mother and father, they can inherit an X-linked dominant disease from either their mother or father.

3.48 Women who were known to be carriers of the X-linked, recessive, hemophilia gene were studied in order to determine the amount of time required for the blood clotting reaction. It was found that the time required for clotting was extremely variable from individual to individual. The values obtained ranged from normal clotting time at one extreme, all the way to clinical hemophilia at the other extreme. What is the most probable explanation for these findings?

Answer: Since hemophilia is an X-linked trait, the most likely explanation is that random inactivation of X chromosomes (lyonization) produces individuals with different proportions of cells with the normal allele. Normal clotting times would be expected in females with a functional h^+ allele, i.e., females whose X^h chromosome was very frequently inactivated. Clinical hemophilia would be expected in females without a functional h^+ allele, i.e., females whose X^{h^+} was very frequently inactivated. Intermediate clotting times would be expected to be proportional to the amount of h^+ function, which is related to the frequency of inactivation of the X^{h^+} chromosome.

3.49 Hurler syndrome is a genetically transmitted disorder of mucopolysaccharide metabolism characterized by short stature, mental retardation and various bony malformations. There are two specific types described with extensive pedigrees in the medical genetics literature. They are:

Type I – recessive autosomal
Type II – recessive X-linked

As a consultant in a hospital ward with several Hurler syndrome patients, several have asked for advice concerning their relatives' offspring. Being aware that both types are extremely rare, and that afflicted individuals virtually never reproduce, state briefly the counsel you would give a Type I Hurler syndrome female patient whose normal brother's daughter is planning marriage concerning the offspring of the proposed marriage. In your answer, state the probabilities of affected offspring and sex differences where relevant.

Answer: Draw out the pedigree of the patient and try to assign genotypes to the relevant individuals. Let *h* represent the Hurler syndrome Type I allele. Then the patient is *hh*. Since we are told that Hurler syndrome patients virtually never reproduce, neither of her parents is expected to be *hh*. Still, in order for the patient to be *hh*, her parents must have been *Hh* x *Hh*. Since her brother is normal, he is *H*–, with a 2/3 chance of being *Hh*. The brother's daughter has a half-chance of receiving either of the brother's (*H* or *h*)alleles. Thus, the chance that the brother's daughter has an *h* allele is 2/3 x 1/2 = 1/3. Since the trait is extremely rare, it is likely she will marry an *HH* individual, and have *H*– children. Therefore there is no chance the brother's daughter will have affected children, as her husband will always provide a dominant, normal *H* allele. Since Type I is autosomal, there will be no sex differences.

CHAPTER 4

EXTENSIONS OF MENDELIAN GENETIC ANALYSIS

I. CHAPTER OUTLINE

MULTIPLE ALLELES
- ABO Blood Groups
- *Drosophila* Eye Color

MODIFICATIONS OF DOMINANCE RELATIONSHIPS
- Incomplete Dominance
- Codominance

GENE INTERACTIONS AND MODIFIED MENDELIAN RATIOS
- Gene Interactions That Produce New Phenotypes
- Epistasis

ESSENTIAL GENES AND LETHAL ALLELES

THE ENVIRONMENT AND GENE EXPRESSION
- Penetrance and Expressivity
- Effects of the Internal Environment
- Effects of the External Environment
- Nature versus Nurture

II. REVIEW OF KEY TERMS, SYMBOLS AND CONCEPTS

Without consulting the text and in your own words, write a brief definition of each term in the groups below. Then, either using a short phrase or a simple diagram, identify the relationship(s) between specific pairs of terms within a set. Finally, consult the text (and perhaps a friend who has also done the exercise) to check your answers.

1	2	3	4
mutant	ABO blood group	epistasis	dihybrid cross
multiple alleles	MN blood group	dominant epistasis	9:3:3:1
allelic series	codominance	recessive epistasis	9:6:1
diploid	blood-typing	duplicate dominant epistasis	9:3:4
wild-type allele	cellular antigen	duplicate recessive epistasis	9:7
	antibody	dominant and recessive epistasis	12:3:1
	agglutination	duplicate gene interaction	13:3
	universal donor	complementary gene action	9:6:1
	universal recipient		15:1

5	6	7	8
complete dominance	dominance	essential gene	penetrance
complete recessiveness	epistasis	lethal gene	expressivity
incomplete dominance	hypostasis	recessive lethal allele	population
partial dominance		dominant lethal allele	individual
codominance			phenocopy
			hereditary trait
			nature vs. nurture
			norm of reaction

9
sex-limited trait
sex-influenced trait
sex-linked gene
autosomal gene

III. THINKING ANALYTICALLY

While the methods of reasoning used in this chapter are not substantively different from those used in Chapters 2 and 3, there is much more information to consider as you tackle problems. Complexity is added when phenotypes are affected by multiple alleles, gene interactions and environmental influences. It is important to gain a solid understanding of how each of these individual factors can contribute to a phenotype. Only then will you understand how a combination of these factors can jointly affect a phenotype.

The action of a gene or the interaction between different genetic loci can be affected in several ways simultaneously, and can therefore be variously labeled. Moreover, alternate phenotypes that are initially studied clinically may exhibit different, or more subtle, relationships when investigated biochemically. As a consequence, a variety of labels may be applied to indicate the kind and degree of interaction that may occur within one set of alleles, between several sets of alleles, and between genetic and environmental influences. Finally, it is important to remember that while a trait results from the action of a set of alleles at one or more loci, genes work within the cellular milieu that has been established genetically and environmentally during an organism's development.

IV. QUESTIONS FOR PRACTICE

A. Multiple Choice Questions

1. In clover leaves, chevron pattern (a light-colored triangular leaf pattern) is controlled by seven different alleles at a single gene. From this information alone, what can be said about this trait?
   a. Alleles at this gene show incomplete dominance.
   b. There is a multiple allelic series with seven alleles that controls chevron pattern.
   c. This gene shows epistasis.
   d. One allele at this gene must be completely dominant.

2. The phenotype associated with a wild-type allele of a particular gene is
   a. the phenotype that is always found in nature.
   b. a reference phenotype used by geneticists for comparison.
   c. a special mutant phenotype.
   d. the most common phenotype seen.

3. A phenocopy is different from a particular heritable mutation in that
   a. a phenocopy cannot have the same phenotype as the mutation.
   b. the biochemical anomaly in a phenocopy and a mutation must be different.
   c. the phenocopy is always dominant.
   d. a phenocopy is not a heritable trait.

4. Neurofibromatosis is a dominantly inherited disease that can show mild, moderate or severe symptoms. Every individual that inherits the dominant allele shows at least mild symptoms. This means that the disease allele shows
   a. variable penetrance and complete expressivity.
   b. variable expressivity and complete penetrance.
   c. variable penetrance and variable expressivity.
   d. complete penetrance and complete expressivity.

5. Two alleles at the *white* locus in *Drosophila melanogaster* are $w^{apricot}$, which has orange colored eyes and w^{coral}, which has pink colored eyes. This is an example of
   a. codominance.
   b. a multiple allelic series.
   c. penetrance.
   d. epistasis.

6. One difference between epistasis and dominance is that
   a. epistasis occurs between two different genes while dominance occurs between alleles at one gene.
   b. only epistasis is influenced by environmental interactions.
   c. dominant traits are completely penetrant, epistatic interactions may not be.
   d. dominant traits may show variable penetrance, epistatic interactions may not.

For questions 7, 8, 9 and 10, choose the kind of inheritance that best explains the data. Explain your choices.
   a. recessive lethal allele
   b. dominant lethal allele
   c. incomplete dominance; two allelic pairs

d. duplicate recessive epistasis; two allelic pairs
e. dominant epistasis; two allelic pairs

7. When crossed, two pure-breeding varieties of white kernel corn yield progeny with purple kernels. When these purple progeny are selfed, a count of the kernels in the F_2 yields an average of 270 purple kernels:210 white kernels per ear.

8. When pure-breeding wheat with red kernels is crossed to the pure-breeding white variety, the progeny have pink kernels. One-hundred F_2 plants exhibit the following phenotypes: 6 red, 6 white, 25 light red, 25 light pink, 38 pink.

9. Yellow mice crossed to pure-breeding agouti mice produce a 50:50 ratio of yellow to agouti offspring. Crosses between yellow mice never produce 100 percent yellow progeny, but rather yield 36 yellow and 15 agouti offspring.

10. At age 52, a man begins to show the symptoms of Huntington's disease. Like his mother before him, he eventually dies from the disease. There is no history of the disease in his wife's family or his father's family. Two of his four children also die from the disease, albeit not until late in their lives.

Answers: 1b; 2b; 3d; 4b; 5b; 6a; 7d; 8c; 9a; 10b.

B. Thought Questions

1. Distinguish between dominance and epistasis, and between epistasis and multiple alleles.
2. How would you distinguish between a gene with pleiotropic effects and the effects resulting from several sets of alleles? (Hint: Review independent assortment.)
3. In Tay-Sachs disease, heterozygotes who are phenotypically normal exhibit a level of the enzyme hexosaminidase A that is midway between that contained in recessive homozygotes, who have no detectable enzyme, and the level in normal, dominant homozygotes. Explain why Tay-Sachs disease could be variously described as an example of incomplete dominance, co-dominance, and/or a recessive lethal.
4. A sex-linked gene controls coat color in cats. One allele determines black coat color while another allele determines orange coat color. A separate set of genes determines where color is produced (i.e., where white patches are). Use this information to explain why most calico (black, orange and white patched) cats are female. How do rare calico males arise? Do you expect them to be fertile?
5. In a very large natural population, could you ever presume to know all the allelic variants in a multiple allelic series? (Hint: Could new variants appear at any time?)
6. Devise a situation in which a multiple allelic series shows elements of co-dominance as well as lethality.
7. Could a lethal allele, either dominant or recessive, be incompletely penetrant? (Hint: Think about small variations in the timing of events during development.)

V. SOLUTIONS TO TEXT PROBLEMS

4.1 In rabbits, c^+ = agouti coat color, c^{ch} = chinchilla, c^h = Himalayan, and c = albino. The four alleles constitute a multiple allelic series. The agouti c^+ is dominant to the three other alleles, c is recessive to all three other alleles, and chinchilla is dominant to Himalayan. Determine the phenotypes of progeny from the following crosses:

a. $c^+/c^+ \times c/c$
b. $c^+/c^{ch} \times c^+/c$
c. $c^+/c \times c^+/c$
d. $c^+/c^h \times c^h/c$
e. $c^+/c^h \times c/c$
f. $c^{ch}/c^h \times c^h/c$
g. $c^h/c \times c/c$
h. $c^+/c^h \times c^+/c$
i. $c^+/c^h \times c^+/c^{ch}$

Answer: a. All agouti
b. 3/4 agouti, 1/4 chinchilla
c. 3/4 agouti, 1/4 albino
d. 1/2 agouti, 1/2 Himalayan
e. 1/2 albino, 1/2 Himalayan
f. 1/2 chinchilla, 1/2 Himalayan
g. 1/2 Himalayan, 1/2 albino
h. 3/4 agouti, 1/4 Himalayan
i. 3/4 agouti, 1/4 chinchilla

4.2 If a given population of diploid organisms contains three, and only three, alleles of a particular gene (say w, $w1$, and $w2$), how many different diploid genotypes are possible in the population? List all possible genotypes of diploids (consider *only* these three alleles).

Answer: Six genotypes are possible: ww, $ww1$, $ww2$, $w1w1$, $w1w2$, $w2w2$.

4.3 The genetic basis of the ABO blood types seems most likely to be:

a. multiple alleles.
b. polyexpressive hemizygotes.
c. allelically excluded alternates.
d. three independently assorting genes.

Answer: a.

4.4 In humans the three alleles I^A, I^B, and i constitute a multiple allelic series that determine the ABO blood group system, as we described in this chapter. For the following problems, state whether the child mentioned can actually be produced from the marriage. Explain your answer.

a. An O child from the marriage of two A individuals.
b. An O child from the marriage of an A to a B.
c. An AB child from the marriage of an A to an O.
d. An O child from the marriage of an AB to an A.
e. An A child from the marriage of an AB to a B.

Answer: a. Yes, if both parents were I^Ai.
b. Yes, if the A parent were I^Ai and the B parent were I^Bi.
c. No, as neither parent has an I^B allele.

d. No, as an O child must obtain an i allele from each parent, and one parent is I^AI^B.
e. Yes, providing that the B parent is I^Bi (and the child is I^Ai).

4.5 A man is blood type O, M. A woman is blood type A, M and her child is type A, MN. The aforesaid man cannot be the father of the child because:
a. O men cannot have type A children.
b. O men cannot have MN children.
c. An O man and an A woman cannot have an A child.
d. An M man and an M woman cannot have an MN child.

Answer: d. If both parents are M, an MN child is impossible because neither has an N allele. An A child is possible however (the mother would contribute her I^A allele and the father an i allele, giving an I^Ai child).

4.6 A woman of blood group AB marries a man of blood group A whose father was group O. What is the probability that
a. their two children will both be group A?
b. one child will be group B and the other group O?
c. the first child will be a son of group AB and the second child a son of group B?

Answer: First determine the genotypes of the parents. The woman is type AB, so must have the genotype I^AI^B. The father is type A, so he must have at least one I^A allele. The grandfather was type O, so only had i alleles. Therefore the cross must be I^AI^B x I^Ai. The four equally likely genotypes (I^AI^A, I^Ai, I^BI^A, I^Bi) give rise to three phenotypes: A (p = 1/2), AB (p = 1/4) or B (p = 1/4).
a. The chance that two children will both be group A is 1/2 x 1/2 = 1/4.
b. There is no chance of producing an O child, so $p = 0$.
c. The chance of a male AB child is 1/2 x 1/4 = 1/8. The chance of a male B child is 1/2 x 1/4 = 1/8. Therefore the chance of both events happening is 1/8 x 1/8 = 1/64. As the first event is independent of the second, this is the chance of the events happening in the order specified.

4.7 If a mother and her child belong to blood group O, what blood group could the father *not* belong to?

Answer: Since the child is group O, its genotype is ii. Therefore the father must have been able to contribute an i allele. The only blood group that would not allow this would be AB, whose genotype is I^AI^B.

4.8 A man of what blood group could not be a father to a child of blood type AB?

Answer: A man who has group O blood, and is genotypically ii, could not contribute either of the I^A and I^B alleles present in an individual of group AB.

4.9 In snapdragons, red flower color (C^R) is incompletely dominant to white (C^W); the C^R/C^W heterozygotes are pink. A red-flowered snapdragon is crossed with a white-flowered one. Determine the flower color of (a) the F_1; (b) the F_2; (c) the progeny of a cross of the F_1 to the red parent; (d) the progeny of a cross of the F_1 to the white parent.

Answer: a. The cross is C^RC^R x C^WC^W, the F_1 are all C^RC^W, and are pink.
b. The F_2 will be 1 C^RC^R:2C^RC^W:1 C^WC^W, and be 1 red:2 pink:1 white.
c. C^RC^W x C^RC^R produces 1/2 C^RC^R (red) and 1/2 C^RC^W (pink) progeny.
d. C^RC^W x C^WC^W produces 1/2 C^RC^W (pink) and 1/2 C^WC^W (white) progeny.

4.10 In Shorthorn cattle the heterozygous condition of the alleles for red coat color (C^R) and white coat color (C^W) is roan coat color. If two roan cattle are mated, what proportion of the progeny will resemble their parents in coat color?

Answer: A cross between two roan cattle can be denoted as C^RC^W x C^RC^W. The progeny will be 1/4 C^RC^R, 1/2 C^RC^W and 1/4 C^WC^W. Thus, half will be roan progeny.

4.11 What progeny will a roan Shorthorn have if bred to (a) red; (b) roan; (c) white?

Answer: In shorthorn cattle, the color roan indicates a heterozygote (C^RC^W) because of incomplete dominance of each allele. Hence, the crosses can be represented as (a) C^RC^W x C^RC^R, (b) C^RC^W x C^RC^W, and (c) C^RC^W x C^WC^W. (a) will give 1/2 red and 1/2 roan, (b) will give 1/4 red, 1/2 roan and 1/4 white, and (c) will give 1/2 roan and 1/2 white.

4.12 In peaches, fuzzy skin (*F*) is completely dominant to smooth (nectarine) skin (*f*), and the heterozygous conditions of oval glands at the base of the leaves (G^O) and no glands (G^N) give round glands. A homozygous fuzzy, no-gland peach variety is bred to a smooth, oval-gland variety.
a. What will be the appearance of the F_1?
b. What will be the appearance of the F_2?
c. What will be the appearance of the offspring of a cross of the F_1 back to the smooth, oval-glanded parent?

Answer: a. The parental cross can be written as $FF\ G^NG^N$ x $ff\ G^OG^O$. The F_1 will be all $Ff\ G^NG^O$ and be fuzzy with round leaf glands.
b. As the alleles at the *O* gene show incomplete dominance, there will be a modified 9:3:3:1 ratio in the F_2. The progeny will be 3/16 fuzzy, oval-glanded ($F\text{–}\ G^OG^O$), 6/16 fuzzy, round-glanded ($F\text{–}\ G^OG^N$), 3/16 fuzzy, no-glanded ($F\text{–}\ G^NG^N$), 1/16 smooth, oval-glanded ($ff\ G^OG^O$), 2/16 smooth, round-glanded ($ff\ G^OG^N$), and 1/16 smooth, no-glanded ($ff\ G^NG^N$).
c. The cross can be diagrammed as $Ff\ G^NG^O$ x $ff\ G^OG^O$. The progeny will be 1/4 fuzzy, oval-glanded ($Ff\ G^OG^O$), 1/4 fuzzy, round-glanded ($Ff\ G^NG^O$), 1/4 smooth, oval glanded ($ff\ G^OG^O$) and 1/4 smooth, round-glanded ($ff\ G^NG^O$).

4.13 In guinea pigs, short hair (*L*) is dominant to long hair (*l*), and the heterozygous condition of yellow coat (C^Y) and white coat (C^W) gives cream coat. A short-haired, cream guinea pig is bred to a long-haired, white-guinea pig, and a long-haired, cream baby guinea pig is produced. When the baby grows up, it is bred back to the short-haired, cream parent. What phenotypic classes and in what proportions are expected from the offspring?

Answer: The initial cross can be diagrammed as $L\text{–}\ C^YC^W$ x $ll\ C^WC^W$. As a long-haired, cream baby ($ll\ C^YC^W$) is produced, one can infer that the short-haired parent must have been *Ll*. The backcross can be diagrammed as $ll\ C^YC^W$ x $Ll\ C^YC^W$. There will be 1/8 short-haired yellow, 1/4 short-haired cream, 1/8 short-haired white, 1/8 long-haired yellow, 1/4 long-haired cream and 1/8 long-haired white progeny.

4.14 The shape of radishes may be long (S^L/S^L), oval (S^L/S^S), or round (S^S/S^S), and the color or radishes may be red (C^R/C^R), purple (C^R/C^W) or white (C^W/C^W). If a long, red radish is crossed with a round, white plant, what will be the appearance of the F_1 and F_2?

Answer: The cross can be diagrammed as S^L/S^L C^R/C^R x S^S/S^S C^W/C^W. The F_1 will be oval and purple (S^L/S^S C^R/C^W) and the F_2 consists of 1/16 long red (S^L/S^L C^R/C^R), 1/8 oval red (S^L/S^S C^R/C^R), 1/16 round red (S^S/S^S C^R/C^R), 1/8 long purple (S^L/S^L C^R/C^W), 1/4 oval purple (S^L/S^S C^R/C^W), 1/8 round purple (S^L/S^S C^R/C^W), 1/16 long white (S^L/S^L C^W/C^W), 1/8 oval white (S^L/S^S C^W/C^W) and 1/16 round white (S^S/S^S C^W/C^W).

4.15 In poultry the genes for rose comb (*R*) and pea comb (*P*), if present together, give walnut comb. The recessive alleles of each gene, when present together in a homozygous state, give single comb. What will be the comb characters of the offspring of the following crosses?

a. *R*/*R P*/*p* x *r*/*r P*/*p*
b. *r*/*r P*/*P* x *R*/*r P*/*p*
c. *R*/*r p*/*p* x *r*/*r P*/*p*
d. *R*/*r P*/*p* x *R*/*r P*/*p*
e. *R*/*r p*/*p* x *R*/*r p*/*p*

Answer:
a. 3/4 walnut (*R*/*r P*/–), 1/4 rose (*R*/*r p*/*p*).
b. 1/2 walnut (*R*/*r P*/–), 1/2 pea (*r*/*r P*/–).
c. 1/4 walnut (*R*/*r P*/*p*), 1/4 rose (*R*/*r p*/*p*), 1/4 pea (*r*/*r P*/*p*), 1/4 single (*r*/*r p*/*p*).
d. 9/16 walnut (*R*/– *P*/–), 3/16 pea (*r*/*r P*/–), 3/16 rose (*R*/– *p*/*p*), 1/16 single (*r*/*r p*/*p*).
e. 3/4 rose (*R*/– *p*/*p*), 1/4 single (*r*/*r p*/*p*).

4.16 For the following crosses involving the comb character in poultry, determine the genotypes of the two parents:

a. A walnut crossed with a single produces offspring 1/4 of which are walnut, 1/4 rose, 1/4 pea, and 1/4 single.
b. A rose crossed with a walnut produces offspring 3/8 of which are walnut, 3/8 rose, 1/8 pea, and 1/8 single.
c. A rose crossed with a pea produces five walnut and six rose offspring.
d. A walnut crossed with a walnut produces one rose, two walnut, and one single offspring.

Answer: To approach solving this problem, first consider the possible genotypes of the parents and their offspring from the phenotypes they present. Then consider the phenotypic ratios in the offspring and determine the precise genotype.

a. *R*/*r P*/*p* (walnut) x *r*/*r p*/*p* (single).
b. *R*/*r p*/*p* (rose) x *R*/*r P*/*p* (walnut).
c. *R*/*r p*/*p* (rose) x *r*/*r P*/*P* (pea).
d. *R*/*r P*/*p* x *R*/*r P*/*p* (note that since a single *r*/*r p*/*p* offspring is produced, both parents must have an *r* and a *p* allele).

4.17 In poultry feathered shanks (*F*) are dominant to clean (*f*), and white plumage of white leghorns (*I*) is dominant to black (*i*).

a. A feathered-shanked, white, rose-combed bird crossed with a clean-shanked white, walnut-combed bird produces these offspring: two feathered, white, rose; four clean, white, walnut; three feathered, black, pea; one clean, black, single; one feathered, white single; two clean, white, rose. What are the genotypes of the parents?

b. A feathered-shanked, white, walnut-combed bird crossed with a clean-shanked, white, pea-combed bird produces a single offspring, which is clean-shanked, black, and single-combed. In further offspring from this cross, what proportion may be expected to resemble each parent, respectively?

Answer: Approach this problem by considering each set of traits separately.

a. A cross between a feathered and a clean bird produces 6 feathered:7 clean offspring (approximately 1:1), indicating the parents are *F/f* x *f/f*. A cross between two white birds produces 9 white and 4 black offspring (approximately 3:1), indicating the parents are *I/i* x *I/i*. Since a cross between a rose and a walnut bird produces single combed (as well as walnut, rose, and pea) offspring, each parent must have an *r* and *p* allele. Thus, the parents must have been *R/r p/p* x *R/r P/p*. The complete genotypes of the parents were *F/f I/i R/r p/p* x *f/f I/i R/r P/p*.

b. The single combed, clean-shanked, black offspring displays the phenotypes associated with the recessive alleles at the loci involved. Its genotype must be *f/f i/i r/r p/p* and each of its parents must have given it a recessive allele. Considering this and the phenotypes of the parents, the parents' genotypes are *F/f I/i R/r P/p* and *f/f i/i r/r p/p*. Since the *f/f i/i r/r p/p* parent will always have *f i r p* gametes, the chance of getting offspring that resemble each parent is the chance of the *F/f I/i R/r P/p* parent giving gametes that are either *F I R P* or *f i r p*. For each, the chance is $(1/2)^4 = 1/16$.

4.18 F_2 plants segregate 9/16 colored:7/16 colorless. If a colored plant from the F_2 is chosen at random and selfed, what is the probability that there will be no segregation of the two phenotypes among its progeny?

Answer: Since there is a modified 9:3:3:1 ratio in the F_2, the F_1 cross must have been of the form *Aa Bb* x *Aa Bb*, and the F_2 colored plant in question must be of genotype *A– B–*, where *A* and *B* are dominant alleles that must be present for colored pigment to be produced. In order for "no segregation of the two phenotypes" to be observed, the colored plant must be true-breeding, i.e., *AA BB*. Thus, the question can be rephrased as: 'what is the chance that if an F_2 colored plant is picked, it is true breeding?' Of all the plants in the F_2, only 1/16 are *AA BB*. Among the 9/16 *A– B–* plants, 1/9 [= (1/16)/(1/9)] chance are true breeding plants that will have no colorless progeny if selfed.

4.19 In peanuts, the plant may be either "bunch" or "runner." Two different strains of peanut, V4 and G2, in which "bunch" occurred were crossed with the following results:

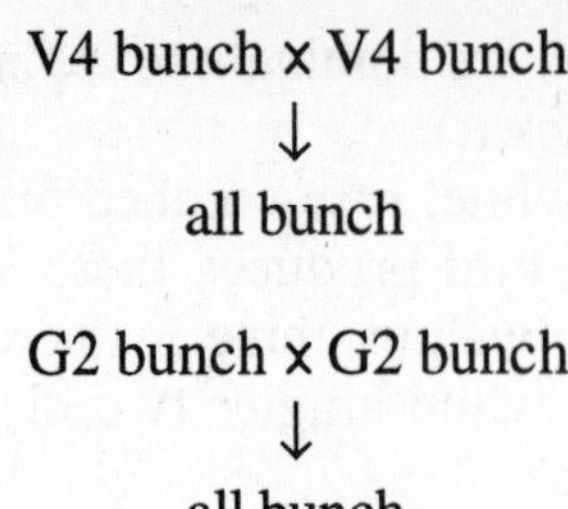

The two true-breeding strains of bunch were crossed in the following way:

V4 bunch x G2 bunch
↓
F_1 runner

$F_1 \times F_1$
↓
F_2 9 runner:7 bunch

What is the genetic basis of the inheritance pattern of runner and bunch in the F_2?

Answer: The 9:7 ratio in the F_2 is a modified 9:3:3:1 ratio, where the *A– B–* genotypes are "runner" and the *A– bb, aa B–* and *aa bb* genotypes are "bunch." This is an example of dominant alleles at each of two genes being necessary to observe a phenotype. When recessive alleles at either of two genes are present, they block (are epistatic to) the "runner" phenotype. Thus, this is an example of duplicate, recessive epistasis.

4.20 In rabbits, one enzyme (the product of a functional gene *A*) is needed to produce a substance needed for hearing. Another enzyme (the product of a functional gene *B*) is needed to produce another substance also required for normal hearing. The genes responsible for the two enzymes are not linked. Individuals homozygous for either one or both of the nonfunctional recessive alleles, *a* or *b*, are deaf.

a. If a large number of matings were made between two double heterozygotes, what phenotypic ratio would be expected in the progeny?
b. This phenotypic ratio is a result of what well-known phenomenon?
c. What phenotypic ratio would be expected if rabbits homozygous recessive for trait A and heterozygous for trait B were mated to rabbits heterozygous for both traits?

Answer:
a. Since individuals homozygous for either one or both recessive alleles are deaf, individuals that are *aa B–*, *A– bb*, and *aa bb* will be deaf. Only *A– B–* individuals are able to hear. Thus, one will get a 9 hearing:7 deaf phenotypic ratio.
b. This is duplicate recessive epistasis. Recessive alleles at either of two genes block hearing, and are epistatic to the dominant alleles at each gene.
c. The cross being described can be written as: *aa Bb* x *Aa Bb*. Use a branch diagram to show that there would be 5/8 deaf progeny (1/8 *Aa bb* + 1/2 *aa* –) and 3/8 hearing (*Aa B–*).

4.21 In Doodlewags, the dominant allele *S* causes solid coat color; the recessive allele *s* results in white spots on a colored background. The black coat color allele *B* is dominant to the brown allele, *b*, but these genes are expressed only in the genotype *a*/*a*. Individuals that are *A*/– are yellow regardless of *B* alleles. Six pups are produced in a mating between a solid yellow male and a solid brown female. Their phenotypes are: 2 solid black, 1 spotted yellow, 1 spotted black, and 2 solid brown.

a. What are the genotypes of the male and female parents?
b. What is the probability that the next pup will be spotted brown?

Answer: Two characteristics controlled by three genes are described here. Spotted coat (*where* pigment is positioned) is controlled by the *S*/*s* alleles. The pigmentation color itself (brown, black or yellow) is controlled by the *B*/*b* and *A*/*a* genes, where *A*– is epistatic to (i.e., blocks the expression of) the *B*/*b* gene. Consider each characteristic separately and remember that when dealing with small numbers of progeny, precise Mendelian ratios are not always obtained.

a. Two solid colored dogs produce 4 solid and 2 spotted pups. In order for any spotted dogs to be produced, both parents must have been heterozygous, i.e., *Ss*. The yellow male parent can initially be assigned the genotype *A*– —, and the brown female must be *aa bb*. Since both black and brown pups are obtained, the yellow parent must be *Aa Bb*. Thus, the parents are *Ss Aa Bb* x *Ss aa bb*.
b. A spotted brown pup has the genotype *ss aa bb*.
   p (*ss aa bb* pup)
   = *p*(*s* from each parent) x *p*(*a* from each parent) x *p*(*b* from each parent)
   = (1/2 x 1/2) x (1/2 x 1) x (1/2 x 1)
   = 1/16.

4.22 The gene *l* in *Drosophila* is a recessive and sex-linked, and lethal when homozygous or hemizygous (the condition in the male). If a female of genotype *L*/*l* is crossed with a normal male, what is the probability that the first two surviving progeny to be observed will be males?

Answer: The cross can be denoted *L*/*l* x *L*/Y, where Y is the Y chromosome. The male progeny of this cross are *L*/Y and *l*/Y, where only *L*/Y males survive. Hence, there is a 1/4 chance of obtaining a viable male offspring (1/2 chance of male, 1/2 chance the male will survive). The probability that the first two surviving progeny will be male is 1/4 x 1/4 = 1/16.

4.23 A locus in mice is involved with pigment production; when parents heterozygous at this locus are mated, 3/4 of the progeny are colored and 1/4 are albino. Another phenotype concerns the coat color produced in the mice; when two yellow mice are mated, 2/3 of the progeny are yellow and 1/3 are agouti. The albino mice cannot express whatever alleles they may have at the independently assorting agouti locus.

a. When yellow mice are crossed with albino, they produce an F_1 consisting of 1/2 albino, 1/3 yellow, and 1/6 agouti. What are the probable genotypes of the parents?
b. If yellow F_1 mice are crossed among themselves, what phenotypic ratio would you expect among the progeny? What proportion of the yellow progeny produced here would be expected to be true breeding?

Answer: Two loci are involved here. The first controls pigment production. From the information that a cross of heterozygous parents gives a 3 colored:1 albino ratio, we can infer that the homozygous recessive condition is that of no pigmentation at all. If we denote this locus as *C/c*, this also means that *C–* is required for pigmentation. Another way to state this is that *cc* blocks the production of pigment.

The second locus determines whether the coat color is yellow or agouti. The 2 yellow:1 agouti progeny ratio seen in a cross between two yellow mice is a modified 3:1 ratio from a monohybrid cross. It indicates recessive lethality. Specifically, it indicates that the yellow allele is dominant for coat color, but that when homozygous, it is lethal. If we denote this locus as *Y/y*, the *YY* condition is lethal, *Yy* individuals are yellow and *yy* individuals are agouti.

Now consider the relationship between the two loci. If *cc* blocks the production of pigment, *cc* is epistatic to *Yy* and *yy* (*cc Yy* and *cc yy* mice will be albino).

a. From the above analysis, the phenotype of a yellow mouse indicates its genotype must be *C– Yy* and the phenotype of an albino mouse indicates its genotype must be *cc y–*. To determine what the unknown alleles are, consider the two traits separately by considering color/no color (the *C/c* locus) and type of color (the *Y/y* locus). There are half albino and half colored offspring, indicating that the parents must have been *Cc* x *cc*. There is a 2:1 ratio of yellow to agouti offspring, indicating that the parents were *Yy* x *Yy*. Therefore, the parental genotypes were *Cc Yy* x *cc Yy*.
b. The yellow F_1 mice have the genotype *Cc Yy*. The progeny that will be obtained from the cross *Cc Yy* x *Cc Yy* are 1/4 albino, 1/2 yellow and 1/4 agouti (remember that the *YY* progeny are inviable). None of the yellow mice will be true breeding, as they are all *Yy*.

4.24 In *Drosophila melanogaster*, a recessive autosomal gene, ebony (*e*), produces a black body color when homozygous, and an independently assorting autosomal gene, black (*b*), also produces a black body color when homozygous. Flies with genotypes e/e $b^+/-$, $e^+/-$ b/b and e/e b/b are phenotypically identical with respect to body color. If true-breeding e/e b^+/b^+ ebony flies are crossed with true-breeding e^+/e^+ b/b black flies,

a. what will be the phenotype of the F_1 flies?
b. what phenotypes and what proportions would occur in the F_2 generation?
c. what phenotypic ratios would you expect to find in the progeny of these backcrosses: (1) F_1 x true-breeding ebony and (2) F_1 x true-breeding black?

Answer: a. The F_1 flies will be b/b^+ e/e^+ and be wild-type in color (gray).
b. The F_2 will show a 9 $b^+/-$ $e^+/-$ (wild-type):3 $b^+/-$ e/e (black):3 b/b $e^+/-$ (black):1 b/b e/e (black) ratio, i.e., 9/16 wild-type and 7/16 black.
c. The F_1 x true-breeding ebony can be denoted b/b^+ e/e^+ x b^+/b^+ e/e, and will give a 1 $b^+/-$ e/e:1 $b^+/-$ e/e^+ progeny ratio, 1/2 black and 1/2 wild-type (gray). The F_1 x true-breeding black can be denoted b^+/b e/e^+ x b/b e^+/e^+, and will give a 1 b^+/b $e^+/-$:1 b/b $e^+/-$ progeny ratio, 1/2 black and 1/2 wild-type (grey).

4.25 In four-o'clock plants, two genes, *Y* and *R*, affect flower color. Neither is completely dominant, and the two interact on each other to produce seven different flower colors:

Y/Y R/R = crimson *Y/y R/R* = magenta
Y/Y R/r = orange-red *Y/y R/r* = magenta-rose
Y/Y r/r = yellow *Y/y r/r* = pale yellow
y/y R/R, *y/y R/r*, and *y/y r/r* = white

a. In a cross of a crimson-flowered plant with a white one (*y/y r/r*), what will be the appearances of the F_1, the F_2, and the offspring of the F_1 backcrossed to the crimson parent?
b. What will be the flower colors in the offspring of a cross of orange-red x pale-yellow?
c. What will be the flower colors in the offspring of a cross of a yellow with a *y/y R/r* white?

Answer: a. *Y/Y R/R* x *y/y r/r* will give a magenta-rose *Y/y R/r* F_1, and an F_2 that is 1/16 crimson (*Y/Y R/R*), 1/8 orange-red (*Y/Y R/r*), 1/16 yellow (*Y/Y r/r*), 1/8 magenta (*Y/y R/R*), 1/4 magenta-rose (*Y/y R/r*), 1/8 pale yellow (*Y/y r/r*) and 1/4 white (*y/y –/–*). A backcross of the F_1 to the crimson parent will give 1/4 crimson (*Y/Y R/R*), 1/4 magenta-rose (*Y/y R/r*), 1/4 magenta (*Y/y R/R*) and 1/4 orange-red (*Y/Y R/r*).
b. The cross can be denoted as *Y/Y R/r* x *Y/y r/r*. The progeny will be 1/4 orange-red (*Y/Y R/r*), 1/4 magenta-rose (*Y/y R/r*), 1/4 yellow (*Y/Y r/r*) and 1/4 pale yellow (*Y/y r/r*).
c. The cross can be denoted as *Y/Y r/r* x *y/y R/r*, and the progeny will be 1/2 *Y/y R/r* magenta-rose and 1/2 *Y/y r/r* pale yellow.

4.26 Two four-o'clock plants were crossed and gave the following offspring: 1/8 crimson, 1/8 orange-red, 1/4 magenta, 1/4 magenta-rose, and 1/4 white. Unfortunately, the person who made the crosses was color-blind and could not record the flower colors of the parents. From the results of the cross, deduce the genotypes and flower colors of the two parents.

Answer: First determine the genotypes of the progeny. They are 1/8 *Y/Y R/R*, 1/8 *Y/Y R/r*, 1/4 *Y/y R/R*, 1/4 *Y/y R/r*, and 1/4 *y/y –/–*. Now consider each gene separately. There are 1/4 *Y/y*, 1/2 *Y/y* and 1/4 *y/y* progeny, a 1:2:1 ratio, so the parental cross must have been *Y/y* x *Y/y*. Among the 3/4 of the progeny whose R/*r* genotypes can be determined, half are *R/R* and half are *R/r*, indicating that the parental genotypes were *R/R* x *R/r*. Thus the parental cross was *Y/y R/R* (magenta) x *Y/y R/r* (magenta-rose).

4.27 Genes *A, B,* and *C* are independently assorting and control production of a black pigment.
a. Assume that *A, B,* and *C* act in a pathway as follows:

$$\text{colorless} \xrightarrow{A} \xrightarrow{B} \xrightarrow{C} \text{black}$$

The alternative alleles that give abnormal functioning of these genes are designated *a, b,* and *c*, respectively. A black *A/A B/B C/C* is crossed by a colorless *a/a b/b c/c* to give a black F_1. The F_1 is selfed. What proportion of the F_2 is colorless? (Assume that the products of each step except the last are colorless, so only colorless and black phenotypes are observed.)
b. Assume that *C* produces an inhibitor that prevents the formation of black by destroying the ability of *B* to carry out its function, as follows:

$$\begin{array}{ccccc} & A & B & & \\ \text{colorless} & \rightarrow & \rightarrow & \text{black} \\ & & \uparrow & & \\ & & C \text{ (inhibitor)} & & \end{array}$$

A colorless *A/A B/B C/C* individual is crossed with a colorless *a/a b/b c/c*, giving a colorless F_1. The F_1 is selfed to give an F_2. What is the ratio of colorless to black in the F_2? (Only colorless and black phenotypes are observed, as in Part a.)

Answer: a. In order for an individual to be black, it must have normal function at each step of the pathway. This is provided by the alleles *A, B,* and *C*. Thus, *A/– B/– C/–* individuals will be black, while all others (those having *a/a* and/or *b/b* and/or *c/c*) will be colorless. The chance of obtaining a black *A/– B/– C/–* individual from a cross of *A/a B/b C/c* x *A/a B/b C/c* is 3/4 x 3/4 x 3/4 = 27/64. The proportion of the F_2 that is colorless is therefore 1 - 27/64 = 37/64.

b. In this situation, an individual is black only if it has the first two steps of the pathway (those provided by *A* and *B*) and lacks the inhibitor provided by *C*. Thus, an organism is black if it is *A/– B/– c/c*. The chance of obtaining this genotype from a cross of *A/a B/b C/c* x *A/a B/b C/c* is 3/4 x 3/4 x 1/4 = 9/64. Thus, the ratio of colorless to black in the F_2 will be 55:9.

4.28 In *Drosophila*, a mutant strain has plum-colored eyes. A cross between a plum-eyed male and a plum-eyed female gives 2/3 plum-eyed and 1/3 red-eyed (wild-type) progeny flies. A second mutant strain of *Drosophila*, called stubble, has short bristles instead of normal long bristles. A cross between a stubble female and a stubble male gives 2/3 stubble and 1/3 normal-bristled flies in the offspring. Assuming that the plum gene assorts independently from the stubble gene, what will be the phenotypes and their relative proportions in the progeny of a cross between two plum-eyed, stubble-bristled flies? (Both genes are autosomal.)

Answer: For either of the stubble or plum mutants, a cross between two mutants gives 2 mutant:1 wild-type progeny ratios. This 2:1 ratio is recognizable as the modified 3:1 ratio of a monohybrid cross, where the 25 percent homozygous mutant individuals are inviable. We can denote the cross (using *P/p* for plum, *S/s* for stubble) as *P/p S/s* x *P/p S/s*. Any *PP* and/or *SS* progeny will be inviable. The kinds of progeny expected can be diagrammed as follows:

2/3 *S/s* < 2/3 *P/p* ⟶ 4/9 *S/s P/p* (stubble, plum)
　　　　 < 1/3 *p/p* ⟶ 2/9 *S/s p/p* (stubble, red)

1/3 *s/s* < 2/3 *P/p* ⟶ 2/9 *s/s P/p* (long, plum)
　　　　 < 1/3 *p/p* ⟶ 1/9 *s/s p/p* (long, red)

4.29 In sheep, white fleece (*W*) is dominant over black (*w*), and horned (*H*) is dominant over hornless (*h*) in males but recessive in females. If a homozygous horned white ram is bred to a homozygous hornless black ewe, what will be the appearance of the F_1 and the F_2?

Answer: One can restate the information given in the problem as following: In males, *H/h* and *H/H* result in horned animals while *h/h* results in hornless animals. Both the notation and phenotypes are that of a typical dominant trait. In females, the fact that the trait is sex-influenced makes the situation is quite different. In females, only *H/H* animals are horned. *H/h* and *h/h* animals are hornless. The white trait shows dominance in both sexes, with *W/–* animals being white and *ww* animals being black.

This information allows one to diagram the cross as *H/H W/W* male x *h/h w/w* female. The F_1 genotype will be *H/h W/w*. Both sex-types will have white fleece, and while males will be horned, females will be hornless. The F_2 genotypes will be 9 *H/– W/–*:3 *H/– w/w*:3 *h/h W/–*:1 *h/h w/w*. Because *H/h* and *H/H* show the dominant horned trait in males, the males will be 9/16 horned white, 3/16 horned black, 3/16 hornless white and 1/16 hornless black. Female phenotypes can be determined with the following branch diagram:

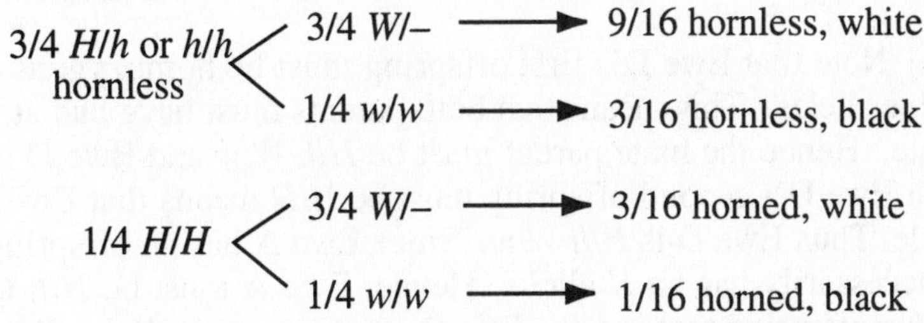

4.30 A horned black ram bred to a hornless white ewe has the following offspring: Of the males 1/4 are horned white; 1/4 are horned, black; 1/4 are hornless, white; and 1/4 are hornless, black. Of the females 1/2 are hornless and black, and 1/2 are hornless and white. What are the genotypes of the parents?

Answer: Start by using the symbols of 4.27 to denote what can be inferred about the genotypes of the parents and the offspring. Remember that in males, the horned condition is dominant, while in females, it is recessive. Hence, *H/–* (=*H/H* or *H/h*) denotes the horned condition in males, while *h/–* (*h/h* or *H/h*) denotes the hornless condition in females. Then the parents are *H/– w/w* (male) x *h/–, W/–* (female). The male progeny are 1/4 *H/– W/–*, 1/4 *H/– w/w*, 1/4 *h/h W/–* and 1/4 *h/h w/w*. The female progeny are 1/2 *h/– w/w* and 1/2 *h/– W/–*. Since homozygous recessive (male) progeny are recovered, we know that each parent must have had (at least) one recessive *h* and one recessive *w* allele. Since none of the female progeny are *H/H*, only one parent must have an *H* allele. Thus, the parents could have been *H/h w/w* (male) x *h/h W/w* (female). This is consistent with the phenotypes and proportions of the progeny.

4.31 A horned white ram is bred to the following four ewes and has one offspring by the first three and two by the fourth: Ewe A is hornless and black; the offspring is a horned white female. Ewe B is hornless and white; the offspring is a hornless black female. Ewe C is horned and black; the offspring is a horned white female. Ewe D is hornless and white; the offspring are one hornless black male and one horned white female. What are the genotypes of the five parents?

Answer: Using the information and symbols of 4.27, start by inferring what you can about genotypes from the phenotypes:

Individual	Phenotype	Initially Inferred Genotype
Male Parent	horned white male	*H/– W/–*
Ewe A	hornless black female	*H/h* or *h/h, w/w*
Ewe A offspring	horned white female	*H/H W/–*
Ewe B	hornless white female	*H/h* or *h/h, W/–*
Ewe B offspring	hornless black female	*H/h* or *h/h, w/w*
Ewe C	horned black female	*H/H w/w*
Ewe C offspring	horned white female	*H/H W/–*
Ewe D	hornless white female	*H/h* or *h/h, W/–*
Ewe D offspring #1	hornless black male	*h/h w/w*
Ewe D offspring #2	horned white female	*H/H W/–*

Note that Ewe D's first offspring must be homozygous for the recessive *h* and *w* alleles. This means that both parents must have had at least one recessive allele. Hence the male parent must be *H/h W/w* and Ewe D is *H/h or h/h, W/w*. That Ewe D's second offspring must be *H/H* means that Ewe D must have an *H* allele. Thus Ewe D is *H/h W/w*. Since Ewe A has an offspring that is *H/H*, Ewe A necessarily has an *H* allele. Hence, Ewe A must be *H/h w/w*. Since Ewe B has an offspring that is *w/w*, Ewe B must have a *w* allele. We cannot tell if Ewe B is *H/h* or *h/h*. So Ewe B is *H/h W/w* or *h/h W/w*. From the genotype of Ewe C, we know she is *H/H w/w*.

4.32 Common pattern baldness is more frequent in males than in females. The appreciable difference in frequency is assumed to be due to

a. Y-linkage of this trait.
b. X-linked recessive mode of inheritance.
c. sex-influenced autosomal inheritance.
d. excessive beer-drinking in males, consumption of gin being approximately equal between the sexes.

Answer: c.

CHAPTER 5

LINKAGE, CROSSING-OVER, AND GENE MAPPING IN EUKARYOTES

I. CHAPTER OUTLINE

DISCOVERY OF GENETIC LINKAGE
- Linkage in the Sweet Pea
- Morgan's Linkage Experiments with *Drosophila*

GENE RECOMBINATION AND THE ROLE OF CHROMOSOMAL EXCHANGE
- Corn Experiments
- *Drosophila* Experiments
- Crossing-Over at the Tetrad (Four-Chromatid) Stage of Meiosis

LOCATING GENES ON CHROMOSOMES: MAPPING TECHNIQUES
- Detecting Linkage Through Testcrosses
- Gene Mapping by Using Two-Point Testcrosses
- Generating a Genetic Map
- Double Crossovers
- Three-Point Crosses
- Mapping Chromosomes by Using Three Point Testcrosses

II. REVIEW OF KEY TERMS, SYMBOLS AND CONCEPTS

Without consulting the text and in your own words, write a brief definition of each term in the groups below. Then, either using a short phrase or a simple diagram, identify the relationship(s) between specific pairs of terms within a set. Finally, consult the text (and perhaps a friend who has also done the exercise) to check your answers.

1	2	3	4
linkage partial linkage linked genes linkage group chromosome	parentals recombinants genetic recombination independent assortment coupling vs. repulsion	crossing-over linkage group chiasma reciprocal exchange	cytological markers genetic markers translocation

5	6	7
genetic recombination	genetic map	two-point testcross
crossing-over	centi-Morgan (cM)	three-point testcross
four-strand stage	map unit	double crossovers
ordered tetrads	recombination frequency	chiasma (chromosomal) interference
ascus	mapping function	coefficient of coincidence
	multiple crossovers	

III. THINKING ANALYTICALLY

When beginning to work on a mapping problem, use a systematic approach. Organize the data to extract important information by inspection. It often helps to consider just two genes at a time and arrange data into more accessible units. Good organization lets you determine whether two genes are linked, and if linked, whether they are sex-linked. A trait appearing in only one sex-type of the progeny should be checked for sex-linkage.

Remember that two genes that show 50 percent recombination in a test-cross may either be on different chromosomes, or be on the same chromosome, but very far apart. The map distances between two genes can exceed 50 map units, but their recombination frequency cannot exceed 50 percent. If two genes showing 50 percent recombination are each linked to a third, intermediate gene (i.e., they each show <50 percent recombination with the third gene), then they will be on the same chromosome but more than 50 map units apart.

In a three-point testcross, be especially methodical.

1. Assign gene symbols to the phenotypes that are seen and rewrite the crosses.
2. Re-organize the progeny genotypes according to the frequency of each class.
3. Assign parental, single-crossover and double-crossover classes—parentals will be most frequent, double-crossovers least frequent.
4. Infer the gene order by comparing the parental and double-crossover classes.
5. Re-write the testcross; infer which single-crossover gives rise to which class.
6. Calculate the recombination frequency and map distance for each gene interval.
7. Calculate the coefficient of coincidence and the interference for the region.

When using the chi-square test to check for linkage, remember that this test, like most statistical tests, only serves as a guide for invalidation, not validation. Carefully consider the hypothesis that you are testing, and what the values of χ^2 and P mean.

IV. QUESTIONS FOR PRACTICE

A. Multiple Choice Questions

Choose the correct answer or answers for the following questions.

1. Evidence that genetic recombination is associated with physical exchange between chromosome homologues was found by
   a. observing the existence of chiasmata during meiosis.
   b. observing the pairing of homologs.

c. using a combined genetic and cytological approach.
d. showing that genetic recombination was accompanied by an exchange of identifiable chromosomal segments.

2. One significant use of ordered tetrads in *Neurospora* was to show that
   a. crossing-over occurs at the 2-chromatid stage.
   b. crossing-over occurs at the 4-chromatid stage.
   c. crossing-over can occur when meiosis does not occur.
   d. crossing-over occurs after meiosis has been completed.

3. Alleles are considered to be linked if recombination is < 50 percent in a (an)
   a. $P \times P \rightarrow F_1$ cross.
   b. $F_1 \times F_1 \rightarrow F_2$ cross.
   c. $F_2 \times F_2 \rightarrow F_3$ cross.
   d. F_1 x homozygous recessive cross.
   e. any F_2 progeny x homozygous recessive cross.

4. If two loci *A/a* and *B/b* are 9 map units apart, and an *A b/a B* individual is testcrossed,
   a. there will be 9 percent *A B* individuals.
   b. there will be 91 percent *A b* individuals.
   c. there will be 4.5 percent *a b* individuals.
   d. there will be 91 percent recombinant individuals.

5. If the genes *a* and *b* show 50 percent recombination, genes *a* and *c* show 35 percent recombination, and genes *b* and *c* show 32 percent recombination,
   a. genes *a* and *b* are on the same chromosome, and 50 map units apart.
   b. genes *a* and *b* are on different chromosomes.
   c. genes *a* and *b* are on the same chromosome, and 67 map units apart.
   d. one cannot tell if genes *a* and *b* are on the same or different chromosomes from this data.

6. A chi-square test is performed with a hypothesis that two genes are unlinked. A chi-square value of 11.35 is obtained, which in turn gives a value of $P = 0.01$ (i.e., 1 percent). Based on this information, one should
   a. accept the hypothesis as being possible and expect that the genes might be unlinked.
   b. accept the hypothesis as being proven true, and know that the genes are unlinked.
   c. reject the hypothesis as being unlikely and expect that the genes might be linked.
   d. reject the hypothesis as being impossible, and know the genes are linked.

7. In a three-point cross, interference is a measure of
   a. how often one chiasma physically impedes the occurrence of a second crossover in a particular region.
   b. the accuracy of the map distances between three genes.
   c. whether the data are considered reliable.
   d. whether the genes are linked.

8. In a region in which interference is 0.85,
   a. there are fewer double crossovers than expected.
   b. there are more double crossovers than expected.
   c. one observed the expected number of double crossovers.
   d. 0.85 percent of testcross progeny show double-crossover phenotypes.

9. If two genes *a* and *b* are 8.3 map units apart, one expects to find
   a. 8.3 percent recombinant gametes from a doubly heterozygous parent.
   b. 91.7 percent recombinant gametes from a doubly heterozygous parent.
   c. A chiasma between the *a* and *b* loci in 8.3% of the meioses.
   d. A chiasma between the *a* and *b* loci in 91.7% of the meioses.

10. Map distances are most accurate when genes are closely linked because as genes become further apart,
   a. they always show independent assortment.
   b. the chance of multiple crossovers increases.
   c. they no longer recombine.
   d. there is chiasma interference.

Answers: 1c & d; 2b; 3d; 4c; 5c; 6c; 7a; 8a; 9a; 10b.

B. Thought Questions

1. Do you think that mapping according to the methods described in this chapter will still be useful if the entire DNA sequence of an organism is known? (Hint: Does genetic and/or cytological mapping narrow the field of possibilities?)
2. Why do you think that sex linkage was described before other examples of linkage in animals? (Hint: Is this an example of a correlation of genetic and cytological data?)
3. Why were the correlations made by Creighton and McClintock in their study of the *waxy* and *colorless* loci in corn important? (Hint: What was the value of cytological data to genetic experiments during that time?)
4. Can one make genetic maps of haploid organisms? (Hint: Review *Neurospora* material.)
5. Does the decision to accept or reject an hypothesis at the 0.05 level mean that one is more likely to discard a correct hypothesis or retain an incorrect one?
6. Do all chromosomes have the same length in terms of map distances?
7. In some organisms, correlations have been made between the physical distance between two closely linked loci and the map distance (i.e., the number of base pairs of DNA) between two loci. If one measures the recombination frequency between two genes in another chromosomal region, can one infer the physical distance between them? (Hint: Consider the basis of interference and the observation that different regions of a chromosome show different amounts of crossing-over.)

V. SOLUTIONS TO TEXT PROBLEMS

5.1 A cross $a^+a^+ \; b^+b^+ \; \times \; aa \; bb$ results in an F_1 of phenotype $a^+ \; b^+$; the following numbers are obtained in the F_2 (phenotypes):

$a^+ \; b^+$	110
$a^+ \; b$	16
$a \; b^+$	19
$a \; b$	15
Total	160

Are genes at the *a* and *b* loci linked or independent? What F_2 numbers would otherwise be expected?

Answer: If two genes are unlinked in a dihybrid cross, one expects a 9:3:3:1 ratio in the F_2. For 160 progeny, one would expect 90:30:30:10. The ratio here is clearly different from this, and the genes must be linked.

5.2 In corn a dihybrid for the recessives *a* and *b* is testcrossed. The distribution of the phenotypes was as follows:

A B	122
A b	118
a B	81
a b	79

Are the genes assorting independently? Test the hypothesis with a χ^2 test. Explain tentatively any deviation from expectation, and tell how you would test your explanation.

Answer: Use the χ^2 test to test the hypothesis that the genes are unlinked. With this hypothesis, one would expect that the cross *Aa Bb* x *aa bb* to produce four equally frequent classes of progeny (i.e., a 1:1:1:1 ratio). With 400 total progeny, one would expect 100 progeny in each phenotypic class. One has:

Phenotype	Observed (= *o*)	Expected (= *e*)	$(o - e)^2/e$
A B	122	100	4.84
A b	118	100	3.24
a B	81	100	3.61
a b	79	100	4.41

$\chi^2 = 16.1$, df = 3

From the χ^2 test, $P < 0.01$. That is, there is less than a 1 percent likelihood of observing this much deviation from expected values by chance alone. This indicates that one can reject the hypothesis of "no linkage" as being unlikely. Note that the chi-square test neither proves that the genes are unlinked, nor proves that they are linked.

Linkage might seem reasonable until it is realized that the minority classes are not reciprocal classes (both carry the *aa* phenotype). If the segregation at each locus is considered, however, the *B/–:b/b* ratio is about 1:1 (203:197), while the *A/–:a/a* ratio is not (240:160). The departure of the observed from expected values is thus due to a lack of *a/a* individuals. This departure should be confirmed in other crosses that test the segregation at the *A/a* locus. It would be particularly informative to check if the lack of *a/a* corn plants is associated with ungerminated seeds or seedlings that die before they are able to mature.

5.3 The F_1 from a cross of *A B/A B* x *a b/a b* is testcrossed, resulting in the following phenotypic ratios:

A B	308
A b	190
a b	292
a B	210

What is the frequency of recombination between genes *a* and *b?*

Answer: Recombination frequency = (# recombinants/total) x 100%
= [(190 + 210)/(308 + 190 + 292 + 210)] x 100%
= 40%

5.4 In rabbits the English type of coat (white-spotted) is dominant over non-English (unspotted), and short hair is dominant over long hair (Angora). When homozygous English, short-haired rabbits were crossed with non-English Angoras and the F_1 crossed back to non-English Angoras, the following offspring were obtained: 72 English and short-haired, 69 non-English and Angora; 11 English and Angora; and 6 non-English and short-haired. What is the map distance between the genes for coat color and hair length?

Answer: Let *E* and *e* represent English and non-English coats, and let *S* and *s* represent short and long hair, respectively. The initial cross can be denoted *EE SS* x *ee ss*, and the cross involving the F_1 can be denoted *Ee Ss* x *ee ss*. The progeny of this test-cross are 72 *Ee Ss*, 69 *ee ss*, 11 *Ee ss* and 6 *ee Ss*. This is clearly not a 1:1:1:1 ratio and the two genes are linked. Recombination frequency = [(11 + 6)/158] x 100% = 10.8%. There are 10.8 map units between the genes.

5.5 In *Drosophila* the mutant black (*b*) has a black body, and the wild type has a grey body; the mutant vestigial (*vg*) has wings that are much shorter and crumpled compared to the long wings of the wild type. In the following cross, the true-breeding parents are given together with the counts of offspring of F_1 females x black and vestigial males:

P black and normal x grey and vestigial
F_1 females x black and vestigial males

Progeny:		
	grey, normal	283
	grey, vestigial	1,294
	black, normal	1,418
	black, vestigial	241

From these data, calculate the map distance between the black and vestigial genes.

Answer: The cross is $b\ vg^+/b\ vg^+ \times b^+\ vg/b^+\ vg$, which gives an F_1 that is $b\ vg^+/b^+\ vg$. An F_1 female is crossed with a homozygous recessive $b\ vg/b\ vg$. The female is used as the doubly heterozygous parent as no crossing-over occurs in male *Drosophila*. The classes can be grouped in reciprocal pairs as follows:

Non-recombinants (parentals):	grey, vestigial	1,294
	black, normal	1,418
Recombinants (non-parentals):	grey, normal	283
	black, vestigial	241
	Total progeny	3,236

Recombination frequency = [(283 + 241)/3,236] x 100% = 16.2%
There are 16.2 map units between the genes.

5.6 A gene controlling wing size is located on chromosome 2 in *Drosophila*. The recessive allele *vg* results in vestigial wings when homozygous; the vg^+ allele determines long wings. A new eye mutation, which we will call "maroon-like," is isolated. Homozygous maroon-like (*m/m*) results in maroon colored eyes; the m^+ allele is bright red. The location of the *m* gene is unknown, and you are asked to design an experiment to determine whether *m* is located on chromosome 2.

You cross true-breeding virgin maroon females to true-breeding *vg/vg* males and obtain all wild-type F_1 progeny. Then you allow the F_1 to interbreed. As soon as the F_2 start to hatch, you begin to classify the flies, and among the first six newly-hatched flies, you find four wild type; one vestigial-winged and red-eyed fly; and one vestigial-winged, maroon-eyed fly. You immediately draw the conclusions that (1) maroon-like is not X-linked and (2) maroon-like is not linked to vestigial. Based on this small sample, how could you tell? On what chromosome is *m* located? (Hint: There is no crossing-over in *Drosophila* males.)

Answer: To eliminate the possibility that *m* is X-linked, consider what kinds of progeny you would expect if it were. If *m* were X-linked, one could write the parental cross as: $X^m/X^m\ vg^+/vg^+ \times X^{m^+}/Y\ vg/vg$. The F_1 males would necessarily be $X^m/Y\ vg^+/vg$, and so be maroon-eyed. Since the F_1 are all wild-type, *m* cannot be X-linked.

To eliminate the possibility that *vg* is linked to *m*, consider the kinds of progeny that could be obtained if it were. If *vg* and *m* are linked, one could write the crosses as:

$$\text{P} \quad \frac{vg^+\ m}{vg^+\ m} \text{ females} \times \frac{vg\ m^+}{vg\ m^+} \text{ males}$$

$$\downarrow$$

$$\text{F}_1 \quad \frac{vg^+\ m}{vg\ \ m^+} \text{ (All wild type)}$$

$$\text{The } F_1 \times F_1 \text{ cross is } \frac{vg^+\ m}{vg\ \ m^+} \text{ females} \times \frac{vg^+\ m}{vg\ \ m^+} \text{ males}$$

Eggs produced by F_1 females may be $vg\ m^+$ (parental), $vg^+\ m$ (parental), $vg\ m$ (recombinant) or $vg^+\ m^+$ (recombinant). Assuming that the two genes are linked, and remembering that there is no crossing-over in *Drosophila* males, the F_1

males' gametes must be *vg* m^+ or vg^+ *m* (parental types). Note that if the two genes are on the same chromosome, no matter how far apart they are, the F_1 males' gametes must be these parental types. Since none of the F_1 males' gametes are *vg m*, it is impossible for the progeny of the F_1 cross to be homozygous *vg m*/*vg m*. (Check this by diagramming a Punnett square.) However, since one of the progeny of the F_1 cross is both vestigial-winged and maroon-eyed, it must have obtained the *m* and *vg* alleles from both parents. Given the genotype of the males, this could only occur through independent assortment, i.e., if the two genes are on different chromosomes.

✯ 5.7 Use the following two-point recombination data to map the genes concerned. Show the order and the length of the shortest intervals.

GENE LOCI	% RECOMBINATION	GENE LOCI	% RECOMBINATION
a, b	50	*b, d*	13
a, c	15	*b, e*	50
a, d	38	*c, d*	50
a, e	8	*c, e*	7
b, c	50	*d, c*	45

Answer: Because of the effects of double crossovers, larger recombination frequencies are less accurate measures of map distances than smaller recombination frequencies between close neighbors. As a consequence, the map distances that can be derived from this data are not strictly additive (e.g., $a-d = 38$ and $a-e = 8$ but $d-e = 45$). The distances between nearest neighbors can be used to infer the following map.

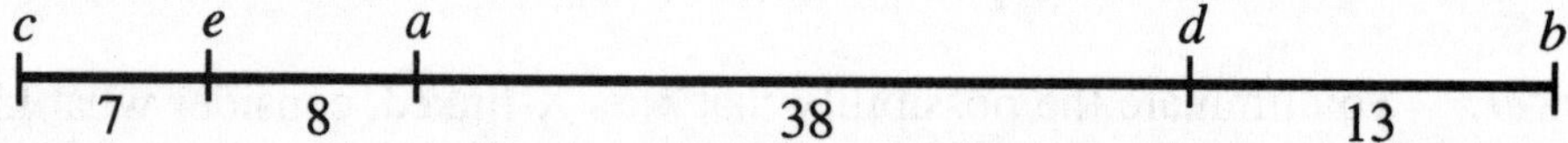

Remember that although recombination frequency cannot exceed 50 percent, map distances can exceed 50 map units. Thus, the map distance between *c* and *b* can be inferred to be 66 map units (= 7 + 8 + 38 + 13), although *c* and *b* will only show 50 percent recombination.

5.8 Use the following two-point recombination data to map the genes concerned. Show the order and the length of the shortest intervals.

LOCI	% RECOMBINATION	LOCI	% RECOMBINATION
a, b	50	*c, d*	50
a, c	17	*c, e*	50
a, d	50	*c, f*	7
a, e	50	*c, g*	19
a, f	12	*d, e*	7
a, g	3	*d, f*	50
b, c	50	*d, g*	50
b, d	2	*e, f*	50
b, e	5	*e, g*	50
b, f	50	*f, g*	15
b, g	50		

Answer: Note that, as discussed in the solution to problem 5.8, the most accurate map distances are obtained by the summing the map distances between nearest neighbors. Also, remember that while the percent recombination between two genes cannot exceed 50 percent, map distances can exceed 50 map units. When the percent recombination between two genes is 50 percent, either they are far apart on the same chromosome or on different chromosomes. In some cases, one of these options can be eliminated: If two genes lie far apart on the same chromosome, they may show 50 percent recombination. If sufficient data are gathered, each gene will show evidence of linkage to other, intervening loci. The map distances between the two distant loci can then be determined by adding up the distances between the intervening loci. Consider two genes that show 50 percent recombination and for which no data exist to show linkage of each of them to intervening loci. For these genes, it will not be possible to distinguish between the genes lying on different chromosomes, and the genes being very far apart on the same chromosome.

Start this problem by initially constructing a map using the smallest map distances, and then checking for linkage with other genes. The solution is:

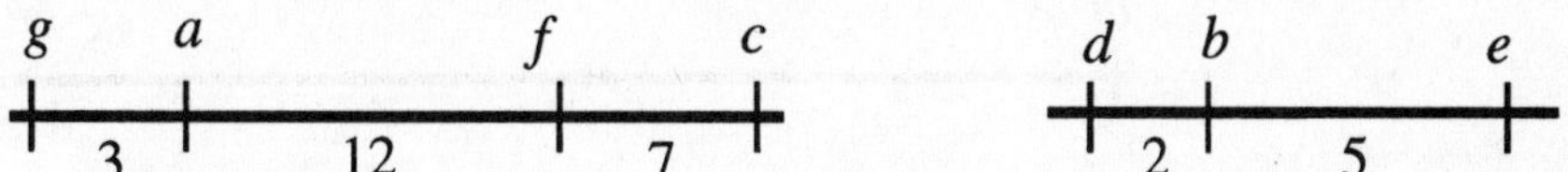

5.9 The following data are from Bridges and Morgan's work on the recombination between the genes black, curved, purple, speck, and vestigial in chromosome 2 of *Drosophila.* On the basis of the data, map the chromosome for these five genes as accurately as possible. Remember that determinations for short distances are more accurate than those for long ones.

GENES IN CROSS	TOTAL PROGENY	NUMBER OF RECOMBINANTS
black, curved	62,679	14,237
black, purple	48,931	3,026
black, speck	695	326
black, vestigial	20,153	3,578
curved, purple	51,136	10,205
curved, speck	10,042	3,037
curved, vestigial	1,720	141
purple, speck	11,985	5,474
purple, vestigial	13,601	1,609
speck, vestigial	2,054	738

Answer: Determine recombination frequencies for each pair of genes by dividing the number of recombinants by the total number of progeny for that pair of genes. One obtains:

GENES IN CROSS	PERCENT OF TOTAL PROGENY
black, curved	22.7
black, purple	6.2
black, speck	47.6
black, vestigial	17.8
curved, purple	20.0
curved, speck	30.2
curved, vestigial	8.2
purple, speck	45.7
purple, vestigial	11.8
speck, vestigial	35.9

Now analyze the data to determine the order of the genes. One approach is to consider that the pair with the greatest recombination frequency are the furthest apart, and then order genes from one or the other end. The genes black and speck are furthest apart:

Now fill in the order by finding the sequence of genes that has an increasing order of recombination with black:

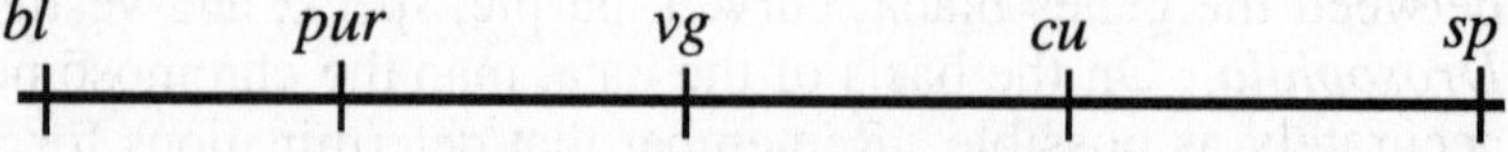

Next, assign map distances based on the recombination frequencies of neighboring genes:

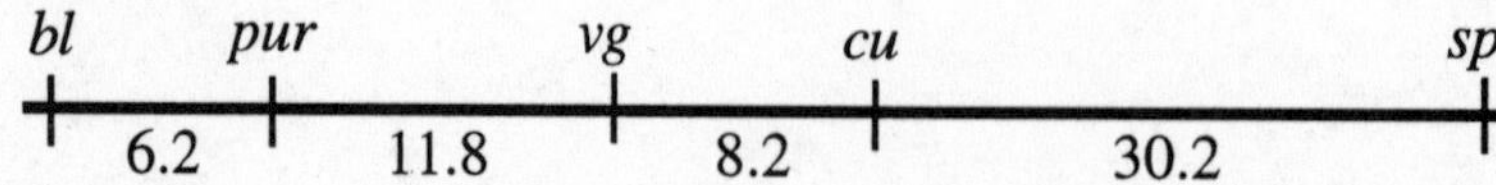

Check gene order by choosing one of the genes (e.g., curved) and verifying that the distances between that gene and the others are approximately correct. Because of multiple crossovers, larger recombination frequencies tend to give underestimates of map distance. The most accurate map distances between two genes can be determined by summing the map distances in all intervals between the genes. Thus, the map distance from *bl* to *sp* is most accurately given by 56.4 (= 6.2 + 11.8 + 8.2 + 30.2) map units, and not 47.6 (= RF x 100%) map units.

5.10 A corn plant known to be heterozygous at three loci is testcrossed. The progeny phenotypes and frequencies are as follows:

+	+	+	455
a	*b*	*c*	470
+	*b*	*c*	35
a	+	+	33
+	+	*c*	37
a	*b*	+	35
+	*b*	+	460
a	+	*c*	475
		Total	2,000

Give the gene arrangement, linkage relations, and map distances.

Answer: First, note that since the offspring result from a testcross, the phenotypes of the offspring reflect the genotypes of the gametes of the heterozygous parent. Consequently, these phenotypes can be used to determine the gene arrangement in the heterozygous parent.

Second, notice that there are eight genotypic classes that appear in two frequencies. Four of the genotypic classes have about 460 individuals, while the other four have about 35 individuals. This would be expected if two of the genes were linked and the third was not. In this case, the chromosome bearing the two linked genes could undergo crossing-over, and give rise to two frequencies of gametes. Because crossing-over between two closely linked genes is infrequent, gametes with non-crossover chromosomes are more frequent. The third gene would assort independently. It is this independent assortment that contributes to the four genotypic classes within each frequency seen.

With this information, one can proceed in either of two ways. First, one can tabulate the data considering only two pairs of genes at one time. This allows one to view this cross as a series of two-point crosses.

a	*b*	*c*	# progeny
a	*b*		470 + 35 = 505
+	*b*		460 + 35 = 495
a	+		475 + 33 = 508
+	+		455 + 37 = 592
a		*c*	470 + 475 = 945
+		*c*	35 + 37 = 72
a		+	33 + 35 = 68
+		+	455 + 460 = 915
	b	*c*	470 + 35 = 505
	+	*c*	475 + 37 = 512
	b	+	460 + 35 = 495
	+	+	455 + 33 = 488

Notice that there is a 1:1:1:1 ratio in the offspring when one considers just the *a* and *b* genes, or just the *b* and *c* genes. This is what one would expect in a dihybrid cross with two independently assorting genes. Thus, *a* assorts independently of *b* and *b* assorts independently of *c*. This is not the case for *a* and *c* however. Here, there are two frequencies of data indicative of two linked genes showing recombination. The recombination frequency between *a* and *c* is RF = [(72 + 68)/(72 + 68 + 945 + 915)] x 100% = 7%, and there are 7 map units between *a* and *c*.

Alternatively, one can use the reasoning presented above to infer that the four progeny classes with about 460 individuals contain parental types of chromosomes for two linked genes, as well as an independently assorting chromosome with the third gene. By inspection of the most frequent phenotypes, one sees that only *a c* and + + chromosomes are found, while the *b*/+ gene assorts independently with these non-crossover chromosomes. Thus, the *a* and *c* loci are linked, and the *b* gene is unlinked. The least frequent progeny classes contain *a* + and + *c* recombinant chromosomes, so that the percent recombination between *a* and *c* is RF = [(35 + 37 + 33 + 35)/2,000] x 100% = 7%. Just as was seen earlier, the map distance between *a* and *c* is 7 map units.

One can diagram these results as:

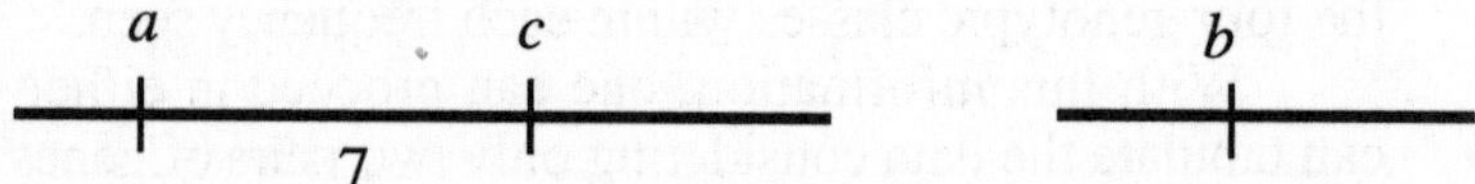

5.11 Genes *a* and *b* are linked, with 10 percent recombination. What would be the phenotypes, and the probability of each, among progeny of the following cross?

$$\frac{a\ b^{+}}{a^{+}\ b} \times \frac{a\ b}{a\ b}$$

Answer: The $a\ b^{+}/a^{+}\ b$ parent will give 90 percent parental type ($a\ b^{+}$ or $a^{+}\ b$) gametes and 10 percent recombinant type gametes ($a\ b$ or $a^{+}\ b^{+}$). The genotypes of these gametes will determine the offspring phenotypes, as the $a\ b/a\ b$ parent will give only $a\ b$ gametes. Therefore there will be 45% $a\ b^+$, 45% $a^+\ b$, 5% $a\ b$ and 5% $a^+\ b^+$ offspring.

5.12 Genes a and b are sex-linked and are located 7 mu apart in the X chromosome of *Drosophila*. A female of genotype $a^+\ b/a\ b^+$ is mated with a wild-type ($a^+\ b^+$) male.

a. What is the probability that one of her sons will be either $a^+\ b^+$ or $a\ b^+$ in phenotype?

b. What is the probability that one of her daughters will be $a^+\ b^+$ in phenotype?

Answer: Since the genes are 7 mu apart, the female will have 93% parental type gametes (46.5% $a^+\ b$ and 46.5% $a\ b^+$) and 7% recombinant type gametes (3.5% $a\ b$ and 3.5% $a^+\ b^+$). The wild-type male will give either an X chromosome bearing a^+b^+ or a Y chromosome. As the genes are X-linked, the sons' phenotype will reflect the chromosome they receive from the mother. The daughters' phenotype will be $a^+\ b^+$, as they receive their father's $a^+\ b^+$ X chromosome.

a. $p = 3.5\%\ (a^+\ b^+) + 46.5\%\ (a\ b^+) = 50\%$.

b. $p = 100\%$

5.13 In *Drosophila*, a and b are linked autosomal genes whose recombination frequency in females is 5 percent; c and d are X-linked genes, located 10 map units apart. A homozygous dominant female is mated to a recessive male, and the daughters are testcrossed. Which of the following would you expect to observe in the testcross progeny?

a. Different ratios in males and females.

b. Nearly equal frequency of $a^+\ b\ c^+\ d^+$, $a^+\ b\ c\ d$, $a\ b^+\ c^+\ d^+$, and $a\ b^+\ c\ d$ classes.

c. Independent segregation of some gene with respect to others involved in the cross.

d. Double-crossover classes less frequent than expected because of interference between the two marked regions.

Answer: The first cross can be represented as $c^+\ d^+/c^+\ d^+$; $a^+\ b^+/a^+\ b^+$ (female) × $c\ d/Y$ $a\ b/a\ b$ (male). The testcross can be represented as $c^+\ d^+/c\ d$; $a^+\ b^+/a\ b$ × $c\ d/Y$; $a\ b/a\ b$. The testcross is set up so that both male and female progeny will have the same ratios, so part (a) is not true. All of the classes shown in part (b) are classes that derive from a single crossover between a and b, so part (b) is true. Genes c and d will segregate independently of genes a and b, so part (c) is true. As interference is only measured between adjacent gene intervals on the same chromosome, part (d) is untrue.

5.14 In maize the dominant genes A and C are both necessary for colored seeds. Homozygous recessive plants give colorless seed, regardless of the genes at the second locus. Genes A and C show independent segregation, while the recessive mutant gene waxy endosperm (wx) is linked with C (20 percent recombination). The dominant Wx allele results in starchy endosperm.

a. What phenotypic ratios would be expected when a plant of constitution $c\ Wx/C\ wx\ A/A$ is testcrossed?

b. What phenotypic ratios would be expected when a plant of constitution $c\ Wx/C\ wx\ A/a$ is testcrossed?

Answer: Use branch diagrams to assess what kind of gametes will arise from the heterozygous plants.

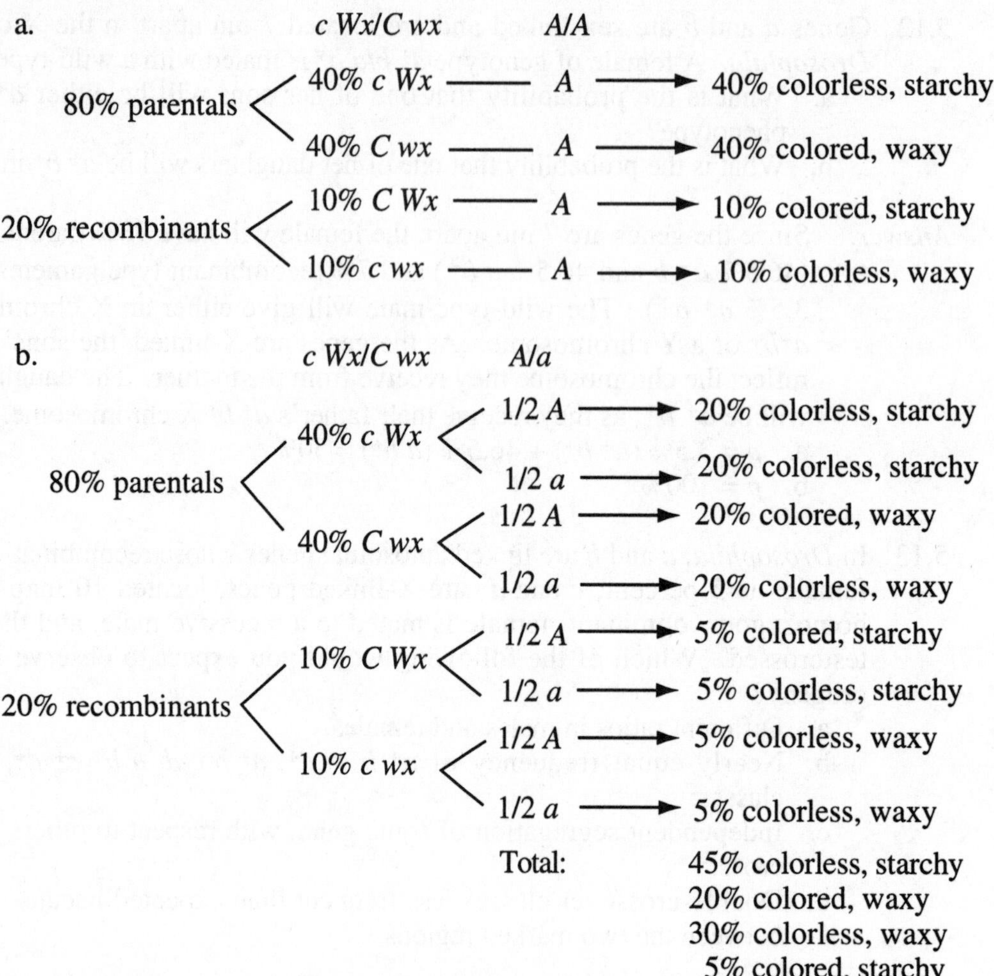

5.15 Assume that genes *a* and *b* are linked and show 20 percent crossing-over.

a. If a homozygous *A B*/*A B* individual is crossed with an *a b*/*a b* individual, what will be the genotype of the F_1? What gametes will the F_1 produce and in what proportions? If the F_1 is testcrossed with a doubly homozygous recessive individual, what will be the proportions and genotypes of the offspring?

b. If, instead, the original cross is *A b*/*A b* x *a B*/*a B*, what will be the genotype of the F_1? What gametes will the F_1 produce and in what proportions? If the F_1 is testcrossed with a doubly homozygous recessive, what will be the proportions and genotypes of the offspring?

Answer:

a. The F_1 will be *A B*/*a b* and produce 80% parental type gametes (40% *A B*, 40% *a b*) and 20% recombinant type gametes (10% *A b*, 10% *a B*). A testcross will produce 40% *Aa B* genotypes with an *A B* phenotype, 40% *aa bb* genotypes with an *a b* phenotype, 10% *Aa bb* genotypes with an *A b* phenotype and 10% *aa Bb* genotypes with an *a B* phenotype.

b. The genotype of the F_1 will be *A b*/*a B* (phenotype = *A B*). The F_1 gametes will be 40% *A b*, 40% *a B*, 10% *a b*, and 10% *A B*. A testcross will produce 40% *Aa bb* genotypes with an *A b* phenotype, 40% *aa Bb* genotypes with an *a B* phenotype, 10% *Aa Bb* genotypes with an *A B* phenotype and 10% *aa bb* genotypes with an *a b* phenotype.

5.16 In tomatoes, tall vine is dominant over dwarf, and spherical fruit shape is dominant over pear shape. Vine height and fruit shape are linked, with a recombinant percentage of 20. A certain tall, spherical-fruited tomato plant is crossed with a dwarf, pear-fruited plant. The progeny are 81 tall, spherical; 79 dwarf, pear; 22 tall, pear; and 17 dwarf, spherical. Another tall and spherical plant crossed with a dwarf and pear plant produces 21 tall, pear; 18 dwarf, spherical; 5 tall, spherical; and 4 dwarf, pear. What are the genotypes of the two tall and spherical plants? If they were crossed, what would their offspring be?

Answer: Let *T* represent tall vine, *t* represent dwarf, *S* represent spherical and *s* represent pear. Then, the initial tall, spherical x dwarf, pear cross can be represented as *T S*/– – x *t s*/*t s*. Since this is a test cross, the genotypes of the gametes of the (potentially) heterozygous plant can be inferred from the progeny phenotypes:

Phenotype	Gamete Genotype	First Cross #	First Cross Type	Second Cross #	Second Cross Type
tall, spherical	*T S*	81	parental	5	recombinant
dwarf, pear	*t s*	79	parental	4	recombinant
tall, pear	*T s*	22	recombinant	21	parental
dwarf, spherical	*t S*	17	recombinant	18	parental

In the first cross, the four classes of gametes, their phenotypes and frequencies indicate that the parental genotypes were *T S*/*t s* x *t s*/*t s*. Similarly, the four classes of progeny and their frequencies in the second cross indicate that it is *T s*/*t S* x *t s*/*t s*. If the two tall and spherical plants were crossed, one would have *T S*/*t s* x *T s*/*t S*. One can diagram the expected progeny using a branch diagram:

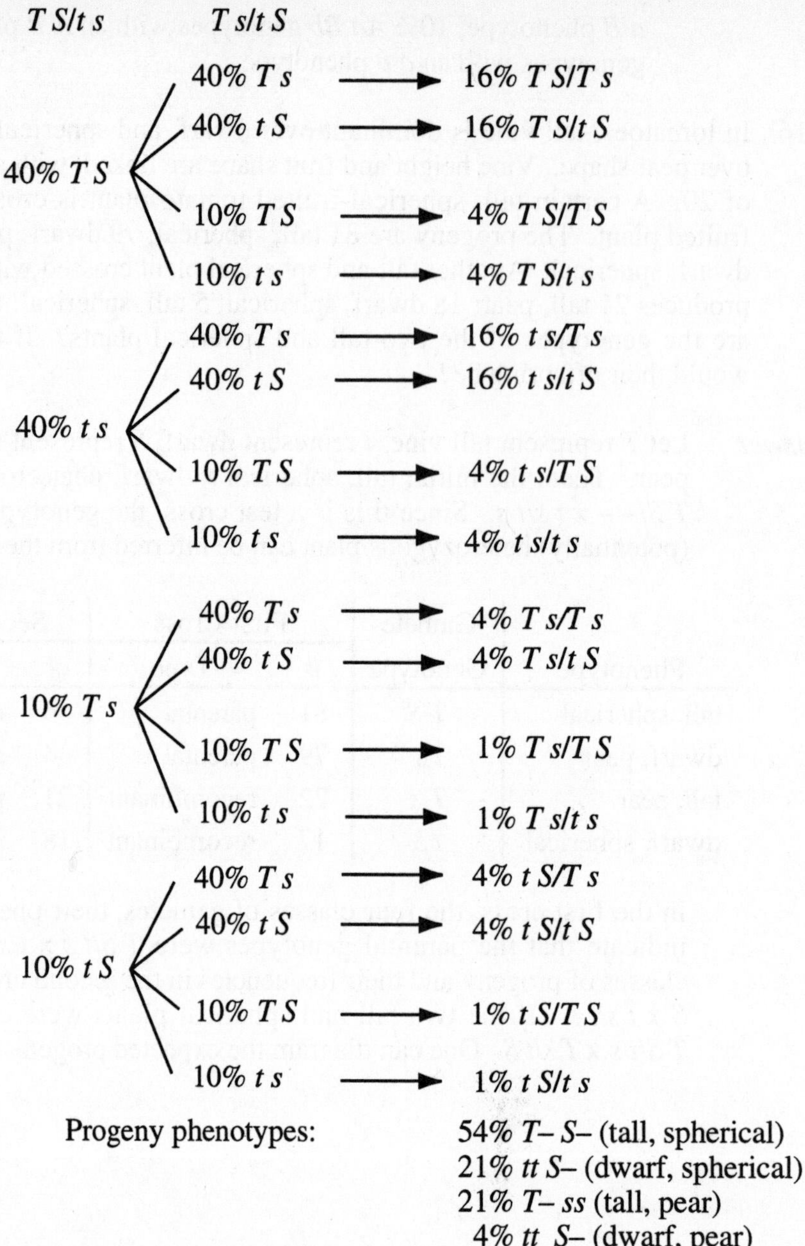

Progeny phenotypes:

54% T– S– (tall, spherical)
21% tt S– (dwarf, spherical)
21% T– ss (tall, pear)
4% tt S– (dwarf, pear)

5.17 Genes *a* and *b* are in one chromosome, 20 mu apart; *c* and *d* are in another chromosome, 10 mu apart. Genes *e* and *f* are in yet another chromosome and are 30 mu apart. Cross a homozygous *A B C D E F* individual with an *a b c d e f* one, and cross the F_1 back to an *a b c d e f* individual. What are the chances of getting individuals of the following phenotypes in the progeny?

a. *A B C D E F*
b. *A B C d e f*
c. *A b c D E f*
d. *a B C d e f*
e. *a b c D e F*

Answer: The frequency of recombination between two genes can be inferred from the map distance between them. For example, the 20 mu between *a* and *b* indicate that in an *AB/ab* individual, gametes will have 80% parental-type chromosomes (40%

each of *AB*, *ab*) and 20% recombinant-type chromosomes (10% each of *Ab*, *aB*). The cross under consideration is *AB/ab; CD/cd; EF/ef* x *ab/ab; cd/cd; ef/ef.*

a. The chance of obtaining the parental-type *AB*, *CD*, and *EF* chromosomes is $p = 0.40 \times 0.45 \times 0.35 = 0.063$ or 6.3%.
b. The chance of obtaining *AB*, *Cd*, and *ef* chromosomes is $p = 0.40 \times 0.05 \times 0.35 = 0.007$ or 0.7%.
c. The chance of obtaining *Ab*, *cD*, and *Ef* chromosomes is $p = 0.10 \times 0.05 \times 0.15 = 0.00075$ or 0.075%.
d. The chance of obtaining *aB*, *Cd*, and *ef* chromosomes is $p = 0.10 \times 0.05 \times 0.35 = 0.00175$ or 0.175%.
e. The chance of obtaining the *ab*, *cD*, and *eF* chromosomes is $p = 0.40 \times 0.05 \times 0.15 = 0.003$ or 0.3%.

5.18 Genes *d* and *p* occupy loci 5 map units apart in the same autosomal linkage group. Gene *h* is a separate autosomal linkage group and therefore segregates independently of the other two. What types of offspring are expected, and what is the probability of each, when individuals of the following genotypes are testcrossed:

a. $\dfrac{D\ P\ h}{d\ p\ h}$ b. $\dfrac{d\ P\ H}{D\ p\ h}$

Answer: a. 95% of the progeny will have *D P* or *d p* parental type chromosomes, and 5% of the progeny will have *D p* or *d P* recombinant type chromosomes. All of the progeny will be *h*. There will be 47.5% *D P h*, 47.5% *d p h*, 2.5% *D p h*, 2.5% *d P h*.
b. 23.75% *d P H*, 23.75% *d P h*, 23.75% *D p H*, 23.75% *D p h*, 1.25% *D P H*, 1.25% *D P h*, 1.25% *d p H*, 1.25% *d p h*.

5.19 A hairy-winged (*h*) *Drosophila* female is mated with a yellow-bodied (*y*), white-eyed (*w*) male. The F_1 are all normal. The F_1 progeny are then crossed, and the F_2 that emerge are as follows:

Females:	wild type	757
	hairy	243
Males:	wild type	390
	hairy	130
	yellow	4
	white	3
	hairy, yellow	1
	hairy, white	2
	yellow, white	360
	hairy, yellow, white	110

Give genotypes of the parents and the F_1, and note the linkage relations and distances, where appropriate.

Answer: Careful inspection of the data shows that the F_2 males, but not the F_2 females, show the yellow and white traits. In contrast, the hairy trait is shown in an approximately 3:1 ratio in both the females and males (females–757 normal:243 hairy; males–757 normal:243 hairy). This indicates that the yellow and white genes are sex-linked, while hairy is not.

The normal phenotype of the F_1 heterozygote indicates that the non-hairy, non-yellow and non-white traits are dominant. Therefore, let h and h^+ represent the hairy and non-hairy traits, respectively, y and y^+ represent the yellow and normal body-color traits, respectively, and w and w^+ represent the white and red-eyed traits, respectively. Then, given the phenotypes of the parents and F_1, and the sex-linkage of y and w, the crosses can be written as:

P: $y^+ w^+/y^+ w^+$; h/h females x $y\ w$/Y; h^+/h^+ males
F_1: $y^+ w^+/y\ w$; h^+/h females x $y^+ w^+$/Y; h^+/h males.

Now determine the map distance between y and w. This can be inferred from the frequency of recombinants among the F_2 males.

PHENOTYPE OF F_2 MALE	GENOTYPE	X-CHROMOSOME RECOMBINANT?	NUMBER
wild type	$y^+ w^+$/Y; h^+/h	no	300
hairy	$y^+ w^+$/Y; h/h	no	130
yellow	$y\ w^+$/Y; h^+/h	yes	4
white	$y^+ w$/Y; h^+/h	yes	3
hairy, yellow	$y\ w^+$/Y; h/h	yes	1
hairy, white	$y^+ w$/Y; h/h	yes	2
yellow, white	$y\ w$/Y; h^+/h	no	360
hairy, yellow, white	$y\ w$/Y; h/h	no	110

The recombination frequency between y and w is [(4 + 3 + 1 + 2)/1,000] x 100% = 1%, so there is 1 map unit between these genes. Note that, since recombination can only be scored in the F_2 males, the number of recombinants is divided by the number of *male* progeny, and not the total number of progeny.

5.20 In the Maltese bippy, amiable (*A*) is dominant to nasty (*a*), benign (B) is dominant to active (*b*), and crazy (*C*) is dominant to sane (*c*). A true-breeding amiable, active, crazy bippy was mated, with some difficulty, to a true-breeding nasty, benign, sane bippy. An F_1 individual from this cross was then used in a testcross (to a nasty, active, sane bippy) and produced, in typical prolific bippy fashion, 4,000 offspring. From an ancient manuscript entitled *The Genetics of the Bippy, Maltese and Other,* you discover that all three genes are autosomal, *a* is linked to *b* but not to *c* and the map distance between *a* and *b* is 20 mu.

a. Predict all the expected phenotypes and the numbers of each type from this cross.
b. Which phenotypic classes would be missing had *a* and *b* shown complete linkage?
c. Which phenotypic classes would be missing if *a* and *b* were unlinked?
d. Again, assuming *a* and *b* to be unlinked, predict all the expected phenotypes of nasty bippies and the frequencies of each type resulting from a self-cross of the F_1.

Answer: The crosses can be denoted as:

Parental: *A b*/*A b*; *C*/*C* x *a B*/*a B*; *c*/*c*
F_1 *A b*/*a B*; *C*/*c* x *a b*/*a b*; *c*/*c*.

a. Since the map distance between *a* and *b* is 20 mu, the F_2 will have 20 percent recombinants (classes indicated by an '*') between *a* and *b*.

Genotype	Phenotype	Percent	Number
A b/a b; C/c	amiable, active, crazy	20	800
A b/a b; c/c	amiable, active, sane	20	800
a B/a b; C/c	nasty, benign, crazy	20	800
a B/a b; c/c	nasty, benign, sane	20	800
a b/a b; C/c	nasty, active, crazy	5	200*
a b/a b; c/c	nasty, active, sane	5	200*
A B/a b; C/c	amiable, benign, crazy	5	200*
A B/a b; c/c	amiable, benign, sane	5	200*

b. If *a* and *b* had shown complete linkage, the recombinant (*) classes would be missing.

c. No phenotypic classes would be missing if *a* and *b* were unlinked. The frequency of each class would be identical (500, or 12.5 percent) however.

d. If *a* and *b* were unlinked, the results of selfing a triple heterozygote (the F_1) are as described previously for a trihybrid cross. The F_2 will have 8 phenotypic classes in a ratio of 27:9:9:9:3:3:3:1. The nasty bippies must be *aa* in genotype, and so their distribution is a subset of these classes. If one is only considering the nasty bippies, the other two pairs of phenotypes will be distributed in the 9:3:3:1 ratio expected from a dihybrid cross. One will see 9 nasty, benign, crazy:3 nasty, benign, sane:3 nasty, active, crazy:1 nasty, active, sane.

5.21 Fill in the blanks. Continuous bars indicate linkage, and the order of linked genes is correct as shown. If all types of gametes are equally probable, write "none" in the right column headed "Least frequent classes." In the right column, show two gamete genotypes, unless all types are equally frequent, in which case write "none."

Parent Genotypes	Number of Different Possible Gametes	Least Frequent Classes	
A b C / a B c (linked: *A b C*)	______	______	______
A b C / a B c (linked: *A b*; *C*)	______	______	______
A b C D / a B c d (linked: *A*; *b C D*)	______	______	______
A b C D E f / a B C d e f (linked: *A b C*; *D E f*)	______	______	______
b D / B d	______	______	______

Answer: Note that the number of classes of gametes is a function of how many different heterozygous loci are present. If there are n heterozygous loci, there will be 2^n possible gametes.

Parent Genotypes	Number of Different Possible Gametes	Least Frequent Classes	
$\frac{A\,b\,C}{a\,B\,c}$	$2^3 = 8$	none	
$\frac{A\ b\ C}{a\ B\ c}$	$2^3 = 8$	$A\,B\,C$	$a\,b\,c$
$\frac{A\ b\,C\,D}{a\ B\ c\,d}$	$2^4 = 16$	$A\,B\,C\,d$	$a\,b\,c\,D$
$\frac{A\ b\ C\,D\,Ef}{a\ B\ C\,d\ ef}$	$2^4 = 16$	$A\,B\,C\,D\,e\,f$	$a\,b\,C\,d\,E\,f$
$\frac{b\,D}{B\,d}$	$2^2 = 4$	$b\,d$	$B\,D$

5.22 For each of the following tabulations of testcross progeny phenotypes and numbers, state which locus is in the middle, and reconstruct the genotype of the tested triple heterozygotes.

a.		b.		c.	
A B C	191	*C D E*	9	*F G H*	110
a b c	180	*c d e*	11	*f g h*	114
A b c	5	*C d e*	35	*F g h*	37
a B C	5	*c D E*	27	*f G H*	33
A B c	21	*C D e*	78	*F G h*	202
a b C	31	*c d E*	81	*f g H*	185
A b C	104	*C d E*	275	*F g H*	4
a B c	109	*c D e*	256	*f G h*	0

Answer: For each of the data sets, there are eight classes of progeny appearing in four frequencies, as is expected for linkage among three genes. Since recombination between closely linked genes is infrequent, the least frequent classes result from crossovers between each of the genes in the triple heterozygote, i.e., double crossovers. The most frequent classes are parental types, where no crossover occurred between the linked genes.

To help you determine which gene is in the middle, consider the example shown below, which shows a double crossover between two non-sister chromatids. Relative to the parental-type chromosomes, the recombinant chromosomes have the middle *A*/*a* alleles switched:

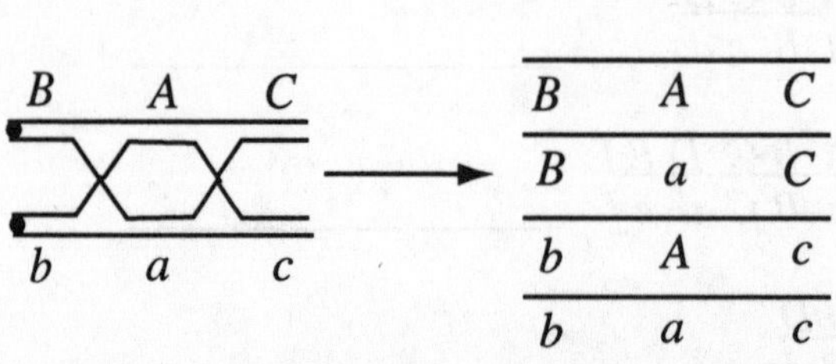

B A C → B A C
B a C
b a c → b A c
b a c

In three-point crosses, comparison of one of the double crossover classes (the least frequent classes) with one of the parental classes (the most frequent classes) can allow you to infer which gene is in the middle. Depending on how they are chosen, one of two scenarios will be found. In one scenario, alleles at two genes will be identical, while alleles at the third gene will be different. In another scenario, alleles at two genes will be different, while alleles at the third gene will be identical. In each case, it is the third gene is the gene in the middle. The middle gene is unlike the other two because its alleles have been "switched."

a. *a* is in the middle. The triple heterozygote's genotype is *BAC*/*bac*.
b. *d* is in the middle. The genotype is *CdE*/*cDe*.
c. *f* is in the middle. The genotype is *GFh*/*gfH*.

5.23 Genes at loci *f*, *m*, and *w* are linked, but their order is unknown. The F_1 heterozygotes from a cross of *FF MM WW* x *ff mm ww* are testcrossed. The most frequent phenotypes in testcross progeny will be *F M W* and *f m w* regardless of what the gene order turns out to be.

a. What classes of testcross progeny (phenotypes) would be least frequent if locus *m* is in the middle?
b. What classes would be least frequent if locus *f* is in the middle?
c. What classes would be least frequent if locus *w* is in the middle?

Answer: The double crossover class will always be the least frequent.

a. *FmW*, *fMw*
b. *MfW*, *mFw*
c. *FwM*, *fWm*

5.24 The following numbers were obtained for testcross progeny in *Drosophila* (phenotypes):

+ *m* +	218
w + *f*	236
+ + *f*	168
w m +	178
+ *m f*	95
w + +	101
+ + +	3
w m f	1
Total	1,000

Construct a genetic map.

Answer: Notice that among the eight phenotypic classes, there are four frequencies of progeny. This would be expected if the three genes were linked, and a triply-heterozygous parent were test-crossed. Since crossovers between closely linked genes occur less often than not, the most frequent classes of progeny will be those having non-recombinant, or parental-type, chromosomes. Thus, the + *m* +, and *w* + *f* chromosomes are parental-type chromosomes, and indicate that the triply heterozygous parent was obtained as follows:

P: *w* + *f*/*w* + *f* x + *m* +/+ *m* +
F_1: *w* + *f*/+ *m* + x *w m f*/*w m f*.

The least frequent classes (+ + +, *w m f*) will be those that resulted from double crossovers. The remaining classes will have resulted from single crossovers.

Comparison of one of the parental-type chromosomes and one of the double-crossover chromosomes (as described in the solution to problem 5.24) indicates that the *m* gene is in the middle:

One parental type chromosomes:	*w*	+	*f*
One double-crossover chromosomes:	*w*	*m*	*f*
Comparison of alleles:	same	different	same

As *m* is unlike the other two, it is in the middle. While the genes are written in the correct order in this problem, this will not always be the case. Always determine the correct order before proceeding!

Now, diagram the crossovers in the triple-heterozygote to infer the progeny classes that are associated with each crossover type:

Single-crossover between *w, m:*

+ *m* + / *w* + *f* → + *m* + ; + + *f* ; *w m* + ; *w* + *f*

Single-crossover between *m, f:*

+ *m* + / *w* + *f* → + *m* + ; + *m f* ; *w* + + ; *w* + *f*

Double-crosover:

+ *m* + / *w* + *f* → + *m* + ; + + + ; *w m f* ; *w* + *f*

With this information, one can re-evaluate the progeny, and determine how much recombination occurred in each gene interval:

+ m +	218	parental
w + f	236	parental
+ + f	168	single crossover between w, m
w m +	178	single crossover between w, m
+ m f	95	single crossover between m, f
w + +	101	single crossover between m, f
+ + +	3	crossover between w, m, and between m, f
w m f	1	crossover between w, m, and between m, f
Total	1,000	

Number of recombinants in w, m interval: $168 + 178 + 3 + 1 = 350$
Recombination Frequency between w, m = (350/1,000) x 100% = 35%

Number of recombinants in m, f interval: $95 + 101 + 3 + 1 = 200$
Recombination Frequency between m, f = (200/1,000) x 100% = 20%

This information can be represented in the following map:

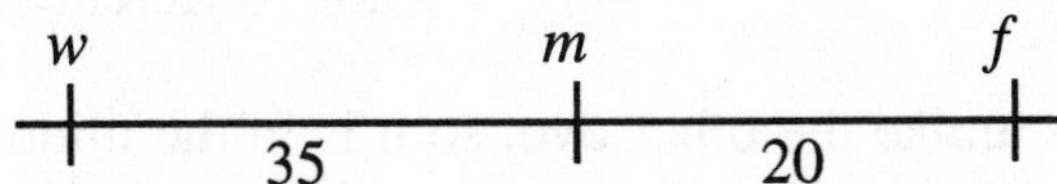

5.25 Three of the many recessive mutations in *Drosophila melanogaster* that affect body color, wing shape, or bristle morphology are black (*b*) body versus grey in the wild type, dumpy (*dp*), obliquely truncated wings versus long wings in the wild type, and hooked (*hk*) bristles at the tip versus not hooked in the wild type. From a cross of a dumpy female with a black and hooked male, all the F_1 were wild type for all three characters. The testcross of an F_1 female with a dumpy, black, hooked male gave the following results:

wild type	169
black	19
black, hooked	301
dumpy, hooked	21
hooked	8
hooked, dumpy, black	172
dumpy, black	6
dumpy	305
Total	1,000

a. Construct a genetic map of the linkage group (or groups) these genes occupy. If applicable, show the order and give the map distances between the genes.

b. (1) Determine the coefficient of coincidence for the portion of the chromosome involved in the cross. (2) How much interference is there?

Answer: a. First, use the symbols for the genes to write out the crosses that were done. It does not matter, at this point, what order you assign to the genes. Just be consistent in the order you choose so that you don't get confused.

P: + + *dp*/+ + *dp* (female) x *b hk* +/*b hk* + (male)
F_1: + + *dp*/*b hk* + (female) x *b hk dp*/*b hk dp* (male)

Now re-write the data to reflect parental type, single crossover (sco) and double crossover (dco) classes (based on frequency of each class).

PHENOTYPE	GAMETE GENOTYPE	NUMBER	CLASS
dumpy	+ + *dp*	305	parental
black, hooked	*b hk* +	301	parental
hooked, dumpy, black	*b hk dp*	172	sco
wild type	+ + +	169	sco
dumpy, hooked	+ *hk dp*	21	sco
black	*b* + +	19	sco
hooked	+ *hk* +	8	dco
dumpy, black	*b* + *dp*	6	dco

Now determine the gene order by comparing one of the double crossovers to one of the parental types (see problem 5.22 solution):

parental type genotype:	+	+	*dp*
dco genotype:	+	*hk*	+
	same	different	different

The *b* gene is unlike the other two, so it is in the middle. The correct gene order is *dp* – *b* – *hk*. Continue to use this gene order. The F_1 female that was testcrossed was *dp* + +/+ *b hk*. Consequently, the *dp b hk* and + + + progeny are single crossovers between *dp* and *b*, while the *dp* + *hk* and + *b* + progeny are single crossovers between *b* and *hk*. Recombination frequencies (RF) can be calculated as:

RF(*dp* – *b*) = [(172 + 169 + 6 + 8)/1,000] x 100% = 35.5%

RF(*b* – *hk*) =[(21 + 19 + 6 + 8)/1,000] x 100% = 5.4%

(Remember to include the double crossovers in this calculation as each double crossover has a crossover in the interval being considered.) The map distance between *dp* and *b* is 35.5 mu and the map distance between *b* and *hk* is 5.4 mu. The map distance between *dp* and *hk* is 35.5 + 5.4 = 40.9 mu.

b. The coefficient of coincidence (c.o.c.) is

c.o.c. = frequency observed doubles / frequency expected doubles
= (14/1,000)/(0.355 x 0.054)
= 0.014/0.01917
= 0.73

Interference = 1 - c.o.c.
= 1 - 0.73
= 0.27

5.26 In Chinese primroses long style (*l*) is recessive to short (*L*), red flower (*r*) is recessive to magenta (*R*), and red stigma (*rs*) is recessive to green (*Rs*). From a cross of homozygous short, magenta flower, and green stigma with long, red flower, and red stigma, the F_1 was crossed back to long, red flower, and red stigma. The following offspring were obtained:

Style	Flower	Stigma	Number
short	magenta	green	1,063
long	red	red	1,032
short	magenta	red	634
long	red	green	526
short	red	red	156
long	magenta	green	180
short	red	green	39
long	magenta	red	54

Map the genes involved.

Answer: Inspection of the data shows eight phenotypic classes appearing in four frequencies. This is expected from three linked genes in a three-point cross. Proceed as in the previous problems: (1) denote the phenotypes with the gene symbols, (2) determine parental, single and double crossover classes, (3) determine the gene order, (4) assess which classes arise from recombination in particular intervals, (5) calculate the recombination frequency and map distance between adjacent genes.

The crosses are: P: *L R Rs*/*L R Rs* x *l r rs*/*l r rs*
F_1: *L R Rs*/*l r rs* x *l r rs*/*l r rs.*

Evaluation of the data (the correct gene order is shown = *L* – *R* – *Rs*; sco = single crossover; dco = double crossover):

Phenotype	Gamete Genotype	Number	Class
short, magenta, green	*L R Rs*	1,063	parental
long, red, red	*l r rs*	1,032	parental
short, magenta, red	*L R rs*	634	sco (*R* – *Rs*)
long, red, green	*l r Rs*	526	sco (*R* – *Rs*)
short, red, red	*L r rs*	156	sco (*L* – *R*)
long, magenta, green	*l R Rs*	180	sco (*L* – *R*)
short, red, green	*L r Rs*	39	dco (*R* – *Rs, L* – *R*)
long, magenta, red	*l R rs*	54	dco (*R* – *Rs, L* – *R*)

Recombination Frequency (RF) = [(# recombinants)/(total)] x 100%
RF (*R* – *Rs*) = [(634 + 526 + 39 + 54)/3,684] x 100% = 34.0%
RF (*L* – *R*) = [(156 + 180 + 39 + 54)/3,684] x 100% = 11.6%

One can therefore draw the following map:

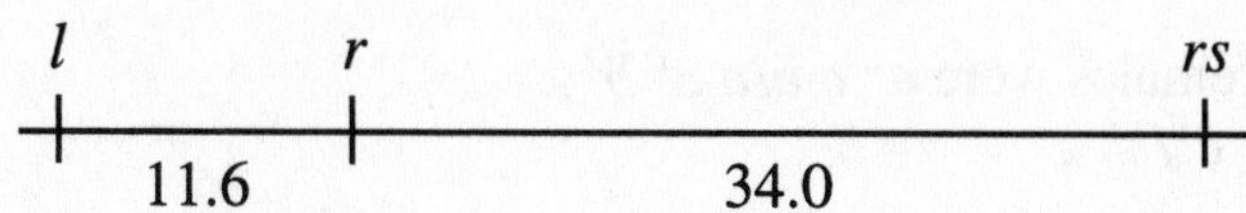

5.27 The frequencies of gametes of different genotypes, determined by testcrossing a triple heterozygote, are as follows:

GAMETE GENOTYPE	%
+ + +	12.9
a b c	13.5
+ + *c*	6.9
a b +	6.5
+ *b c*	26.4
a + +	27.2
a + *c*	3.1
+ *b* +	3.5
Total	100.0

a. Which gametes are known to have been involved in double crossovers?
b. Which gamete types have not been involved in any exchanges?
c. The order shown is not necessarily correct. Which gene locus is in the middle?

Answer: a. *a* + *c* and + *b* + (least frequent).
b. + *b c* and *a* + + (most frequent).
c. locus *c*.

5.28 Genes *a, b,* and *c* are recessive. Females heterozygous at these three loci are crossed to phenotypically wild-type males. The progeny are phenotypically shown:

Daughters: all + + +

Sons:		
	+ + +	23
	a b c	26
	+ + *c*	45
	a b +	54
	+ *b c*	427
	a + +	424
	a + *c*	1
	+ *b* +	0
	Total:	1,000

a. What is known of the genotype of the females' parents with respect to these three loci? Give gene order and the arrangement in the homologs.
b. What is known of the genotype of the male parents?
c. Map the three genes.

Answer: a. The females were $a^+ \; c \; b / a \; c^+ \; b^+$.
b. $a^+ \; b^+ \; c^+/Y$.

c.

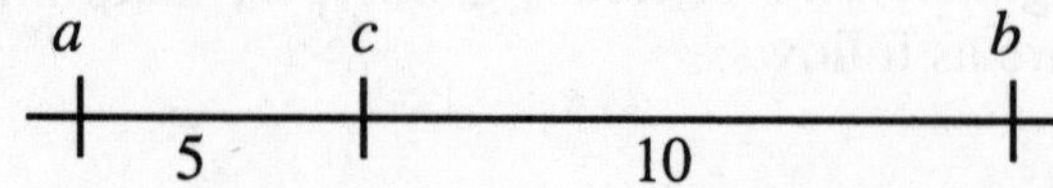

5.29 Two normal-looking *Drosophila* are crossed and yield the following phenotypes among the progeny:

Females:	+ + +	2,000
Males:	+ + +	3
	a b c	1
	+ *b c*	839
	a + +	825
	a b +	86
	+ + *c*	90
	a + *c*	81
	+ *b* +	75
	Total:	4,000

Give parental genotypes, gene arrangement in the female parent, map distances, and the coefficient of coincidence.

Answer: The differential appearance of the traits in males and females indicates that the traits are sex-linked, i.e., they are X-linked. To explain why all of the female progeny are normal, consider that they received their father's X chromosome, which was + + +/Y. As they are phenotypically normal, they cannot be used in any mapping analysis. The male progeny received their mother's X chromosome, and therefore can be used to analyze the results of recombination in the mother.

The most frequent progeny classes in the males are + *b c* and *a* + +. These are the parental type chromosomes. The least frequent classes (the double crossover classes) are + + + and *a b c*. Comparison of these two classes (see the solution to problem 5.22) indicates that gene *a* is in the middle, and the correct order is *b* – *a* – *c*. The parental genotypes (with the correct gene order) were *c* + *b*/+ *a* + (female) x + + +/Y (male). Now analyze the data to determine which progeny classes result from crossing over in each gene interval and calculate the recombination frequencies in each interval. (sco = single crossover; dco = double crossover).

GAMETE GENOTYPE	NUMBER	CLASS
c + *b*	839	parental
+ *a* +	825	parental
+ *a b*	86	sco (*a* – *b*)
c + +	90	sco (*a* – *b*)
c a +	81	sco (*c* – *a*)
+ + *b*	75	sco (*c* – *a*)
c a b	1	dco (*a* – *b*, *c* – *a*)
+ + +	3	dco (*a* – *b*, *c* – *a*)
	Total 2,000	

Note that, since recombination cannot be scored in the female progeny, and only the male progeny are being considered, the total that is used as the divisor in the calculation of recombination frequency (RF) is 2,000.

$$\text{RF } (a - b) = [(86 + 90 + 1 + 3)/2{,}000] \times 100\% = 9.0\%$$
$$\text{RF } (c - a) = [(81 + 75 + 1 + 3)/2{,}000] \times 100\% = 8.0\%$$

From this information, one has following map:

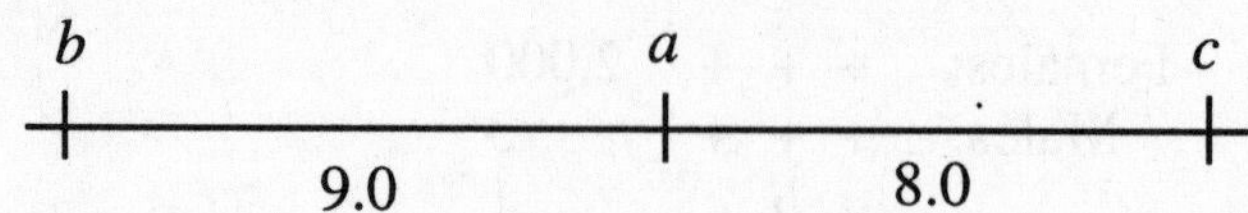

To calculate the coefficient of coincidence (c.o.c.), compare the frequency of actual double crossovers to that expected based on the crossover frequency observed in each interval.

c.o.c. = observed dco frequency/expected dco frequency
= [(3 + 1)/2,000]/(0.09 x 0.08)
= 0.002/0.0072
= 0.28

5.30 The questions below make use of this genetic map:

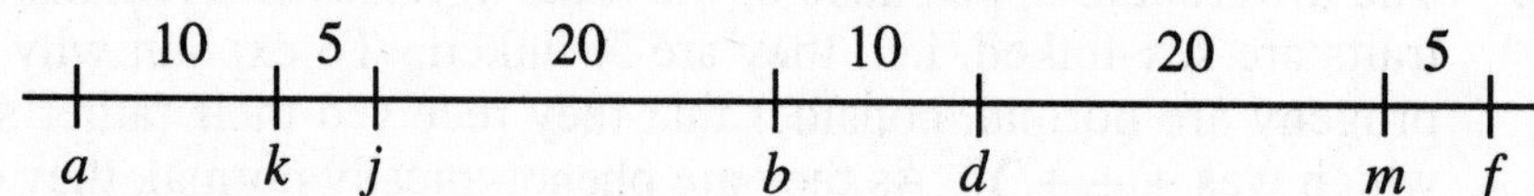

Calculate:

a. the frequency of *j b* gametes from a *J B*/*j b* genotype.
b. the frequency of *A M* gametes from an *a M*/*A m* genotype.
c. the frequency of *J B D* gametes from a *j B d*/*J b D* genotype.
d. the frequency of *J B d* gametes from a *j B d*/*J b D* genotype.
e. the frequency of *j b d*/*j b d* genotypes in a *j B d*/*J b D* x *j B d*/*J b D* mating.
f. the frequency of *A k F* gametes from an *A K F*/*a k f* genotype.

Answer: a. There are 20 map units between *j* and *b*, so there will be 20 percent recombinant-type and 80 percent parental-type gametes. In a *J B*/*j b* parent, *j b* gametes are half of the parental-type gametes, or 40 percent of the total gametes.

b. There are 65 map units between *A* and *M*, so these loci will show independent assortment, as the frequency of recombinant type gametes cannot exceed 50 percent. In an *a M*/*A m* parent, the *A M* gametes are half of the recombinant-type gametes. There will be 25 percent *A M* gametes.

c. In a *j B d*/*J b D* individual, *J B D* gametes are produced by a double crossover: a crossover in the interval *j* – *b*, and a crossover in the interval *b* – *d*. As there are 20 map units between *j* and *b*, and 10 map units between *b* and *d*, the frequency of double crossovers is expected to be 0.20 x 0.10 x 100% = 2%. *J B D* gametes are half of the double crossovers produced, so will be seen 1 percent of the time.

d. By the reasoning in (c), *j b d* gametes will be seen 1 percent of the time in a *j B d*/*J b D* individual. To obtain a *j b d*/*j b d* genotype, one must obtain such gametes from both parents. The frequency of this is (0.01 x 0.01) x 100% = 0.01%.

e. Based on the map distances, one expects 10% recombination between *a* and *k*, and 50 percent recombination between *k* and *f*. (Again, note that even though *k* and *f* are more than 50 map units apart, one observes only 50 percent recombination between these two genes.) In an *A K F*/*a k f* individual, an *A k F* gamete results from a double crossover: one crossover between *a* and *k* (10 percent chance) and one crossover between *k* and *f* (50 percent chance). The chance of both crossovers occurring simultaneously is 0.10 x 0.50 = 0.05, or 5%. Since *A k F* gametes are half of the double crossovers produced, they will be 2.5% of the total.

5.31 A female *Drosophila* carries the recessive mutations *a* and *b* in repulsion on the X-chromosome (she is heterozygous for both). She is also heterozygous for an X-linked recessive lethal allele, *l*. When she is mated to a true-breeding, normal male, she yields the following progeny:

Females:	1,000	+ +
Males:	405	*a* +
	44	+ *b*
	48	+ +
	2	*a b*

Draw a chromosome map of the three genes in the proper order, and with map distances as nearly as you can calculate them.

Answer: First, consider what happens when a female that is heterozygous for an X-linked lethal is crossed to a normal male. One can diagram such a cross as:

P: *l*/+ female x +/Y male
F_1: females: *l* /+ and +/+ (phenotypically normal)
males: +/Y (normal) and *l*/Y (dead, not recovered).

In such a cross, one-half of the male progeny are not recovered due to the presence of the lethal gene. Only progeny bearing the + allele are recovered. Thus, the 500 males that are recovered in this cross are half of the expected male progeny.

Second, consider that the lethal-bearing chromosome can be recovered in the female progeny. Since the lethal allele is a recessive mutation, it is "rescued" by the normal + allele contributed to the female progeny by the father's X chromosome. However, as the father's X chromosome bears + alleles at each gene, the females are phenotypically normal and cannot be used for recombination analysis.

Now, analyze the phenotypes of the male progeny that are recovered to infer map distances and the gene order. Since one-half of the male progeny are not recovered, each of the four classes seen represents one of the two reciprocal classes of progeny recovered in a three point cross. Include the third (*l*/+) locus and assign each genotype to a progeny class based on its frequency:

Male progeny:	405	*a*	+	+	parental type
	44	+	*b*	+	single crossover
	48	+	+	+	single crossover
	2	*a*	*b*	+	double crossover

Comparison of the parental and double crossover classes indicates that *b* is in the middle, and the correct order is *a* – *b* – *l*. Since one of the parental type chromosomes was *a* + +, the other must have been + *b c*, the reciprocal. The heterozygous female was therefore *a* + +/+ *b l*.

The 44 + *b* + progeny are obtained from gametes arising from single crossovers between *b* and *l*. They are half of the total single crossovers in that interval. The other half were not recovered as they bore the *l* allele. Similarly, the 48 + + + progeny are half of the single crossovers in the interval *a* – *b*. This information can be used to calculate recombination frequencies (RF) and draw a map.

RF $(a - b) = [(48 + 2)/500] \times 100\% = 10\%$
RF $(b - l) = [(44 + 2)/500] \times 100\% = 9.2\%$

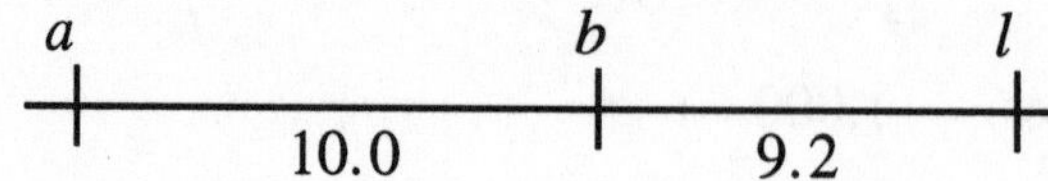

5.32 The following *Drosophila* cross is done:

$$\frac{a + b}{+ c +} \times \frac{a\, c\, b}{\longrightarrow}$$

Predict the numbers of phenotypes of male and female progeny which will emerge if the gene arrangement is as shown, the distance between *a* and *c* is 14 map units, the distance between *c* and *b* is 12 map units, the coefficient of coincidence is 0.3, and the number of progeny is 2,000.

Answer: Since the male parent is triply recessive, the phenotypes associated with the female parent's gametes will be evident equally in males and females. One expects 14% single crossover recombinants between *a* and *c* (7% each of *a c* + and + + *b*), and 12% single crossover recombinants between *c* and *b* (6% each of *a* + + and + *c b*). Since the coefficient of coincidence is 0.3, one observes 30% of the double crossovers that would be expected based on the recombination frequency in each of the two gene intervals. The expected double crossover frequency is (0.12 x 0.14) x 100% = 1.68%. The observed double crossover frequency will be 1.68% x 30% = 0.50% (0.25% each of *a c b* and + + +). The remaining progeny [100% - (14% + 12% + 0.50%) = 73.5%] will be parental types (36.75% each of *a* + *b* and + *c* +). Thus, one predicts:

Genotype	Percent	Number
a + *b*	36.75	735
+ *c* +	36.75	735
a c +	7.00	140
+ + *b*	7.00	140
a + +	6.00	120
+ *c b*	6.00	120
a c b	0.25	50
+ + +	0.25	50

5.33 A farmer who raises rabbits wants to break into the seasonal Easter market. He has stocks of two true-breeding lines. One is hollow, and long-eared, but not chocolate, while the second is solid, short-eared and chocolate. Hollow (*h*), long ears (*le*), and chocolate (*ch*) are all recessive, autosomal and linked as in the following map:

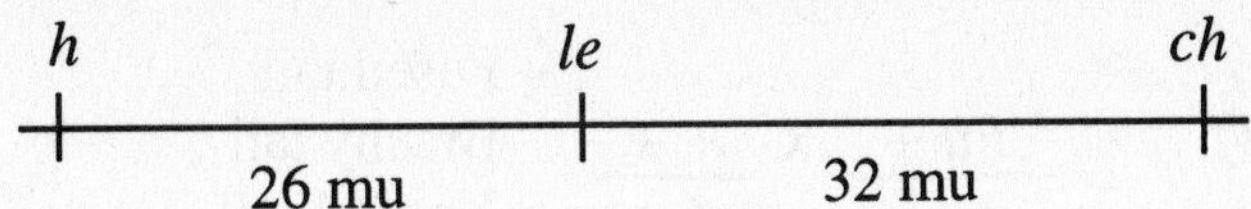

The farmer can generate a trihybrid by crossing his two lines and at great expense he is able to obtain the services of a male homozygous recessive at all three loci to cross with his F_1 females.

The farmer has buyers for both solid and hollow bunnies; however, all must be chocolate and long eared. Assuming that interference is zero, if he needs 25 percent of the progeny of the desired phenotypes to be profitable, should he continue with his breeding? Calculate the percent of the total progeny that will be the desired phenotypes.

Answer: The crosses under consideration are:

P: *h le Ch*/*h le Ch* x *H Le ch*/*H Le ch*
F_1: *h le Ch*/*H Le ch* x *h le ch*/*h le ch*

To be profitable, the farmer needs 25 percent of the progeny to be either *h le ch*/*h le ch* or *H le ch*/*h le ch*. Hence, he is concerned with the probability of obtaining a *h le ch* or a *H le ch* gamete from an *h le Ch*/*H Le ch* individual. Gametes that are *h le ch* arise from a single crossover between *le* and *ch*. Since there are 32 mu between *le* and *ch*, there will be 16 percent *h le ch* individuals (one-half of the recombinants between *le* and *ch*). Gametes that are *H le ch* arise from a crossover in each interval. Since there is no interference, their frequency will be half of the expected double recombinant frequency, or (0.5 x 0.32 x 0.26) x 100% = 4.16%. The expected frequency of *h le ch* and *H le ch* gametes is 4.16% + 16% = 20.16%. The farmer should stop breeding and cut his losses.

5.34 Three different semidominant mutations affect the tail of mice. They are linked genes, and all three are lethal in the embryo when homozygous. Fused-tail (*Fu*) and kinky-tail (*Ki*) mice have kinky-appearing tails, while brachyury (*T*) mice have short tails. A fourth gene, histocompatibility-2 (*H-2*), is linked to the three tail genes and is concerned with tissue transplantation. Mice that are *H-2*/+ will accept tissue grafts, whereas +/+ mice will not. In the following crosses, the normal allele is represented by a +. The phenotypes of the progeny are given for four crosses.

Cross	Parents	Phenotype	Number
(1)	*Fu* + / + *Ki* x + + / + +	Fused tail	106
		Kinky tail	92
		Normal tail	1
		Fused-kinky tail	1
(2)	*Fu H-2* / + + x + + / + +	Fused tail, accepts graft	88
		Normal tail, rejects graft	104
		Normal tail, accepts graft	5
		Fused tail, rejects graft	3
(3)	*T H-2* / + + x + + / + +	Brachy tail, accepts graft	1,048
		Normal tail, rejects graft	1,152
		Brachy tail, rejects graft	138
		Normal tail, accepts graft	162

(4)	$\frac{Fu\ +}{+\ T}$ x $\frac{+\ +}{+\ +}$	Fused tail	146
		Brachy tail	130
		Normal tail	14
		Fused-brachy tail	10

Make a map of the four genes involved in these crosses, giving gene order and map distances between the genes.

Answer: Since all of the genes have a dominant phenotype, and crosses are made to a homozygous recessive strain, the data can be analyzed as a series of two-point testcrosses as follows:

CROSS	GENES INVOLVED	# RECOMBINANTS / TOTAL	RECOMBINATION FREQUENCY (%)	APPARENT MAP DISTANCE
1	*Fu, Ki*	2/200	1	1 mu
2	*Fu, H-2*	8/200	4	4 mu
3	*T, H-2*	300/2,500	12	12 mu
4	*Fu, T*	24/300	8	8 mu

Two maps are consistent with these data.

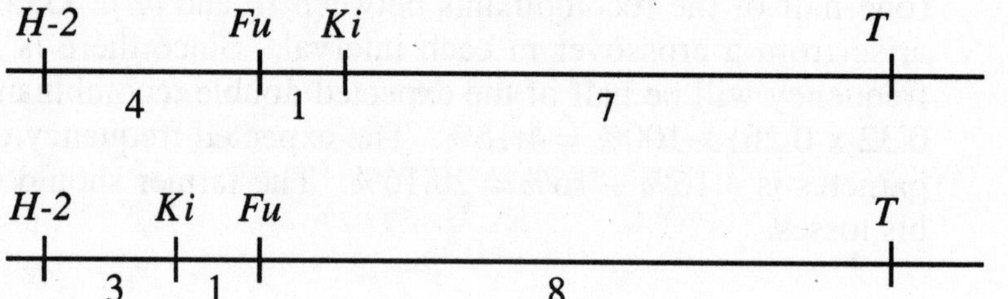

5.35 The cross in *Drosophila* of

$$\frac{a^+\ b^+\ c\ d\ e}{a\ b\ c^+\ d^+\ e^+} \times \frac{a\ b\ c\ d\ e}{a\ b\ c\ d\ e}$$

gave 1,000 progeny of the following 16 phenotypes:

	Genotype	Number		Genotype	Number
(1)	$a^+\ b^+\ c\ d\ e$	220	(9)	$a\ b^+\ c^+\ d\ e^+$	14
(2)	$a^+\ b^+\ c\ d\ e^+$	230	(10)	$a\ b^+\ c^+\ d\ e$	13
(3)	$a\ b\ c^+\ d^+\ e$	210	(11)	$a^+\ b\ c\ d^+\ e^+$	8
(4)	$a\ b\ c^+\ d^+\ e^+$	215	(12)	$a^+\ b\ c\ d^+\ e$	8
(5)	$a\ b^+\ c^+\ d^+\ e$	12	(13)	$a^+\ b^+\ c^+\ d\ e^+$	7
(6)	$a\ b^+\ c^+\ d^+\ e^+$	13	(14)	$a^+\ b^+\ c^+\ d\ e$	7
(7)	$a^+\ b\ c\ d\ e^+$	16	(15)	$a\ b\ c\ d^+\ e^+$	6
(8)	$a^+\ b\ c\ d\ e$	14	(16)	$a\ b\ c\ d^+\ e$	7

a. Draw a genetic map of the chromosome, indicating the linkage of the five genes and the number of map units separating each.
b. From the single-crossover frequencies, what would be the expected frequency of $a^+ b^+ c^+ d^+ e^+$ flies?

Answer: a. Approach this problem by considering two genes at a time. Then one has

GENE PAIR	# PARENTAL TYPE PROGENY	# RECOMBINANT TYPE PROGENY	RECOMBINATION FREQUENCY (%)	LINKED?
a, b	902	98	9.8	Yes
a, c	973	27	2.7	Yes
a, d	957	43	4.3	Yes
a, e	497	503	50.0	No
b, c	875	125	12.5	Yes
b, d	945	55	5.5	Yes
b, e	497	503	50.0	No
c, d	930	70	7.0	Yes
c, e	498	502	50.0	No
d, e	496	504	50.0	No

The genes *a, b, c,* and *d* are linked, while *e* is unlinked to the other four genes. Use the smallest distances as the most accurate map distances. One has:

b — 5.5 — d — 4.3 — a — 2.7 — c   e

b. Rewrite the parents using the correct gene order:

$$\frac{b^+\ d\ a^+\ c}{b\ d^+\ a\ c^+}\ \frac{e}{e^+} \times \frac{a\ b\ c\ d}{a\ b\ c\ d}\ \frac{e}{e}$$

To obtain a $b^+ d^+ a^+ c^+$ fly, there must be crossovers between *b* and *d*, *d* and *a*, and *a* and *c*. Two reciprocal crossovers will be found: one-half of the progeny will be $b^+ d^+ a^+ c^+$ and one-half of the progeny will be *b d a c*. Among these progeny, one-half will be e^+ and one-half will be *e*.

$$p(b^+ d^+ a^+ c^+ e^+) = p(\text{triple crossover}) \times 1/2 \times 1/2$$
$$= (0.055 \times 0.043 \times 0.027) \times 0.25$$
$$= 1.6 \times 10^{-5}.$$

5.36 Complete all diagrams in the figure by correctly showing the centromeres, chromosome strands, and alleles on each strand.

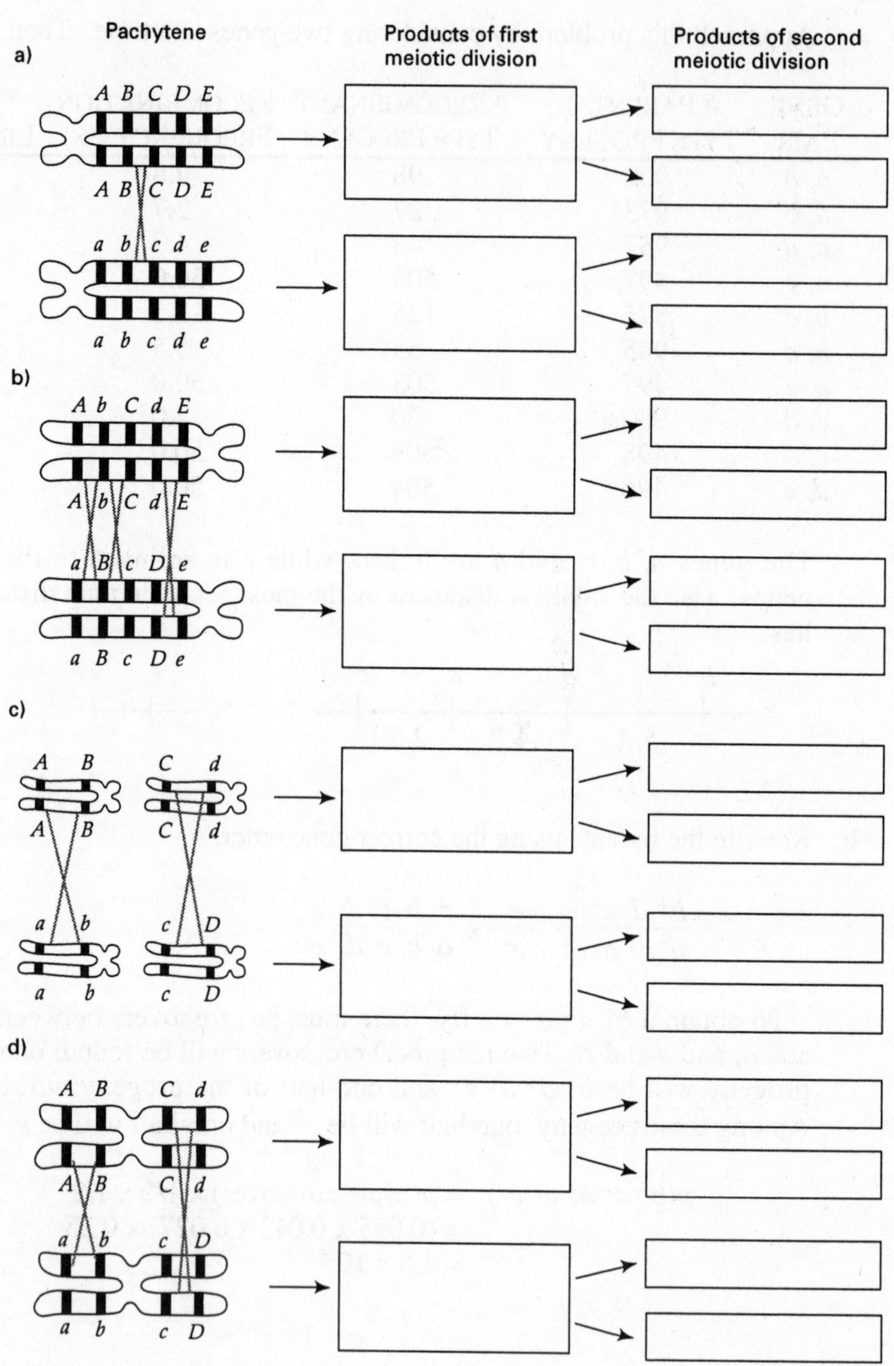

Answer:

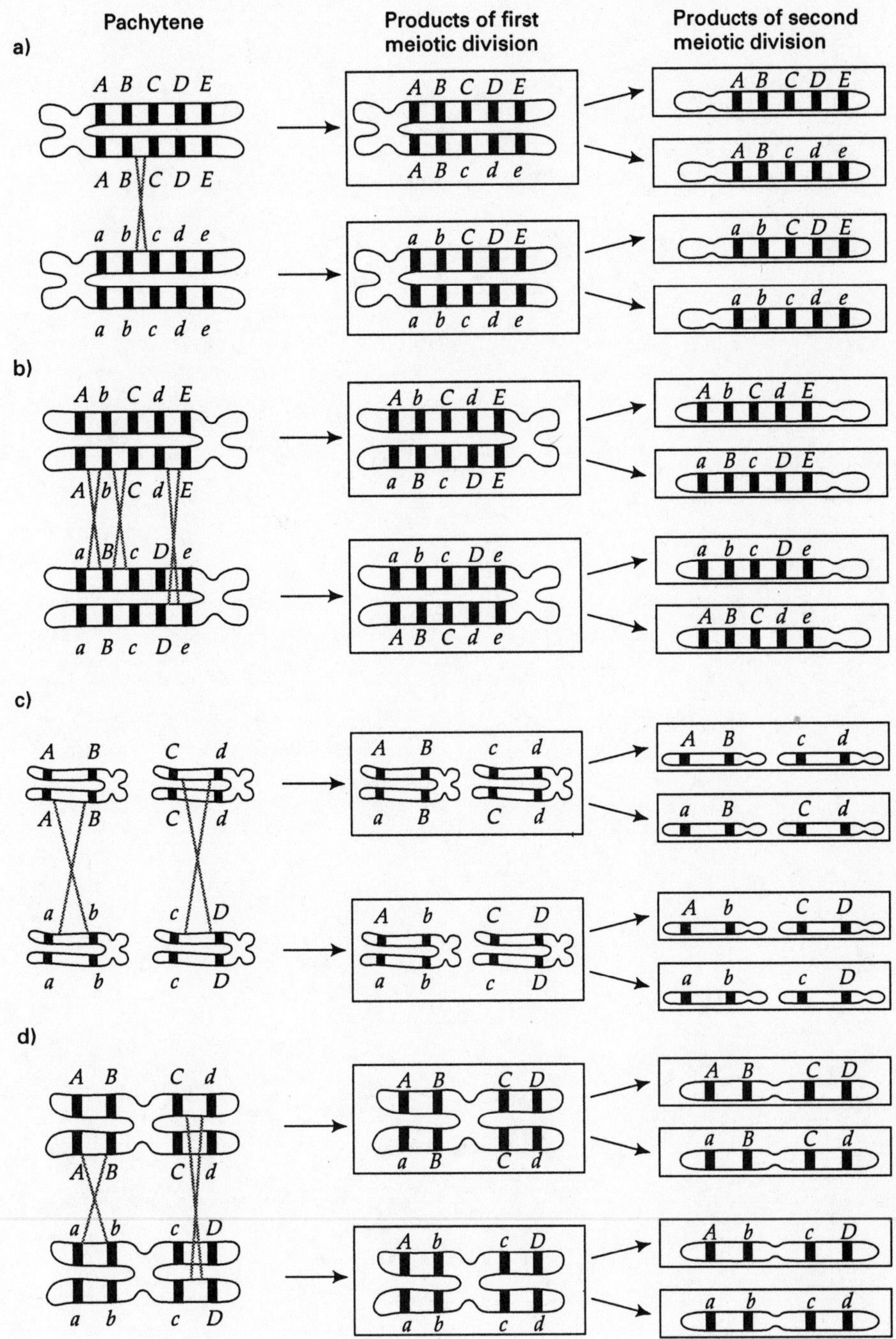

Pachytene
Products of first meiotic division
Products of second meiotic division
a)
A B C D E
A B C D E
a b c d e
a b c d e
A B C D E
A B c d e
A B C D E
A B c d e
a b C D E
a b c d e
a b C D E
a b c d e
b)
A b C d E
A b C d E
a B c D e
a B c D e
A b C d E
a B c D E
A b C d E
a B c D E
a b c D e
A B C d e
a b c D e
A B C d e
c)
A B C d
A B C d
a b c D
a b c D
A B c d
a B C d
A B c d
a B C d
A b C D
a b c D
A b C D
a b c D
d)
A B C d
A B C d
a b c D
a b c D
A B C D
a B C d
A B C D
a B C d
A b c D
a b c d
A b c D
a b c d

CHAPTER 6

ADVANCED GENETIC MAPPING IN EUKARYOTES

I. CHAPTER OUTLINE

TETRAD ANALYSIS
- Life Cycle of Yeast
- Life cycle of *Chlamydomonas reinhardtii*
- Using Random-Spore Analysis to Map Genes in Haploid Eukaryotes
- Using Tetrad Analysis to Map Two Linked Genes
- Calculating Gene-Centromere Distance in Organisms with Linear Tetrads

MITOTIC RECOMBINATION
- Discovery of Mitotic Recombination
- Mechanism of Mitotic Crossing-Over
- Examples of Mitotic Recombination

MAPPING GENES IN HUMAN CHROMOSOMES
- Mapping Human Genes by Somatic Cell Hybridization Techniques

II. REVIEW OF KEY TERMS, SYMBOLS AND CONCEPTS

Without consulting the text and in your own words, write a brief definition of each term in the groups below. Then, either using a short phrase or a simple diagram, identify the relationship(s) between specific pairs of terms within a set. Finally, consult the text (and perhaps a friend who has also done the exercise) to check your answers.

1	2	3	4
haploid	tetrad analysis	PD, T, NPD	four-strand stage
diploid	ordered tetrads	first-division segregation	mitotic recombination
ascospore	unordered tetrads	second-division segregation	meiotic recombination
vegetative cell	linear tetrads	ordered tetrad	mosaic
mitosis	random-spore analysis	unordered tetrad	twin spot frequency
ascus			

5	6	7	8
deletion translocation p, q synkaryon	heterokaryon diploidization haploidization parasexual system	hereditary retinoblastoma sporadic retinoblastoma mitotic recombination tumor-suppressor	somatic cell hybridization HAT technique syntenic synkaryon HGPRT, TK

III. THINKING ANALYTICALLY

Many students find the analysis of tetrads, mitotic recombination and somatic cell hybrids quite challenging. Step back for a minute though and ask yourself, "Can I recognize the *principles* behind these methods?" Tetrad analysis is based on the normal events of meiosis in organisms where all four products of a meiosis remain together. Mitotic recombination is based on the occurrence of a crossover in the context of a normal mitosis. Somatic cell hybrids are derived from mitotically dividing cells that result from the fusion of mouse and human cells. Underlying each of these processes are the fundamental processes of cell division: meiosis and mitosis.

A solid understanding of chromosome behavior in meiosis and mitosis will enable you to master these advanced genetic mapping techniques. Review how chromosomes behave during meiosis. Consider *how* T, PD and NPD type tetrads are produced. Ask yourself what physical basis underlies first- and second-division segregation. Review how sister and non-sister chromatids are distributed to daughter cells during mitosis, and then diagram what happens if a crossover between non-sister chromatids occurs between the centromere and a particular locus. Approaching the material this way will provide you with a solid, *mechanistic* understanding of these processes.

Once you understand how meiosis and mitosis relate to these mapping techniques, you will be able to more readily apply them. Analyzing the results of crosses in which tetrads are produced is in many regards much more straightforward than analyzing the results of crosses where they are not produced. Try employing the following systematic approach:

1. Assign gene symbols to the traits and write out the cross, considering just two loci at a time.
2. Decide which are parental-type spores and which are recombinant-type spores.
3. If you have unordered asci, calculate the recombination frequency (RF).
   RF = (# recombinant spores/total # spores) x 100%.
4. If you have ordered tetrads, identify PD, T and NPD tetrads.
   a. If PD = NPD (or is close), the genes are not linked.
   b. If PD >> NPD, the genes are linked.
   RF = [(1/2 T + NPD)/total # asci] x 100%.
6. If you are analyzing linear (ordered) tetrads, consider each locus separately and determine the number of first-division (MI; 2:2 or 4:4) and second division (MII; 1:1:1:1 or 2:2:2:2) segregation patterns.
   RF (gene–centromere) = [1/2(MII patterns)/total # asci] x 100%.
7. Draw a map.

In problems involving mitotic recombination, it often helps to remind yourself of one of the basic results of mitosis: Normally, parent cells give rise to identical daughter cells. The following diagram shows the results of normal chromatid segregation in the absence or presence of a mitotic crossover (arrows trace the movement of the centromeres).

No mitotic recombination:

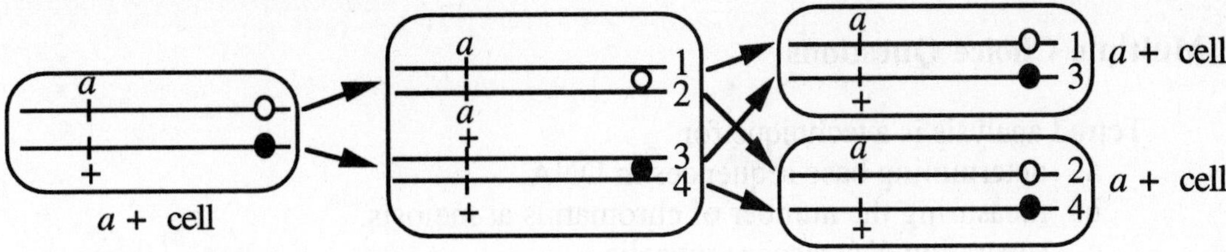

Mitotic recombination between *a* and its centromere leads to the formation of a twin spot:

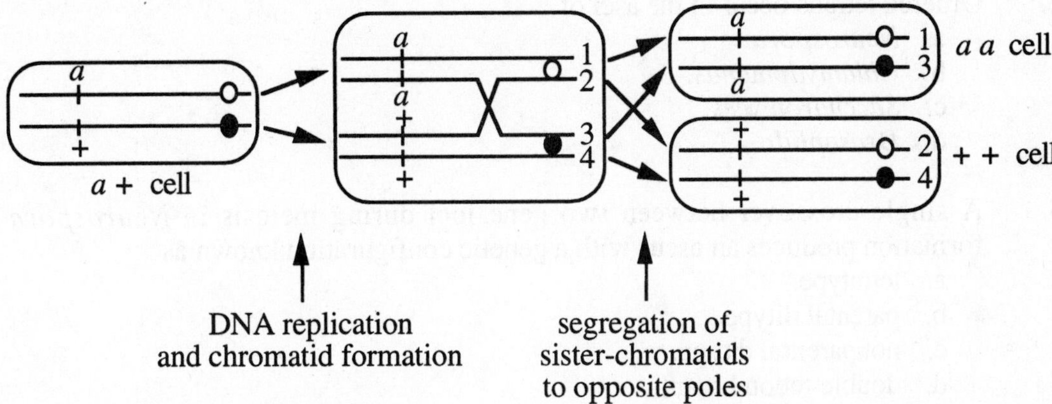

Chromatids segregate according to their centromere, not according to the loci attached to that centromere. By following how centromeres normally segregate, one can see how mitotic division normally gives rise to identical cells, and in the presence of mitotic recombination, how it gives rise to twin spots.

In the above example, members of a pair of homologues are identified by white and black centromeres. DNA replication occurs, chromatids form and then, during anaphase of mitosis, the centromeres split and go to opposite poles. The white centromere splits to form new white centromeres 1 and 2 and the black centromere splits to form new black centromeres 3 and 4. Since the centromeres normally move to opposite poles of the cell at anaphase, one possible outcome is that the even-numbered centromeres will go to one pole, and the odd-numbered centromeres will go to the other pole. This is the result that is shown. (What other possible outcome(s) exist?)

Normally, the movement of chromosomes to opposite poles in mitosis insures that two identical daughter cells will be produced. A rare mitotic crossover leads to the production of a twin spot in which a cell becomes homozygous at all loci distal to the cross-over event. Note that the tissue surrounding the twin spot will remain heterozygous. Since the organism is now composed of tissues having different genotypes, it is referred to as a mosaic.

Additional hints on dealing with questions involving somatic cell genetics are given in the "Solutions to Text Questions" section below.

IV. QUESTIONS FOR PRACTICE

A. Multiple Choice Questions

1. Tetrad analysis is a technique for
   a. determining base sequences in DNA.
   b. measuring the number of chromatids at meiosis.
   c. separating *Neurospora* mycelia.
   d. mapping genes in organisms retaining the four meiotic products in one structure.

2. Ordered tetrads occur in the asci of
   a. *Neurospora.*
   b. *Chlamydomonas.*
   c. *Saccharomyces.*
   d. *Drosophila.*

3. A single crossover between two gene loci during meiosis in *Neurospora* spore formation produces an ascus with a genetic configuration known as
   a. tetratype.
   b. parental ditype.
   c. nonparental ditype.
   d. double recombinant.

4. The formula (1/2 T + NPD)/(total # of tetrads) x 100% is used to calculate
   a. the length of a chromosome.
   b. the location of the centromere.
   c. the number of crossovers.
   d. the percentage of recombinants.

For questions 5 and 6, refer to the following ordered tetrad resulting from the *Neurospora* cross *a* + x *b* + .

a +
a +
a b
a b
+ *b*
+ *b*
+ +
+ +

5. What configuration does this represent?
   a. NPD
   b. PD
   c. T
   d. NPD + PD

6. For the *a* locus, what segregation pattern is shown?
   a. equal
   b. random
   c. first-division
   d. second-division

7. In a tetrad in which a single cross-over between two loci has occurred, what percentage of the spores will be recombinant-types?
   a. 0%
   b. 25%
   c. 50%
   d. 100%

8. A heterokaryon cell produced by the fusion of two mycelia of *Aspergillus* contains
   a. a single, fused, diploid nucleus.
   b. two fused, diploid nuclei.
   c. two unfused, haploid nuclei.
   d. two fused haploid nuclei.

9. The HAT-technique
   a. causes selected human and mouse cells to fuse.
   b. segregates hybridized human/mouse cells from unfused human and mouse cells.
   c. causes certain human chromosomes to be eliminated from hybridized cells.
   d. identifies the genotypes of selected human chromosomes.

Answers: 1d; 2a; 3a; 4d; 5c; 6c; 7c; 8c; 9b

B. Thought Questions

1. Tests for the presence of the gene for Human Clotting Factor 7 (HCF-7) in certain somatic cell hybrid lines gave the results shown in the table below. A + or – indicates the presence or absence of a chromosome in each cell line, and the presence or absence of the clotting factor. In which chromosome is the gene located?

	Chromosome Number							
Cell line	1	2	4	6	8	14	15	20	HCF-7
1	+	–	–	+	+	+	–	–	+
2	+	–	+	+	–	–	+	+	–
3	–	+	–	+	+	–	+	+	+
4	+	–	–	+	+	+	+	+	+
5	–	+	+	+	–	–	+	+	–

2. *Drosophila* normally have deep red eyes. A large number of alleles at the X-linked *white* locus cause lighter colored eyes, with w^l causing a completely white eye, w^a causing apricot (light orange) colored eyes, and w^c causing coral (pink) colored eyes. In females heterozygous for these two *white* alleles the amount of pigmentation contributed by each allele is additive, so that w^l/w^a females have apricot colored eyes and w^a/w^c females have reddish eyes. Diagram the appearance of a twin spot that

results from a single mitotic crossover between the centromere and the *white* locus in a w^a/w^c *Drosophila* female.

3. How might one use ordered tetrad analysis to show that recombination occurs at the four-strand stage of meiosis? Hint: Use diagrams to compare the results of a single crossover between *a* and *b* in a cross of *a b* × + + that occur before and after DNA replication.
4. How could you use mitotic recombination to map linked loci relative to each other? Would you expect the map distances obtained from mitotic recombination to be the same as those derived using meiotic recombination?
5. In *Drosophila*, meiotic recombination does not occur during spermatogenesis. Might mitotic recombination occur in *Drosophila* males? For all chromosomes? If the answer is yes, what might this tell you about the "normal" role of mitotic recombination in cells?

V. SOLUTIONS TO TEXT PROBLEMS

6.1 In *Saccharomyces*, *Neurospora*, and *Chlamydomonas*, what meiotic events give rise to PD, NPD and T tetrads?

Answer: It is useful here to visualize the chromosomes in a tetrad at the time of crossing over, and refer to text Figure 6.6. PD, or parental ditypes, can arise from tetrads in which no crossovers have taken place between the markers being observed, as well as from tetrads in which a double crossover has taken place between the *same* two non-sister chromatids, a so-called two-strand double crossover. NPD, or non-parental ditypes, also arise from tetrads with double crossovers between the two markers. With NPD tetrads, each of the two crossovers occurs on a separate pair of non-sister chromatids, a so-called 4-strand double. T, or tetratype tetrads, occur from either single crossovers between any pair of non-sister chromatids, as well as from double crossovers involving three chromatids, so-called 3-strand doubles.

6.2 What important item of information regarding crossing-over can be obtained from tetrad analysis (as in *Neurospora*) but not from single-strand analysis (as in *Drosophila*)?

Answer: Because one can follow *all* of the products of a single meiosis in a tetrad analysis, experiments could be designed to determine how and when crossing-over takes place. These experiments have shown that crossing over takes place at the four-strand stage of meiosis.

6.3 A cross was made between a pantothenate-requiring (*pan*) strain and a lysine-requiring (*lys*) strain of *Neurospora crassa*, and 750 random ascospores were analyzed for their ability to grow on a minimal medium (a medium lacking pantothenate and lysine). Thirty colonies subsequently grew. Map the *pan* and *lys* loci.

Answer: First, write out the strains that were crossed and the kinds of progeny that are possible.

pan + x + *lys*	→	*pan* +	parental type spore
		+ *lys*	parental type spore
		pan lys	recombinant type spore
		+ +	recombinant type spore

Notice that, since spores that are *pan* or *lys* require supplements to grow, the only spores that can grow on minimal medium are wild-type, or + + spores. All parental type spores and half of the recombinants cannot grow. Thus, the 30 spores that grew are half of the recombinants. Therefore, there were a total of 60 recombinants in 750 progeny tested, giving a map distance of (60/750) x 100% = 8%.

6.4 In *Neurospora* the following crosses yielded the progeny as shown:

Cross		Number	Progeny
$a^+ b$ x $a b^+$	→	981	a^+ b
		1,000	a b^+
		10	a^+ b^+
		9	a b
		2,000	
$a^+ c$ x $a c^+$	→	850	a^+ c
		833	a c^+
		169	a^+ c^+
		148	a c
		2,000	
$b^+ c$ x $b c^+$	→	850	b^+ c
		850	b c^+
		140	b^+ c^+
		160	b c
		2,000	

What is the probable gene order and what are the approximate map distances between adjacent genes?

Answer: Note that one is not really utilizing tetrad analysis in this problem, and that it can be solved by calculating the map distances between each pair of genes, and then ordering them relative to one another. For each cross, one sees that the parental types of spores are much greater in number than the recombinant types of spores, and therefore each pair of loci appear to be linked. The map distance between any two loci is given by the (number of recombinant spores/total) x 100%. The distances are:

$$a - b = [(9 + 10)/2{,}000] \text{ x } 100\% = 0.95\%$$
$$a - c = [(169 + 148)/2{,}000] \text{ x } 100\% = 15.8\%$$
$$b - c = [(140 + 160)/2{,}000] \text{ x } 100\% = 15.0\%$$

Therefore, it appears that *b* is closest to *a*, and *b* is in the middle between *a* and *c*. A map is:

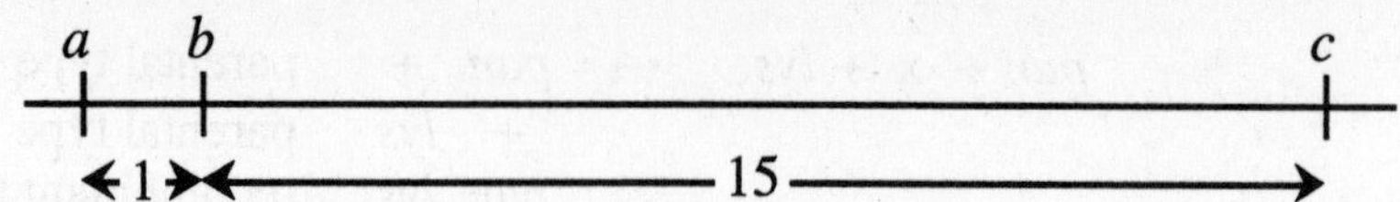

6.5 Four different albino strains of *Neurospora* were each crossed to the wild-type. All crosses resulted in half wild type and half albino progeny. Crosses were made between the first strain and the other three with the following results:

1 x 2: 975 albino, 25 wild type

1 x 3: 1,000 albino

1 x 4: 750 albino, 250 wild type

Which mutations represent different genes, and which genes are linked? How did you arrive at your conclusions?

Answer: First, consider the results of the set of crosses to wild-type.

albino x + → 50% albino, 50% +

The observation of equal frequencies of parental type spores in the progeny indicates that the albino strains have an allele that behaves in a Mendelian fashion. That is, for each of the four strains, the albino and + alleles segregate from each other as would be expected of alleles at one gene. The question now becomes whether each of the albino strains results from a mutation in the same or a different gene. This issue can be resolved by considering the second set of crosses and the possibility that the different strains are mutant at different genes. With this idea in mind, it becomes clear that any wild-type spores that are found represent half of the recombinant-types of spores, as follows:

1 x 2: a^1 + x + a^2 → 975 albino (a^1 a^2, a^1 + or + a^2)
25 wild-type (+ +)

1 x 3: a^1 + x + a^3 → 1,000 albino (a^1 a^2, a^1 + or + a^2)
0 wild-type (+ +)

1 x 4: a^1 + x + a^4 → 750 albino (a^1 a^2, a^1 + or + a^2)
250 wild-type (+ +)

In the cross of 1 x 2, the appearance of 25 wild-type spores indicates that there are a total of 50 recombinant spores (0.5%). Hence, genes 1 and 2 are linked, and 0.5 map units apart. In the cross of 1 x 3, there are no recombinant spores, and so these albino alleles appear to be at the same gene, and may be identical. In the cross of 1 x 4, there are a total of 500 recombinant spores (50%) and so genes 1 and 4 appear to be unlinked. There are therefore three different genes, with genes 1 (= 3) and 2 being linked.

6.6 Genes *met* and *thi* are linked in *Neurospora crassa*; we wish to locate *arg* with respect to *met* and *thi*. From the cross *arg* + + x + *thi met*, the following random ascospore isolates were obtained. Map these three genes.

arg	*thi*	*met*	26	*arg*	+	+	51
arg	*thi*	+	17	+	*thi*	+	4
arg	+	*met*	3	+	+	*met*	14
+	*thi*	*met*	56	+	+	+	29

Answer: Notice that there are 8 classes of progeny which appear in four frequencies, reminiscent of a three-point cross with three-linked loci. Therefore, treat this as you would a three-point cross. Comparison of one of the parental types (+ *thi met*) to one of the double recombinants (*arg* + *met*) tells us that *met* is in the middle, between *arg* and *thi*. Rewrite the progeny types as (sco = single crossover):

parental	+	*met*	*thi*	56
parental	*arg*	+	+	51
sco between *arg* – *met*	*arg*	*met*	*thi*	26
sco between *arg* – *met*	+	+	+	29
sco between *met* – *thi*	*arg*	+	*thi*	17
sco between *met* – *thi*	+	*met*	+	14
double crossover	*arg*	*met*	+	3
double crossover	+	+	*thi*	4
				200

The map distance between *arg* and *met* is [(26 + 29 + 3 + 4)/200] x 100% = 31 mu, and the map distance between *met* and *thi* is (17 + 14 + 3 + 4)/200 = 19 mu. Therefore, we have the following map:

6.7 Given a *Neurospora* zygote of the constitution shown in the figure below, diagram the significant events producing an ascus where the *A* alleles segregate at the first division and the *B* alleles segregate at the second division.

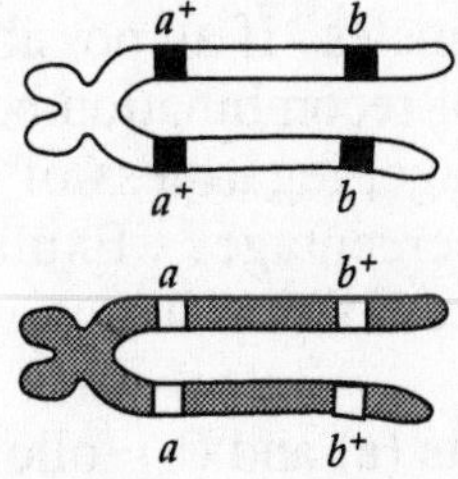

Answer: Recall what happens when there is a single crossover between an allele and its centromere. Alleles will segregate from each other at the first meiotic division if there is no crossover between them and the centromere and at the second meiotic division if there is a crossover between them and the centromere. A single crossover between *a* and *b* will satisfy the conditions of the problem. This is diagrammed below.

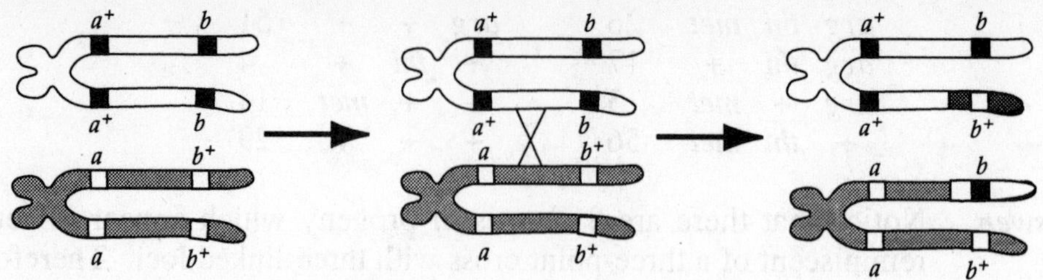

6.8 Double exchanges between two loci can be of several types, called two-strand, three-strand, and four-strand doubles.

a. Four recombination gametes would be produced from a tetrad in which the first of two exchanges is depicted in the figure below. Draw in the second exchange.

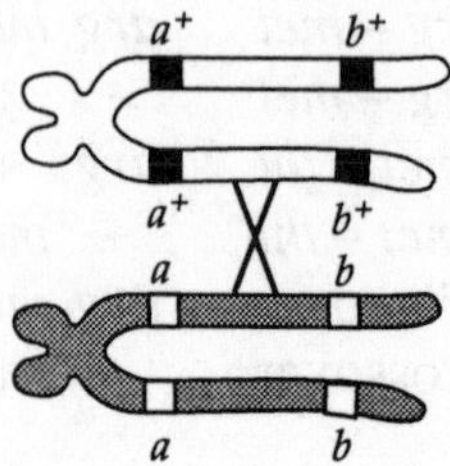

b. In the following figure, draw in the second exchange so that four non-recombination gametes would result.

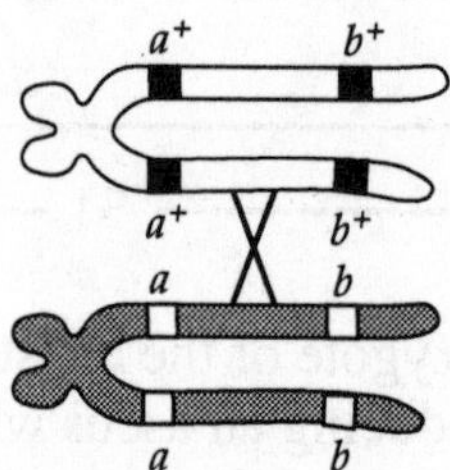

c. Other possible double-crossover types would result in two recombination and two parental gametes. If all possible multiple-crossover types occur at random, the frequency of recombination between genes at two loci will never exceed a certain percentage, regardless of how far apart they are on the chromosome. What is that percentage? Explain the reason why the percentage cannot theoretically exceed that value.

Answer: A solution to parts (a) and (b) follows from trial and error. Try diagramming a number of crossovers and their resolution (by following a chromosome "through" the crossover from left to right). A diagram of solutions for (a) and (b), with resolution for the crossover events, is shown below. The solution to (a) is a four-strand double crossover, and the solution to (b) is a two-strand double crossover. Note that the diagram shown in (b) is equivalent to a two-strand double-crossover between the outer two chromatids.

a.

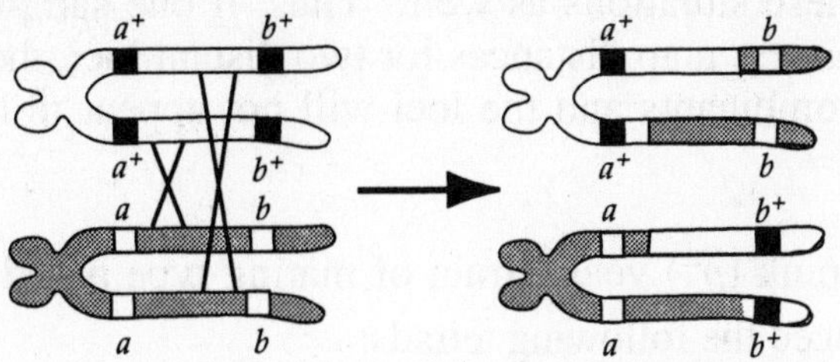

b.

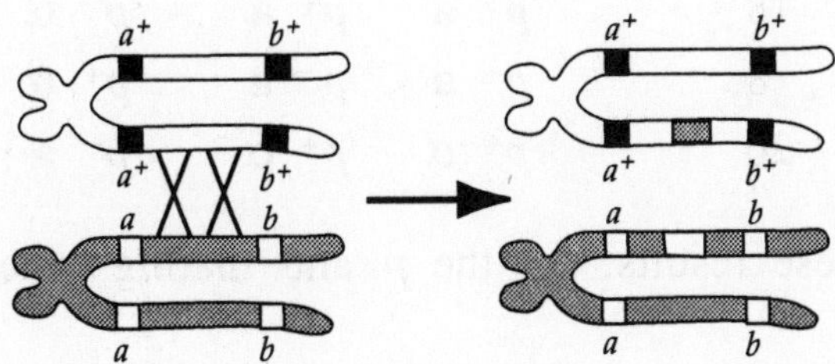

c. The remaining possible double-crossovers are three-strand double-crossovers, that is, involving three of the chromatids. There are two types diagrammed below:

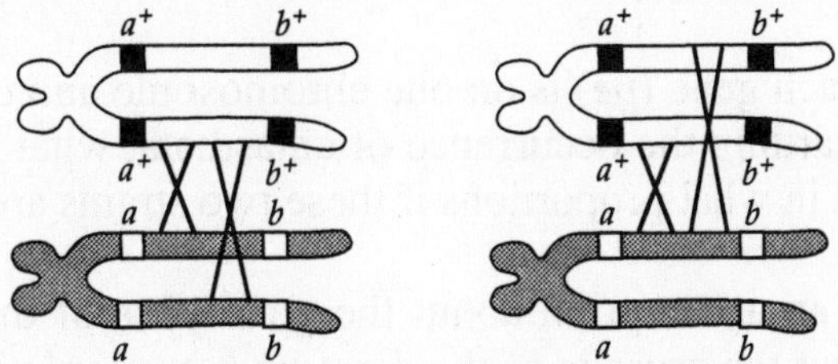

Resolution of these double-crossovers will result in a tetratype tetrad, with two recombinant-type spores and two parental-type spores. To determine the maximum frequency of recombination that is possible if double-crossover types occur at random, note the following:

Type of Event	Recombinant types	Parental-types
two-strand double	0	4
four-strand double	4	0
three-strand double, first way	2	2
three-strand double, second way	2	2
Total	8	8

This tabulation shows that if many double-crossovers occur, and the type of double-crossover that occurs is chosen randomly (i.e., there is no non-sister chromatid interference), then on average, there will be an equal number of recombinant and parental-type spores produced. Note that when a single crossover occurs, a tetratype tetrad is produced that contains two recombinant and two parental type spores. Analysis of all of the possible tetrads that can be produced when a higher-order number of crossovers occur at random (e.g., 3, 4, 5 etc.) shows that there will be at most 50 percent recombinant spores in these situations as well. Thus, if one samples progeny spores at random to assess map distances for two distant loci, there will be at most 50 percent recombinants and the loci will not appear to be more than 50 map units apart.

6.9 A cross between a pink (p^-) yeast strain of mating type **a** and a grey strain (p^+) of mating type α produced the following tetrads:

Number of Tetrads	Kinds of Tetrads			
18	p^+ **a**	p^+ **a**	p^- α	p^- α
8	p^+ **a**	p^- **a**	p^+ α	p^- α
20	p^+ α	p^+ α	p^- **a**	p^- **a**

On the basis of these results, are the p and mating type genes on separate chromosomes?

Answer: As the initial cross is p^- **a** x p^+ α, there are 20 PD tetrads, 18 NPD tetrads and 8 T tetrads. Since the number of PD and NPD tetrads is approximately equal, these two genes appear to assort independently. (Note that once one realizes which tetrads are of which type, determination of linkage is much easier than in non-tetrad analysis. One simply asks if PD = NPD!)

6.10 In *Neurospora* the peach gene (*pe*) is on one chromosome and the colonial gene (*col*) is on another. Disregarding the occurrence of chiasmata, what kinds of tetrads (asci) would you expect and in what proportions if these two strains are crossed?

Answer: One must make an assumption about the genotypes of the parent strains here. Let us assume that the cross is of the form *pe* + x + *col*. Then we can diagram the possible outcomes (ignoring chiasmata) as follows:

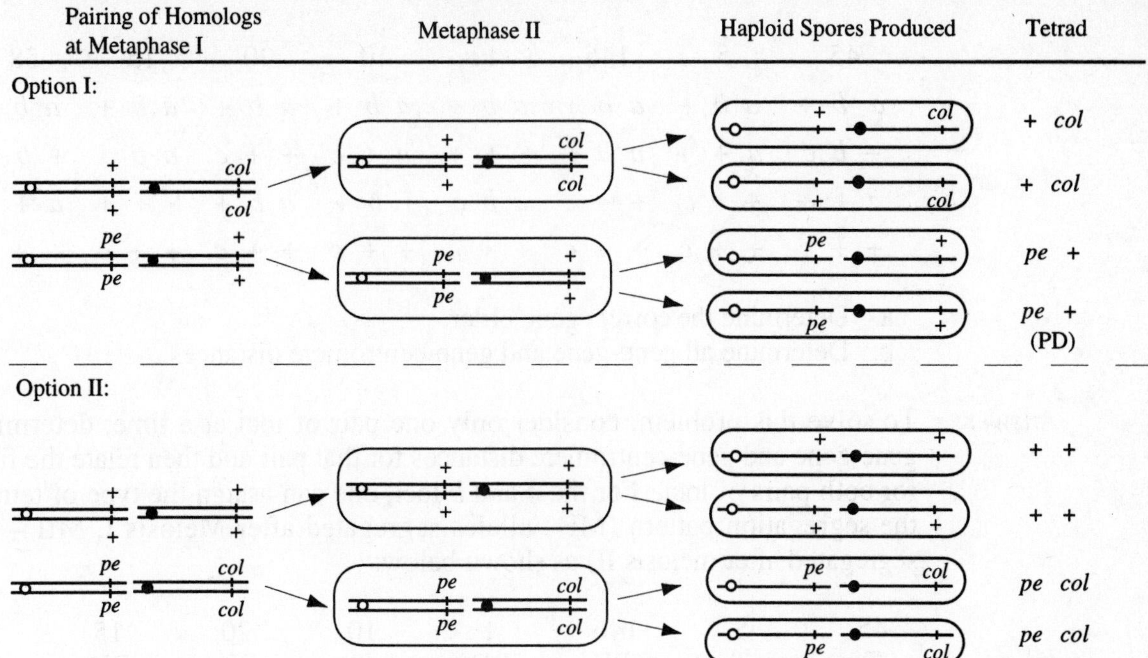

The type of tetrad that is formed (barring chiasmata) is solely dependent on the alignment of homologues in Meiosis I. Hence, one expects to find NPD and PD tetrads equally frequently for unlinked loci.

6.11 The following unordered asci were obtained from the cross *leu* + x + *rib* in yeast. Draw the linkage map and determine the map distance.

110	45	6	39
leu +	*leu rib*	+ +	*leu* +
+ *rib*	*leu* +	*leu rib*	+ *rib*
leu +	+ +	*leu rib*	+ +
+ *rib*	+ *rib*	+ +	*leu rib*

Answer: There are 110 PD, 6 NPD and 45 + 39 = 84 T tetrads. Since PD>> NPD, the loci are linked. The percent recombination between them is given by:

RF = [(1/2 T + NPD)/(total # asci)] x 100%
= [[(1/2 x 84) + 6]/(110 + 6 + 45 + 39)] x 100%
= (48/200) x 100%
= 24%

The map distance between *leu* and *rib* is 24 mu.

6.12 The genes *a*, *b*, and *c* are on the same chromosome arm in *Neurospora crassa*. The following ordered asci were obtained from the cross *a b* + x + + *c*.

45	5	146	1	10	20	15	58
a b +	*a b* +	*a b* +	*a b* +	*a b* +	*a b* +	*a b* +	*a b* +
+ *b c*	*a* + +	*a b* +	+ + +	*a* + *c*	+ + *c*	*a b c*	+ *b* +
a + +	+ *b c*	+ + *c*	*a b c*	+ *b* +	*a b* +	+ + +	*a* + *c*
+ + *c*	+ + *c*	+ + *c*	+ + *c*	+ + *c*	+ + *c*	+ + *c*	+ + *c*

a. Determine the correct gene order.
b. Determine all gene-gene and gene-centromere distances.

Answer: To solve this problem, consider only one pair of loci at a time, determine the gene-gene and gene-centromere distances for that pair and then relate the findings for both pairs of loci. For the *a* and *b* loci, one can assign the type of tetrad and the segregation pattern (MI – alleles segregated after Meiosis I, MII – alleles segregated after meiosis II) as shown below:

45	5	146	1	10	20	15	58
T	T	PD	PD	T	PD	PD	T
a b	*a b*	*a b*	*a b*	*a b*	*a b*	*a b*	*a b*
+ *b*	*a* +	*a b*	+ +	*a* +	+ +	*a b*	+ *b*
a +	+ *b*	+ +	*a b*	+ *b*	*a b*	+ +	*a* +
+ +	+ +	+ +	+ +	+ +	+ +	+ +	+ +
MII MI	MI MII	MI MI	MII MII	MI MII	MII MII	MI MI	MII MI

For the cross *a b* x + + then, there are (146 + 1 + 20 + 15) = 182 PD, (45 + 5 + 10 + 58) = 118 T, and 0 NPD tetrads. Since PD>>NPD, the loci are linked. The recombination frequency between them is given by:

RF = [(1/2T + NPD)/total # asci] x 100%
= [(1/2 x 118)/300] x 100%
= 19.7%

The recombination frequency of *a* and *b* with the centromere is given by:

RF (gene – centromere) = (1/2 MII type patterns/total) x 100%
RF (*a* – centromere) = [1/2(45 + 1 + 20 + 58)/300] x 100%
= 20.7%
RF (*b* – centromere) = [1/2(5 + 1 + 10 + 20)/300] x 100%
= 6%

For the cross *b* + x + *c*, we have:

45	5	146	1	10	20	15	58
T	T	PD	T	PD	PD	T	PD
b +	*b* +	*b* +	*b* +	*b* +	*b* +	*b* +	*b* +
b c	+ +	*b* +	+ +	+ *c*	+ *c*	*b c*	*b* +
+ +	*b c*	+ *c*	*b c*	*b* +	*b* +	+ +	+ *c*
+ *c*	+ *c*	+ *c*	+ *c*	+ *c*	+ *c*	+ *c*	+ *c*
MI MII	MII MI	MI MI	MII MI	MII MII	MII MII	MI MII	MI MI

There are (146 + 10 + 20 + 58) = 234 PD, (45 + 5 + 1 + 15) = 66 T and 0 NPD tetrads. Since PD>>NPD, *b* and *c* are linked. The recombination frequency between them is given by:

RF = [(1/2 x 66)/300] x 100%
= 11%

Their recombination frequency of *c* with the centromere is given by:

RF (*c* – centromere) = [1/2(45 + 10 + 20 + 15)/300] x 100%
= 15%

For the cross *a* + x + *c*, we have:

45	5	146	1	10	20	15	58
PD	PD	PD	T	T	PD	T	T
a +	*a* +	*a* +	*a* +	*a* +	*a* +	*a* +	*a* +
+ *c*	*a* +	*a* +	+ +	*a c*	+ *c*	*a c*	+ +
a +	+ *c*	+ *c*	*a c*	+ +	*a* +	+ +	*a c*
+ *c*	+ *c*	+ *c*	+ *c*	+ *c*	+ *c*	+ *c*	+ *c*

There are (45 + 5 + 146 + 20) = 216 PD, (1 + 10 + 15 + 58) = 84 T and 0 NPD tetrads. Since PD>>NPD, *a* and *c* are linked. The recombination frequency between them is given by:

RF = [(1/2 x 84)/300] x 100%
= 14%

Using the calculations above, *a* – *b* is 19.7 map units, *a* – *c* is 14 map units, *b* – *c* is 11 map units, and *b* – centromere is 6 map units. Thus, the gene order is centromere – *b* – *c* – *a*, and one has the following map (taking into consideration that, due to increasing numbers of multiple-crossovers, more accurate RFs are obtained over smaller intervals):

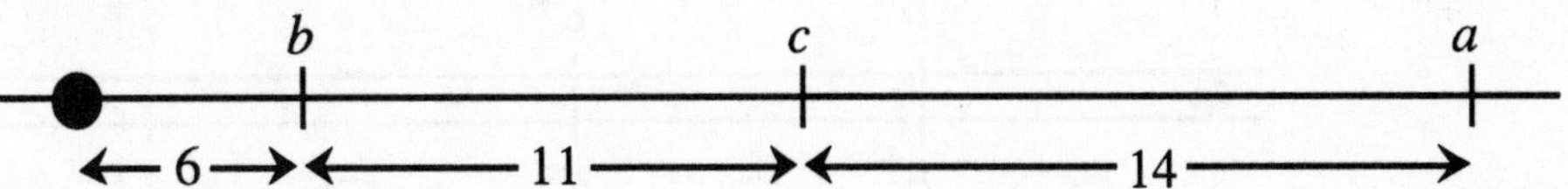

6.13 If 10 percent of the asci analyzed in a particular two-point cross of *Neurospora crassa* show that an exchange has occurred between two loci (i.e., exhibit second division segregation), what is the map distance between the two loci?

Answer: Recall that a centromere can be mapped from knowledge of the frequency of second division segregation patterns of a locus in ordered tetrads (MII patterns) by the formula:

RF (gene – centromere) = 1/2(# of MII patterns)/total number of asci.

Thus, if 10 percent of the asci show second division pattern of segregation of the two loci, they show a 1/2(10%) = 5% recombination frequency with each other and are 5 map units apart.

6.14 An *Aspergillus* diploid was obtained from a heterokaryon made by fusing together the haploid strains $a^+ \; b \; c \; i$ and $a \; b^+ \; c^+ \; i^+$, where the recessive allele a determines a requirement for growth substance A, the recessive allele b determines a requirement for growth substance B, the recessive allele c determines a requirement for growth substance C, and the recessive allele i determines *resistance* to an inhibitor I. When diploid asexual spores were plated on a medium lacking substance A, but containing the inhibitor, I, 43 diploid colonies resistant to the inhibitor grew. Analysis of the colonies indicated that 17 required neither B nor C, 14 required C but not B, and 11 required both B and C. What is the relationship between the b, c, and i genes and the centromere?

Answer: The initial heterokaryon was $a^+ \; b \; c \; i/a \; b^+ \; c^+ \; i^+$. In order for segregant colonies to grow without substance A and with the inhibitor I, they must be $a^+ – \; ii$. If, in addition, they require:

neither B nor C, they are	$a^+ – \;\; b^+ – \;\; c^+ – \;\; ii$	(17 colonies);
C but not B, they are	$a^+ – \;\; b^+ – \;\; c \;\; c \;\; ii$	(14 colonies);
both B and C, they are	$a^+ – \;\; b \;\; b \;\; c \;\; c \;\; ii$	(11 colonies).

Notice also that there are no colonies that are $c^+ – \; bb$.

Consider how these events could have come about. Since the original heterokaryon was heterozygous for all four loci, it would be able to grow without substance A, but not in the presence of I. In order to grow in the presence of I, mitotic recombination is required to make the i allele homozygous. Since mitotic recombination is relatively rare, that some of the ii segregants are homozygous for c and/or b suggests that these loci were made homozygous simultaneously with i. Therefore the i, c and b loci appear to be linked.

To order the loci on a chromosome relative to their centromere, consider how mitotic recombination in different regions of a chromosome can give rise to the observations. A single crossover will render all loci distal to the crossover point homozygous. To be consistent with the types of colonies seen, and in particular, to be consistent with the absence of $c^+ – \; bb$ segregants, the order of the loci must be:

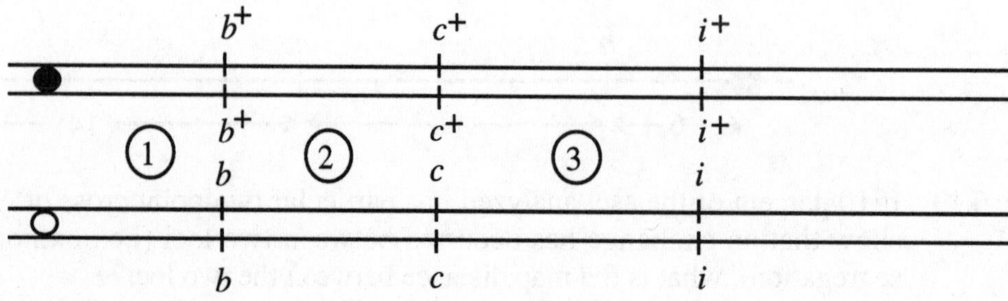

Single crossovers in region 1 could result in *bb cc ii* segregants, single crossovers in region 2 could result in b^+b *cc ii* segregants and single crossovers in region 3 could result in b^+b c^+c *ii* segregants.

6.15 The accompanying table shows the only human chromosomes present in stable human-mouse cell hybrid lines.

		Human Chromosomes			
		2	4	10	19
	A	−	+	+	−
Hybrid lines	*B*	+	−	+	+
	C	−	+	+	+
	D	+	+	−	−

The presence of four enzymes, I, II, III, and IV, was investigated: I was present in *A*, *B*, and *C* but absent in *D*; II was in *B* and *D* but absent in *A* and *C*; III was in *A*, *C*, and *D* but not in *B*; and IV was in *B* and *C* but not in *A* and *D*. On what chromosomes are the genes for the four enzymes?

Answer: To solve this problem, one needs to determine a correlation between the presence, or absence, or a chromosome and the presence, or absence, of an enzyme activity. This can be done by trial and error, or by tabulating the data, to help relationships become apparent. The data given can be tabulated as follows:

		Human Chromosomes			
		2	4	10	19
	A	−	+	**+**	−
Hybrid lines	*B*	+	−	**+**	+
	C	−	+	**+**	+
	D	+	+	**−**	−
		Enzyme			
		I	II	III	IV
	A	**+**	−	+	−
Hybrid lines	*B*	**+**	+	−	+
	C	**+**	−	+	+
	D	**−**	+	+	−

Notice that the pattern of being present or absent is the same for enzyme I and chromosome 10 (read the columns of pluses and minuses from top to bottom; note the emboldened and enlarged symbols), enzyme II and chromosome 2, enzyme III and chromosome 4, and enzyme IV and chromosome 19. Therefore, these genes are on those chromosomes.

		2	[illegible]	10	19
Hybrid lines	A	[illegible]	[illegible]	[illegible]	[illegible]
	B	[illegible]	[illegible]	[illegible]	[illegible]
	C	[illegible]	[illegible]	[illegible]	[illegible]
	D	[illegible]	[illegible]	[illegible]	[illegible]

Human Chromosome

		I	II	III	IV
	A	[illegible]	[illegible]	[illegible]	[illegible]
Hybrid lines	B	[illegible]	[illegible]	[illegible]	[illegible]
	C	[illegible]	[illegible]	[illegible]	[illegible]
	D	[illegible]	[illegible]	[illegible]	[illegible]

CHAPTER 7

CHROMOSOMAL MUTATIONS

I. CHAPTER OUTLINE

TYPES OF CHROMOSOMAL MUTATIONS
VARIATIONS IN CHROMOSOME STRUCTURE
- Deletion
- Duplication
- Inversion
- Translocation
- Position Effect
- Fragile Sites and Fragile X Syndrome

VARIATIONS IN CHROMOSOME NUMBER
- Changes in One of a Few Chromosomes
- Changes in Complete Sets of Chromosomes

CHROMOSOME REARRANGEMENTS THAT ALTER GENE EXPRESSION
- Amplification or Deletion of Genes
- Inversions That Alter Gene Expression
- Transpositions That Alter Gene Expression

II. REVIEW OF KEY TERMS, SYMBOLS AND CONCEPTS

Without consulting the text and in your own words, write a brief definition of each term in the groups below. Then, either using a short phrase or a simple diagram, identify the relationship(s) between specific pairs of terms within a set. Finally, consult the text (and perhaps a friend who has also done the exercise) to check your answers.

1	2	3	4
chromosomal mutation	chromosomal aberrations	deletion	duplication
chromosomal aberration	karyotype	pseudodominance	terminal tandem duplication
deletions	polytene chromosomes	*cri-du-chat* syndrome	reverse tandem duplication
duplications	bands & interbands	*Prader-Willi* syndrome	*Drosophila Bar* mutation
translocations			multigene family
inversions			unequal crossing-over
position affect			

5	6	7	8
inversion pericentric inversion paracentric inversion dicentric bridge acentric fragment	translocation intrachromosomal translocation interchromosomal translocation reciprocal translocation nonreciprocal translocation alternate segregation adjacent-1, -2 segregation semisterility	translocation Philadelphia chromosome Burkitt's lymphoma chronic myelogenous leukemia oncogenes	fragile sites fragile X syndrome triplet repeat amplification

9	10	11	12
euploid aneuploid polyploid diploid haploid	nullisomy monosomy trisomy tetrasomy	autopolyploidy allopolyploidy allotetraploid	yeast mating type genes transposition programmed amplification polytenization programmed deletion

13
Down syndrome Patau syndrome Edwards syndrome Turner syndrome Klinefelter syndrome familial Down syndrome Robertsonian translocation

III. THINKING ANALYTICALLY

This chapter requires you to visualize the physical structure of chromosomes in the context of various cellular processes. It is particularly important to be able to visualize the effects of internal alterations of chromosome structure such as inversions, transpositions, translocations, duplications and deletions.

In order to fully understand the impact of chromosomal mutations on gene expression and inheritance of traits, it is essential to have a solid grasp on the fundamental cellular processes that involve chromosome movement. Review how chromosomes move and align during meiosis. This will give you a solid foundation to visualize how chromosomal aberrations and mutations affecting chromosome number impact on this process, and thereby impact on the inheritance of traits. The disruptions in meiosis that are common sequellae of certain chromosomal aberrations are best understood when examined structurally, as opposed to purely conceptually. When doing the problems, take the time to draw out the consequences of crossing-over in inversion and translocation heterozygotes.

Profound changes in gene expression often result from alterations in the physical positions of genes relative to each other, even without any other kind of change being involved. The importance of gene dosage for a "normal" phenotype also becomes apparent when the effects of aneuploidy and polyploidy are examined. Keep these issues in mind as you consider the spectrum of abnormal traits that are presented in this chapter.

IV. QUESTIONS FOR PRACTICE

A. Multiple Choice Questions

1. The number of polytene chromosomes in dipterans, such as *Drosophila*, is characteristically
   a. twice the diploid number.
   b. hundreds or more times the diploid number.
   c. half the diploid number.
   d. the same as the diploid number.

2. The chromosomes that become polytene chromosomes are
   a. somatically paired.
   b. genetically inert.
   c. uniformly pycnotic.
   d. acentric.

3. Nonhomologous chromosomes that have exchanged segments are the products of a
   a. double deletion.
   b. reciprocal translocation.
   c. pericentric inversion.
   d. paracentric inversion.

4. The abbreviated karyotype, 2N - 1, describes
   a. nullisomy.
   b. monosomy.
   c. trisomy.
   d. haploidy.

5. The abbreviated karyotype, 2N + 1 + 1, describes
   a. double trisomy.
   b. tetrasomy.
   c. double monosomy.
   d. none of these.

6. All known monosomics in humans have been
   a. lethal.
   b. semi-lethal.
   c. treatable.
   d. deletion heterozygotes.

7. The cultivated bread wheat, *Triticum aestivum,* although a polyploid, is fertile because it
   a. has an odd number of chromosome sets.
   b. has an even multiple of chromosome sets.
   c. is a double hybrid.
   d. is propagated by grafting.

8. Cultivated bananas and Baldwin apples are
   a. double diploids.
   b. tetraploids.
   c. double haploids.
   d. triploids.

9. Down Syndrome is associated with
   a. an inversion.
   b. a Robertsonian translocation.
   c. trisomy 21.
   d. both b and c.

10. Gene expression can be altered by
   a. an inversion.
   b. a duplication.
   c. a transposition.
   d. any of the above.

Answers: 1b; 2a; 3b; 4b; 5a; 6a; 7b; 8d; 9d; 10d.

B. Thought Questions

1. Identify and describe five examples of changes in chromosome structure that alter gene expression.
2. A chromosomal inversion in the heterozygous condition seems to act as a suppressor of crossing over. Why might this be the case? How might such chromosomes be important for keeping a laboratory stock of a three linked recessive mutations?
3. What is genetically and immunologically noteworthy about the protozoan parasite, *Trypanosoma*, the causative agent of African sleeping sickness?
4. What are two modes by which Down syndrome can be caused? Why might one be more frequent than the other? If a normal female has a sibling with Down syndrome, what information would be critical in her assessment of the chances that she might have a Down syndrome child? How would the assessment be different if the individual were a normal male with a Down syndrome sibling?
5. How do you account for the fact that polyploidy is significantly more common in plants that in animals? (Hint: What basic difference between sexual and asexual reproduction might be significant here?)
6. What might be the significance of the following in the evolutionary process of speciation? (1) autopolyploidy, (2) allopolyploidy, (3) inversions, (4) translocations. Provide an explanation that considers how animals might be differently affected than plants.

7. Chromosomal rearrangements such as inversions and translocations are sometimes associated with dominant or recessive phenotypes. How might such phenotypes be related to a "position effect"?
8. Consider the nature of a position effect, and offer an hypothesis as to why triplet repeats, such as those found in Fragile X syndrome and Huntington's disease, might only result in an abnormal phenotype when a threshold number of triplet repeats is exceeded?

V. SOLUTIONS TO TEXT PROBLEMS

7.1 A normal chromosome has the following gene sequence:

A B C D ○ *E F G H*

Determine the chromosome mutation illustrated by each of the following chromosomes:

a. *A B C F E* ○ *D G H*
b. *A D* ○ *E F B C G H*
c. *A B C D* ○ *E F E F G H*
d. *A B C D* ○ *E F F E G H*
e. *A B D* ○ *E F G H*

Answer: a. pericentric inversion [inversion of *D* – (centromere) – *E* – *F*]
b. nonreciprocal translocation [*B* – *C* moved from left to right arm]
c. tandem duplication [*E* – *F* duplicated]
d. reverse tandem duplication [*E* – *F* duplicated]
e. deletion [*C* deleted]

7.2 Distinguish between pericentric and paracentric inversions.

Answer: A pericentric inversion is an inversion that includes the centromere, while a paracentric inversion is an inversion that is wholly within one chromosomal arm. See text pages 201-202 and Figure 7.9.

7.3 Very small deletions behave in some instances like recessive mutations. Why are some recessive mutations known not to be deletions?

Answer: Deletions, whether just a few bases or large segments of DNA, are unable to be reverted. Point mutations that affect only a single base can be reverted. If a recessive mutation is able to be reverted to a wild-type phenotype, it cannot be a deletion.

7.4 What would be the effect on protein structure, if a small inversion were to occur within the coding region of a structural gene?

Answer: Review the material presented in Chapter 1 on DNA structure, transcription, and translation. DNA is composed of two strands that have opposite polarity (one is

5' to 3', the other 3' to 5'). The transcription of DNA into RNA always proceeds so that the RNA is transcribed in the 5' to 3' direction using a 3' to 5' template strand. The sequence of bases in the mRNA transcript are then read as a triplet code and translated into an amino acid sequence.

There are several possible consequences of an inversion in the DNA of a single gene, all of them serious. Since the polarity of DNA must be maintained, within the region of an inversion, the bases of the non-template strand of DNA would be inserted in the template strand (Try drawing this out). After the DNA is transcribed, it will be decoded during translation. The region of the transcript originating from the inverted region will encode an amino acid sequence that is quite different than the normal, uninverted region, and may have possible nonsense codons (translation termination signals) within it. Consequently, a non-functional or truncated protein may be produced.

7.5 Inversions are known to affect crossing-over. The following homologs with the indicated gene order are given:

● *A B C D E*

○ *A D C B E*

a. Diagram the alignment of these chromosomes during meiosis.
b. Diagram the results of a single crossover between homologous genes B and C in the inversion.
c. Considering the position of the centromere, what is this sort of inversion called?

Answer: a.

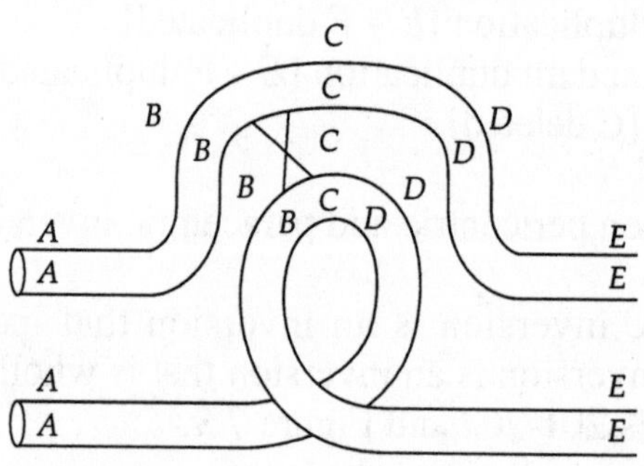

b. A crossover between B and C results in the following chromosomes:

● *A B C D E* (normal order)

● *A B C D A* ○ (dicentric, duplication for *A* deletion for *E*)

E B C D E (acentric, duplication for *E* deletion for *A*)

○ *A D C B E* (inverted order)

c. Paracentric inversion, because the centromere is not included in the inverted DNA segment.

7.6 Single crossovers within the inversion loop of inversion heterozygotes give rise to chromatids with duplications and deletions. What happens when, within the inversion loop, there is a two-strand double crossover in such an inversion heterozygote when the centromere is outside the inversion loop?

Answer: If a two-strand double crossover occurs within a paracentric inversion, the four products of meiosis have a complete set of genes, without duplications or deletions. No acentric or dicentric fragments are formed. Consequently, all four meiotic products are viable. This is illustrated in the following example:

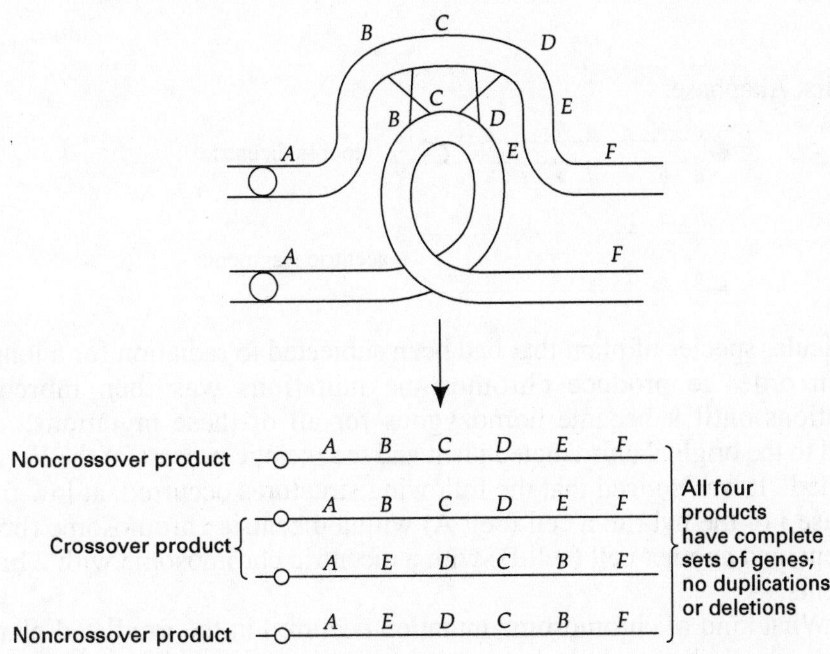

7.7 An inversion heterozygote possesses one chromosome with genes in the normal order:

a b c d e f g h

It also contains one chromosome with genes in the inverted order:

a b f e d c g h

A four-strand double crossover occurs in the areas *e* – *f* and *c* – *d*. Diagram and label the four strands at synapsis (showing the crossovers) and at the first meiotic anaphase.

Answer: Diagrammed below is a four-strand double crossover, where the crossover between *c* and *d* involves strands 2 and 4, and the crossover between *e* and *f* involves strands 1 and 3.

Synapsis:

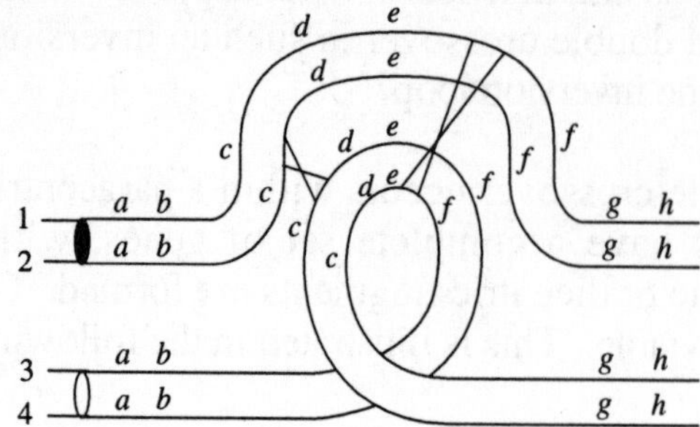

First Anaphase:

1 a b c d e f b a 3
2 a b c d e f b a 4 double dicentric

h g f e d c g h

h g f e d c g h acentric fragments

7.8 A particular species of plant that had been subjected to radiation for a long period of time in order to produce chromosome mutations was then inbred for many generations until it became homozygous for all of these mutations. It was then crossed to the original unirradiated plant and the meiotic process of the F_1 hybrids was examined. It was noticed that the following structures occurred, at low frequency, in anaphase I of the hybrid: a cell (cell A) with a dicentric chromosome (bridge) and a fragment, and another cell (cell B) with a dicentric chromosome with 2 bridges and 2 fragments.

a. What kind of chromosome mutation occurred in the irradiated plant? In your answer, indicate where the centromeres are.

b. Explain in words and with a clear diagram where crossover(s) occurred, and how the bridge chromosome of cell A arose.

c. Explain in words and with a clear diagram where crossover(s) occurred, and how the double bridge chromosome of cell B arose.

Answer: a. The irradiated chromosome has a paracentric inversion, with an inversion within one of its arms. Dicentric chromosomes and fragments arise as the result of single crossovers within paracentric inversions, and dicentric chromosomes with two bridges and two fragments result from a four-strand double crossover within paracentric inversions. An example of such a chromosome is diagrammed below.

a b c ● *d e f g h i* normal order

a b c ● *d h g f e i* paracentric inversion

b. The bridge chromosome would arise by a single crossover within an inversion loop during meiosis. In the above example, the crossover would occur between *d* and *i*. See text Figure 7.10b for an illustration of such a crossover.

c. The double bridge chromosome would arise by a four-strand double crossover within an inversion loop during meiosis. See the figure in the solution to problem 7.7 for an illustration of such a crossover.

7.9 Mr. and Mrs. Lambert have not yet been able to produce a viable child. They have had two miscarriages and one severely defective child who died soon after birth. Studies of banded chromosomes of father, mother, and child showed all chromosomes were normal except for pair number 6. The number 6 chromosomes of mother, father, and child are shown in the following figure.

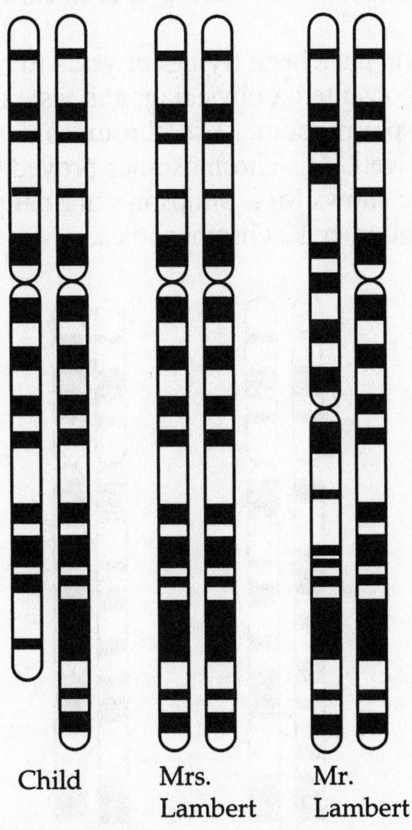

a. Does either parent have an abnormal chromosome? If so, what is the abnormality?
b. How did the chromosomes of the child arise? Be specific as to what events in the parents gave rise to these chromosomes.
c. Why is the child not phenotypically normal?
d. What can be predicted about future conceptions by this couple?

Answer: a. Mr. Lambert is heterozygous for a pericentric inversion of chromosome 6. One of the breakpoints is within the fourth light band up from the centromere, while the other is in the sixth dark band below the centromere. Mrs. Lambert's chromosomes are normal.

b. When Mr. Lambert's number 6 chromosomes paired during meiosis, they formed an inversion loop which included the centromere. Crossing over occurred within the loop, and gave rise to the partially duplicated, partially deficient chromosome 6 which the child received.

c. The child's abnormalities stem from having three copies of some, and only one copy of other, chromosome 6 regions. The top part of the short arm is duplicated, and there is a deficiency of the distal part of the long arm in this case.

d. The inversion appears to cover more than half of the length of chromosome 6, so crossing over will occur in this region in the majority of meioses. In the minority of meioses where crossing over occurs outside the loop, and in the cases where it has occurred within the loop but the child receives an uncrossed-over chromatid, the child can be normal. There is significant risk for abnormality, so monitoring of fetal chromosomes should be done.

7.10 Mr. and Mrs. Simpson have been trying for years to have a child but have been unable to conceive. They consulted a physician, and tests revealed that Mr. Simpson had a markedly reduced sperm count. His chromosomes were studied, and a testicular biopsy was done as well. His chromosomes proved to be normal, except for pair 12. The following figure shows Mrs. Simpson's normal pair of number 12 chromosomes and Mr. Simpson's number 12 chromosomes.

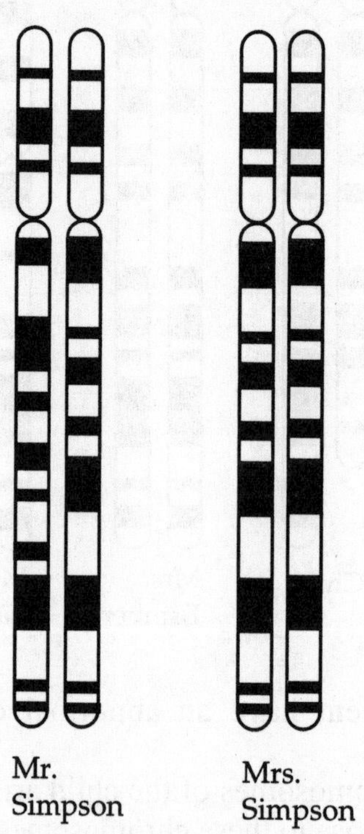

a. What is the nature of the abnormality in Mr. Simpson's chromosomes of pair number 12?
b. What abnormal feature would you expect to see in the testicular biopsy (cells in various stages of meiosis can be seen).
c. Why is Mr. Simpson's sperm count low?
d. What can be done about it?

Answer: a. Mr. Simpson has a paracentric inversion in the long arm of one of his number 12 chromosomes. Moving downwards (distally) from the centromere, the breakpoints are after the first band and after the sixth band.

b. Crossing over within the inversion loop will produce dicentric chromatids which would form anaphase bridges. These chromatin bridges would be visible joining the two chromatin masses at anaphase I.

c. The inversion is large, so the frequency of crossing over within it will be significant. Consequently, bridges will be formed in the majority of meioses. Cells which have bridges form do not complete meiosis or form sperm in mammals.

d. Nothing can be done to increase Mr. Simpson's sperm count. This might be an instance to consider *in vitro* fertilization.

7.11 Irradiation of *Drosophila* sperm produces translocations between the X chromosome and autosomes, and between the Y chromosome and autosomes, and between different autosomes. Translocations between the X and Y chromosomes are not produced. Explain the absence of X-Y translocations.

Answer: Mature sperm either bear an X or a Y chromosome, not both. For X-Y translocations to be obtained, both chromosomes must be present in the same sperm cell.

7.12 The following gene arrangements in a particular chromosome are found in *Drosophila* populations in different geographical regions. Assuming the arrangement in part a is the original arrangement, in what sequence did the various inversion types most likely arise?

a. *ABCDEFGHI*
b. *HEFBAGCDI*
c. *ABFEDCGHI*
d. *ABFCGHEDI*
e. *ABFEHGCDI*

Answer: One can envision a series of inversions occurring in the following sequence:

a → c → e → d
↓ (from e)
b

The regions inverted in each step are illustrated below.

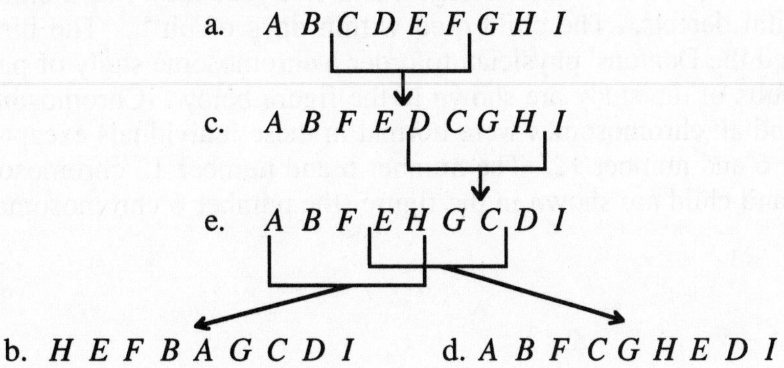

7.13 Chromosome I in maize has the gene sequence *ABCDEF*, whereas chromosome II has the sequence *MNOPQR*. *A* reciprocal translocation resulted in *ABCPQR* and

MNODEF. Diagram the expected pachytene configuration of the F_1 of a cross of homozygotes of these two arrangements.

Answer:

F F
E E
D D
A B C ONM

A B C ONM
P P
Q Q
R R

7.14 Diagram the pairing behavior at prophase of meiosis I of a translocation heterozygote that has normal chromosomes of gene order *abcdefg* and *tuvwxyz* and has the translocated chromosomes *abcdvwxyz* and *tuefg*. Assume that the centromere is at the left end of all chromosomes.

Answer:

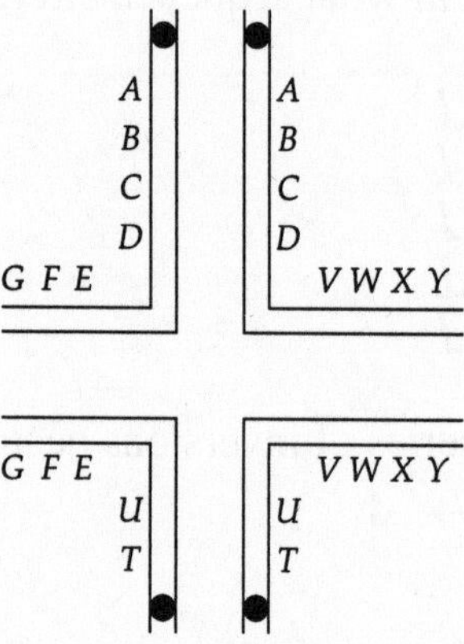

7.15 Mr. and Mrs. Denton have been trying for several years to have a child. They have experienced a series of miscarriages, and last year they had a child with multiple congenital defects. The child died within days of birth. The birth of this child prompted the Dentons' physician to order a chromosome study of parents and child. The results of the study are shown in the figure below. Chromosome banding was done, and all chromosomes were normal in these individuals except some copies of number 6 and number 12. The number 6 and number 12 chromosomes of mother, father, and child are shown in the figure (the number 6 chromosomes are the larger pair).

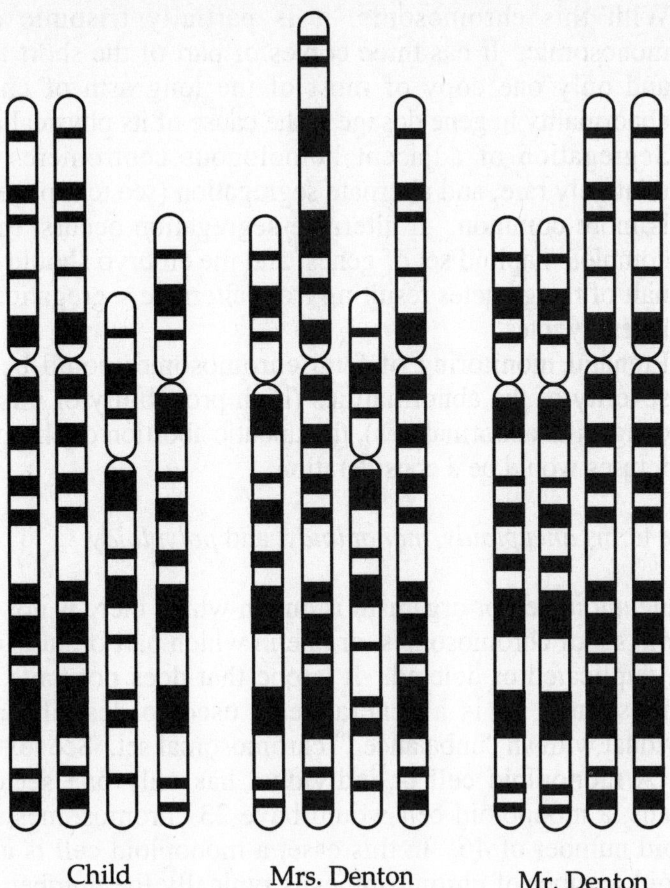

a. Does either parent have an abnormal karyotype? If so, which parent has it, and what is the nature of the abnormality?
b. How did the child's karyotype arise (what pairing and segregation events took place in the parents)?
c. Why is the child phenotypically defective?
d. What can this couple expect to occur in subsequent conceptions?
e. What medical help, if any, can be offered to them?

Answer:

a. Mr. Denton has normal chromosomes. Mrs. Denton is heterozygous for a balanced reciprocal translocation between chromosomes 6 and 12. Most of the short arm of chromosome 6 has been reciprocally translocated onto the long arm of chromosome 12. The breakpoints appear to be in the thick, dark band just above the centromere of 6 and in the third dark band below the centromere of 12.
b. The child received a normal 6 and a normal 12 from his father. In prophase I of meiosis in Mrs. Denton, chromosome 6 and 12 and the reciprocally translocated 6 and 12 paired to form a cruciform-like structure. Segregation of adjacent, nonhomologous centromeres to the same pole ensued, so that the child received a gamete containing a normal 6 and one of the translocation chromosomes. See text Figure 7.13.
c. The child has a normal 6 and a normal 12 chromosome from Mr. Denton. The child also has a normal 6 chromosome from Mrs. Denton. However, the child also has one of the translocation chromosomes from Mrs. Denton.

With this chromosome, it is partially trisomic as well as partially monosomic. It has three copies of part of the short arm of chromosome 6 and only one copy of most of the long arm of chromosome 12. This abnormality in gene dosage is the cause of its physical abnormality.

d. Segregation of adjacent homologous centromeres to the same pole is relatively rare, and alternate segregation (see text page 205 and Figure 7.13) is more common. If alternate segregation occurs, the gamete will have a complete haploid set of genes, and the embryo should be normal. However, half of the gametes resulting from alternate segregation will be translocation heterozygotes.

e. Prenatal monitoring of fetal chromosomes could be done, and given the severity of the abnormalities (high probability of miscarriage and multiple congenital abnormalities), therapeutic abortion of chromosomally unbalanced fetuses would be a consideration.

7.16 Define the terms *aneuploidy*, *monoploidy*, and *polyploidy*.

Answer: An aneuploid cell or organism is one in which there is not an exact multiple of a haploid set of chromosomes, or one in which part or parts of chromosomes have been duplicated or deleted. It is one that does not have a *euploid* number of chromosomes. It is a general term used to describe a typically abnormal individual with an "unbalanced" chromosomal set. See text pages 211-216.

A monoploid cell or individual has only one set of chromosomes. In humans, a monoploid cell would have 23 chromosomes instead of the normal diploid number of 46. In this case, a monoploid cell is also a haploid cell. A haploid number of chromosomes is typically the number of chromosomes in a gamete. Thus, in diploid individuals, a haploid gamete is also a monoploid cell. This is not always the case for non-diploid individuals. For example, in a hexaploid plant that has 36 chromosomes, a gamete will have 18 chromosomes (haploid number = 18), but the monoploid number will be 6. A polyploid cell or individual has multiple sets of chromosomes. It is euploid, having multiple *complete* sets of chromosomes. See text pages 216-217.

7.17 If a normal diploid cell is 2N, what is the chromosome content of the following:

a. a nullisomic
b. a monosomic
c. a double monosomic
d. a tetrasomic
e. a double trisomic
f. a tetraploid
g. a hexaploid

Answer:
a. 2N - 2 (two copies of the same chromosome are missing)
b. 2N - 1 (missing one chromosome)
c. 2N - 1 - 1 (one copy of each of two different chromosomes are missing)
d. 2N + 2 (two extra copies of one chromosome)
e. 2N + 1 + 1 (an extra copy of each of two different chromosomes)
f. 4N
g. 6N

7.18 In humans, how many chromosomes would be typical of nuclei of cells that are

a. monosomic
b. trisomic
c. monoploid
d. triploid
e. tetrasomic

Answer: a. 45
b. 47
c. 23
d. 69
e. 48

7.19 An individual with 47 chromosomes, including an additional chromosome 15, is said to be

a. triplet.
b. trisomic.
c. triploid.
d. tricycle.

Answer: b. trisomic

7.20 A color-blind man marries a homozygous normal woman and after four joyful years of marriage they have two children. Unfortunately, both children have Turner syndrome, although one has normal vision and one is color-blind. The type of color blindness involved is a sex-linked recessive trait.

a. For the color-blind child with Turner syndrome, did nondisjunction occur in the mother or the father? Explain your answer.
b. For the Turner child with normal vision, in which parent did nondisjunction occur? Explain your answer.

Answer: a. The cross can be written as $X^+X^+ \times X^cY$. Turner syndrome children are XO. If the child is color-blind, it received its father's X^c via a chromosomally normal X-bearing sperm. Therefore, the egg that was fertilized must have lacked an X, and non-disjunction occurred in the mother.
b. If the child has normal vision, it must have received its mother's normal X^+. To be XO, this must be the only X the embryo received, so that the egg must have been fertilized by a nullo-X, nullo-Y sperm. Nondisjunction occurred in the father.

7.21 Assume that *x* is a new mutant gene in corn. A female *x/x* plant is crossed with a triplo-10 individual (trisomic for chromosome 10) carrying only dominant alleles at the *x* locus. Trisomic progeny are recovered and crossed back to the *x/x* female plant.

a. What ratio of dominant to recessive phenotypes is expected if the *x* locus is *not* on chromosome 10?
b. What ratio of dominant to recessive phenotypes is expected if the *x* locus *is* on chromosome 10?

Answer: a. If the new mutant is not on chromosome 10, the cross can be written as *x/x* × +/+ (considering only the chromosome carrying the *x* gene). The progeny will be $x/+$, and the backcross will be $x/+ \times x/x$. A 1:1 ratio of x/x (affected) to $x/+$ (normal) individuals will be seen.

b. If the new mutant is on chromosome 10, the cross can be written as $x/x \times +/+/+$. Trisomic progeny will be $x/+/+$, and the backcross can be written as $x/+/+ \times x/x$. In the trisomic $x/+/+$ individual, there will be four kinds of gametes produced depending on how chromosome 10 segregates during meiosis. In 1/3 of the meioses, x and $+/+$ gametes will be produced, giving 1/6 x and 1/6 $+/+$ gametes. In 2/3 of the meioses, $+$ and $+/x$ gametes will be produced, giving 1/3 $+/x$ and 1/3 $+$ gametes. When such gametes fuse with a x-bearing gamete (from the xx parent) at fertilization, 1/6 of the progeny will have the mutant phenotype (i.e., be x/x) and 5/6 will be normal (i.e., $+/+/x$, $+/x/x$, $+/x$).

7.22 Why are polyploids with even multiples of the chromosome set generally more fertile than polyploids with odd multiples of the chromosome set?

Answer: Polyploids with even multiples of the chromosome set can better form chromosome pairs in meiosis than can polyploids with odd multiples. Triploids, for example, will generate an unpaired chromatid pair for each chromosome type in the genome, so that chromosome segregation to the gametes is irregular and the resulting zygotes will not be euploid.

7.23 Select the correct answer from the key below for the following statement: One plant species (N = 11) and another (N = 19) produced an allotetraploid.

I. The chromosome number of this allotetraploid is 30.
II. The number of nuclear linkage groups of this allotetraploid is 30.

Key:

A. Statement I is true and Statement II is true.
B. Statement I is true, but Statement II is false.
C. Statement I is false, but Statement II is true.
D. Statement I is false and Statement II is false.

Answer: To form the initial alloploid, gametes from each species fused, so that the initial alloploid had 11 + 19 = 30 chromosomes. A fertile, allotetraploid plant was produced by the doubling of this chromosome set, so that the allotetraploid plant has 60 chromosomes, two sets of 11 and two sets of 19. Thus statement II is true, but statement I is not. The allotetraploid has 60 chromosomes and 30 linkage groups (30 pairs of homologs). C is correct.

7.24 According to Mendel's first law, genes *A* and *a* segregate from each other and appear in equal numbers among the gametes. But Mendel did not know that his plants were diploid. In fact, since plants are frequently tetraploid, he could have been unlucky enough to have started with peas that were 4N rather than 2N. Let us assume that Mendel's peas were tetraploid, that every gamete contains two alleles, and that the distribution of alleles to the gamete is random. Suppose we have a cross of *AA AA* × *aa aa,* where *A* is dominant, regardless of the number of *a* alleles present in an individual.

a. What will be the genotype of the F_1?
b. If the F_1 is selfed, what will be the phenotypic ratios in the F_2?

Answer: a. The F_1 will be *AA aa*.

b. Label the four alleles in the F_1 as A^1, A^2, a^1 and a^2. These four gametes can segregate into gametes in six ways: A^1A^2, A^1a^1, A^1a^2, A^2a^1, A^2a^2, a^1a^2. 1/6 of the gametes will be *AA*, 4/6 of the gametes will be *Aa*, and 1/6 will be *aa*. Selfing the F_1 will give the following pairings:

	1/6 *AA*	4/6 *Aa*	1/6 *aa*
1/6 *AA*	1/36 *AAAA*	4/36 *AAAa*	1/36 *AAaa*
4/6 *Aa*	4/36 *AAAa*	16/36 *AAaa*	4/36 *Aaaa*
1/6 *aa*	1/36 *AAaa*	4/26 *Aaaa*	1/36 *aaaa*

This would give a phenotypic ratio of 35/36 *A–* :1/36 *aa*.

7.25 What phenotypic ratio of *A* to *a* is expected if *AA aa* plants are testcrossed against *aa aa* individuals? (Assume that the dominant phenotype is expressed whenever at least one *A* is present, no crossing-over occurs, and each gamete receives two chromosomes.)

Answer: Label the four alleles of the *AAaa* plant as A^1, A^2, a^1 and a^2. As in problem 7.24, these four gametes can segregate into gametes in six ways: A^1A^2, A^1a^1, A^1a^2, A^2a^1, A^2a^2, a^1a^2. 1/6 of the gametes will be *AA*, 4/6 of the gametes will be *Aa*, and 1/6 will be *aa*. When these gametes fuse with *aa* gametes at fertilization, the testcross progeny will be 1/6 *AAaa,* 4/6 *Aaaa,* 1/6 *aaaa*, or 5/6 *A–* :1/6 *aa.*

7.26 The root tip cells of an autotetraploid plant contain 48 chromosomes. How many chromosomes were contained by the gametes of the diploid from which this plant was derived?

Answer: The somatic cells of an autotetraploid plant have four identical sets of chromosomes. As root tip cells are mitotically dividing, somatic cells, they too have four identical sets of chromosomes. Therefore, the gametes of the diploid from which this plant was derived had 12 chromosomes.

7.27 A number of species of the birch genus have a somatic chromosome number of 28. The paper birch is reported as occurring with several different chromosome numbers; *fertile* individuals with the somatic numbers 56, 70, and 84 are known. How should the 28-chromosome individuals be designated with regard to chromosome number?

Answer Plants with 56, 70 and 84 chromosomes have gametes that have 28, 35 and 42 chromosomes, respectively. If these plants are fertile polyploids, they should have an even number of chromosome sets. This would be the case if the monoploid number of these plants is 7, and the 56, 70 and 84 chromosome plants are have eight, ten and twelve times the monoploid number of chromosomes. A plant with 28 chromosomes would therefore be tetraploid, and should be fertile.

7.28 How many chromosomes would be found in somatic cells of an allotetraploid derived from two plants, one with N = 7 and the other with N = 10?

Answer: The initial alloploid will have 17 chromosomes. After doubling, the somatic cells will have 34 chromosomes.

7.29 Plant species A has a haploid complement of 4 chromosomes. A related species B has 5. In a geographical region where A and B are both present, C plants are found that have some characters of both species and somatic cells with 18 chromosomes. What is the chromosome constitution of the C plants likely to be? With what plants would they have to be crossed in order to produce fertile seed?

Answer: The C plants are allotetraploids, containing a diploid chromosome set from each of species A and species B. These plants should be fertile, as they will have no abnormal chromosome pairing or unpaired chromosomes. They should be able to produce gametes with 9 chromosomes (the four of species A and the five of species B), and can either be selfed or crossed to other C plants to produce fertile seed.

CHAPTER 8

GENETIC MAPPING IN BACTERIA AND BACTERIOPHAGES

I. CHAPTER OUTLINE

GENETIC ANALYSIS OF BACTERIA
BACTERIAL TRANSFORMATION
CONJUGATION IN BACTERIA
- Discovery of Conjugation in *E. coli*
- The Sex Factor *F*
- High-Frequency Recombination Strains
- *F′* factors
- Using Conjugation and Interrupted Mating to Map Bacterial Genes
- Circularity of the *E. coli* Map

TRANSDUCTION IN BACTERIA
- Bacteriophages: An Introduction
- Transduction Mapping of Bacterial Chromosomes

MAPPING GENES IN BACTERIOPHAGES
FINE-STRUCTURE ANALYSIS OF A BACTERIOPHAGE GENE
- Recombination Analysis of *rII* Mutants
- Deletion Mapping
- Defining Genes by Complementation Tests

II. REVIEW OF KEY TERMS, SYMBOLS AND CONCEPTS

Without consulting the text and in your own words, write a brief definition of each term in the groups below. Then, either using a short phrase or a simple diagram, identify the relationship(s) between specific pairs of terms within a set. Finally, consult the text (and perhaps a friend who has also done the exercise) to check your answers.

1	2	3	4
transformation conjugation transduction	plating colony titer minimal medium complete medium auxotroph prototroph replica plating	transduction phage vectors transductants phage lysate virulent phage lysogenic pathway prophage lysogeny temperate phage lytic cycle	conjugation transconjugants sex factor F^-, F', *Hfr*, F^+ strains plasmid F^- duction (sexduction) merodiploid interrupted mating donor vs. recipient *F* pili (sex pili) *F* factor origin exconjugant

5	6	7	8
transformation competent cells heteroduplex DNA cotransformation "natural" vs. "engineered"	generalized transduction specialized transduction transducing phage selected marker unselected marker cotransduction	bacteriophage cross plaque permissive host nonpermissive host	intergenic mapping "beads on a string" view intragenic mapping fine-structure mapping homoallelic mutations heteroallelic mutations

9	10
deletion mapping point mutant deletion mutant hot spot	cis-trans test complementation test cistron a gene as a unit of function

III. THINKING ANALYTICALLY

While this chapter focuses on the genetics of bacteria and bacteriophages, many of the core concepts introduced here are used throughout the study of genetics. Of particular importance are the genetic principles that underlie selection, recombination and gene structure, and complementation and gene function. Geneticists working in many different research areas are concerned with how to *select* for or against specific phenotypes. Analysis of recombination within a gene (*intragenic* recombination) formed a foundation for our current understanding of gene structure. The concept of a gene as a unit of function was developed by using complementation tests to measure whether two mutants affect the same function. It is therefore essential to thoroughly understand these genetic principles as they are developed using bacterial and phage systems in this chapter.

Having stressed the importance of the concepts presented in this chapter, it is also important to advise you that much of the material used to develop these concepts is complicated, and will not necessarily be intuitive. For example, there are many refinements of bacteria and phage types and special characteristics that are associated with one or another aspect of bacterial or phage growth and mating. Carefully go over the chapter sections that describe these characteristics until you thoroughly understand them. The names are quite descriptive, and once you use them a few times in solving problems, you will become adept at using them. Approach using the terms with more than rote memorization: use them in context until you fix upon what is practically a visual image of them.

As with the proceeding chapters, carefully read the statements and conditions of a particular problem so that you gain a clear understanding of it. Then organize and re-organize the data, and sometimes, re-organize again, using scratch paper and pencil until you can see how to use the data to solve the problem. If you still get stuck, try representing the processes that are used in the problem with diagrams, and ask yourself how the data fit into the biological process you've drawn out.

The problems in this chapter are especially well-organized to help you build-up your understanding of the material. Work through each of them sequentially to fully develop your understanding and analytical skills.

IV. QUESTIONS FOR PRACTICE

A. Multiple Choice Questions

1. A strain, such as a strain of *E. coli*, that requires nutritional or other kind of supplements for growth and/or survival is known as
   a. a prototroph.
   b. a heterotroph.
   c. a pleiotroph.
   d. an auxotroph.

2. A conjugating bacterial cell that typically transfers only part of its *F* factor and some chromosomal genes is
   a. F^+.
   b. F'.
   c. *Hfr*.
   d. F^-.

3. The process by which cells take up genetic material from the extracellular milieu and incorporate it into their genetic complement is called
   a. transduction.
   b. transformation.
   c. translocation.
   d. DNA transfusion.

4. In bacteria, conjugation involves
   a. the union of two bacterial genomes.
   b. the fusion of two cells of opposite mating types.
   c. a virus mediated exchange of DNA.
   d. the transfer of DNA from one cell into another.

5. A self-replicating genetic element found in the cytoplasm of bacteria is a
   a. sex-pilus.
   b. plasmid.
   c. viral capsid.
   d. cotransductant.

6. Which of the following is true about cells having *F* factors?
   a. In both F^+ and *Hfr* cells, fertility genes are transferred first.
   b. In both F^+ and *Hfr* cells, genes are transferred through sex-pili.
   c. In *Hfr* cells, bacterial genes closest to the origin are transferred first.
   d. In *F'* cells, only *F* factor genes are transferred.

7. Phage that can be in either a lytic or lysogenic pathway are called
   a. virulent.
   b. temperate.
   c. intemperate.
   d. transductant.

8. Recombination that occurs between two alleles at one gene is
   a. unheard of.
   b. intergenic recombination.
   c. intragenic recombination.
   d. more frequent that recombination between two closely linked genes.

9. A complementation test measures
   a. whether two mutants undergo intergenic recombination.
   b. whether two mutants undergo intragenic recombination.
   c. whether two mutants are linked.
   d. whether two mutants affect the same function.

10. *E. coli K12(λ)* cells will only support the growth of T4 phage that are rII^+. If two different mutant rII^- phage co-infect an *E. coli K12(λ)* cell and lysis occurs,
   a. intragenic recombination must have occurred.
   b. complementation must have occurred.
   c. the two rII^- phage must have mutations in different cistrons.
   d. both a and b are true.
   e. both b and c are true.
   f. a, b, and c are true.

Answers: 1d; 2c; 3b; 4d; 5b; 6c; 7b; 8c; 9d; 10e.

B. Thought Questions

1. Explain why two genes that are close together on a chromosome show a high frequency of cotransduction and cotransformation.

2. Compare the life cycles of T4, P1 and λ. Which is (are) temperate? Which is (are) virulent? Which can undergo lysogeny?

3. Contrast the features and/or conditions of generalized transduction and specialized transduction.

4. What is the significance of bacterial transformation and conjugation in relation to the rapidity with which certain infectious organisms adapt to changing environmental conditions or factors, such as antibiotics. (Hint: Sexual reproduction promotes genetic variability. How do parasexual systems compare?)

5. Explain how intragenic recombination is different from complementation.

6. Deletion mapping as used by Benzer in T4 was important in defining the location of point mutations at the *rII* locus. As was discussed in Chapter 7, recessive mutations can be

"uncovered" by deletions in eukaryotes. Diagram how you might use a set of nested deletions to localize a *Drosophila* point mutant. (Recall that *Drosophila* have polytene chromosomes in which deletion breakpoints can be mapped cytologically).

7. Consider how we may select for prototrophic and drug-resistant bacterial strains. Can you think of clever (or even not so clever) ways to attempt to select for (a) a pesticide-resistant strain of beetles, (b) bacteria that are able to degrade a toxic chemical, (c) very sweet sweet-corn, and (d) a strain of cats that doesn't respond to catnip?

V. SOLUTIONS TO TEXT PROBLEMS

8.1 If an *E. coli* auxotroph *A* could grow only on a medium containing thymine, and an auxotroph *B* could grow only on a medium containing leucine, how would you test whether DNA from *A* could transform *B*?

Answer: Strain *A* is thy^- leu^+, while strain *B* is thy^+ leu^-. To test if DNA from *A* can transform *B*, one is interested in whether one can transform the leu^- allele of *B* to leu^+. To do this, add DNA from strain *A* to a leucine–fortified culture of *B*. Incubate long enough for transformation to occur, then plate out these potentially transformed *B* cells on minimum medium or on medium supplemented only with thymine. As shown in the table below, one can select for transformants by plating on such media, as growth on such media requires leu^+.

	MEDIUM TYPE			
Strain	Minima l	Plus leucine	Plus thymine	Plus thymine & leucine
leu^+ thy^+	+	+	+	+
leu^+ thy^-	–	–	+	+
leu^- thy^+	–	+	–	+
leu^- thy^-	–	–	–	+

8.2 Distinguish among F^-, F^+, F', and *Hfr* strains of *E. coli.*

Answer: An F^- strain lacks an *F* factor, and is a recipient strain. F^+, F' and *Hfr* strains possess *F* factors, are donor strains and differ from each other based on the location of the *F* factor relative to the bacterial chromosome or a plasmid. See text pp. 234-238.

8.3 In $F^+ \times F^-$ crosses the F^- recipient is converted to a donor with very high frequency. However, it is rare for a recipient to become a donor in $Hfr \times F^-$ crosses. Explain why.

Answer: The frequency in which a recipient is converted to a donor reflects the frequency that a complete *F* factor is transferred. In $F^+ \times F^-$ crosses, only the *F* factor is transferred, and this occurs relatively quickly. In an $Hfr \times F^-$ crosses, transfer starts at the origin within the *F* element, and then must proceed through the bacterial chromosome before reaching the *F* factor. In order for the entire *F* factor to be transferred, the whole chromosome would have to be transferred. This would take about 100 minutes, and usually, the conjugal unions break apart before then.

8.4 With the technique of interrupted mating four *Hfr* strains were tested for the sequence in which they transmitted a number of different genes to an F^- strain. Each *Hfr* strain was found to transmit its genes in a unique sequence, as shown in the accompanying table (only the first six genes transmitted were scored for each strain).

ORDER OF TRANSMISSION	*Hfr* STRAIN 1	2	3	4
First	*O*	*R*	*E*	*O*
	F	*H*	*M*	*G*
	B	*M*	*H*	*X*
	A	*E*	*R*	*C*
	E	*A*	*C*	*R*
Last	*M*	*B*	*X*	*H*

What is the gene sequence in the original strain from which these *Hfr* strains derive? Indicate on your diagram the origin and polarity of each of the four *Hfr*'s.

Answer: The diagram below shows the gene sequence in the original F^+ strain, and the locations of the four different *F* factor insertions. (Note that only one insertion exists in a particular *Hfr* strain.)

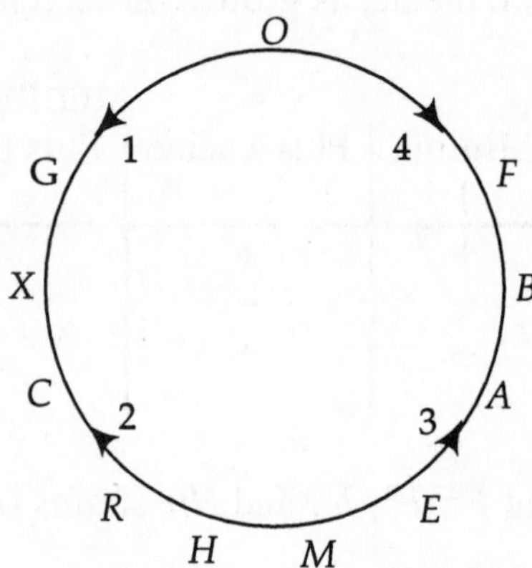

8.5 When *Hfr* donors conjugate with F^- recipients that are lysogenic for phage λ, the recipients (the transconjugant zygotes) usually survive. However, when *Hfr* donors that are lysogenic for λ conjugate with F^- cells that are nonlysogens, the transconjugant zygotes produced from matings that have lasted at least 100 minutes usually lyse, releasing mature phage particles. This event is called zygotic induction of λ.

a. Explain zygotic induction.
b. Explain how the locus of the integrated λ prophage can be determined.

Answer: a. In a λ lysogen, repressor molecules are normally present. These prevent the induction of the prophage and a lytic cycle. In a *Hfr(λ)* x F^- mating, no repressor molecules are present in the recipient (F^-) cell. When the prophage enters the recipient via conjugal transfer, it is induced, and the cell enters the lytic cycle. This is called zygotic induction since the initiation of the lytic cycle depends on the formation of a zygote (the conjugal union) between the donor and recipient. As long as the prophage (i.e., the λ gene set) is still within the donor, there are enough repressor molecules present to keep the genome in the prophage state.

b. The insertion site of the λ prophage can be mapped by an interrupted mating experiment. Prior to its transfer to the recipient, cells will not have the λ prophage and will not undergo lysis. After its transfer, cells will have the λ prophage and will undergo lysis. To map the prophage then, interrupt a *Hfr(λ)* x F^- mating at various times, and plate exconjugants on suitable plates that will select for those F^- that have received donor markers, that will select against the parentals, and that contain a lawn of sensitive bacteria on which plaques of λ can be detected. Up to a certain time, no plaques will be seen. Once the λ is transferred, plaques will be seen.

8.6 At time zero an *Hfr* strain (strain 1) was mixed with an F^- strain, and at various times after mixing, samples were removed and agitated to separate conjugating cells. The cross may be written as:

Hfr 1: $a^+\ b^+\ c^+\ d^+\ e^+\ f^+\ g^+\ h^+\ str^s$
F^-: $a^-\ b^-\ c^-\ d^-\ e^-\ f^-\ g^-\ h^-\ str^r$

(No order is implied in listing the markers.)

The samples were then plated onto selective media to measure the frequency of $h^+\ str^r$ recombinants that had received certain genes from the *Hfr* cell. The graph of the number of recombinants against time is shown in the following figure:

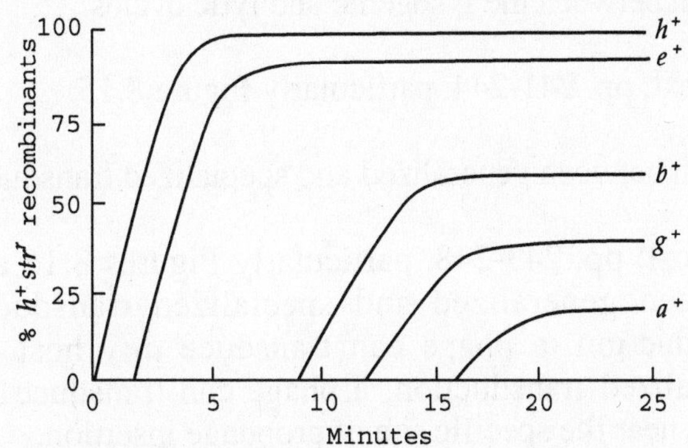

a. Indicate whether each of the following statements is true or false.
   i. All F^- cells which received a^+ from the *Hfr* in the chromosome transfer process must also have received b^+.
   ii. The order of gene transfer from *Hfr* to F^- was a^+ (first), then g^+ then b^+, then e^+, then h^+.
   iii. Most $e^+\ str^r$ recombinants are likely to be *Hfr* cells.
   iv. None of the $b^+\ str^r$ recombinants plated at 15 minutes are also a^+.
b. Draw a linear map of the *Hfr* chromosome indicating the
   i. point of insertion, or origin;
   ii. the order of the genes a^+, b^+, e^+, g^+, and h^+;
   iii. the shortest distance between consecutive genes on the chromosomes.

Answer: a.

i. True. The graph indicates that a^+ is transferred last, starting about 16 minutes after conjugal pairing. The genes are transferred linearly in the order h^+ (starting at 0 minutes), e^+ starting at 2 minutes), b^+ (starting at 9 minutes), g^+ (starting at 12 minutes) and then a^+.

ii. False. The data show that a^+ took the longest time to be transferred and so was the last gene to be received.

iii. False. In order for the recipient cell to be transformed to F^+ or *Hfr*, it must receive a complete copy of the *F* factor. Since only part of the *F* factor is transferred at the beginning of conjugation, and e^+ is transferred quite quickly after conjugal pairing, most e^+ str^r recombinants will have only a small part of the donor chromosome. The remaining part of the *F* factor is only transferred after the entire bacterial chromosome is transferred. Because of turbulence, the conjugal pairing is almost always disrupted before transfer of an entire chromosome is completed. Hence, the chance that complete transfer will occur is quite remote.

iv. True. The graph shows that a^+ is not transferred until after about 16 minutes has elapsed.

b.

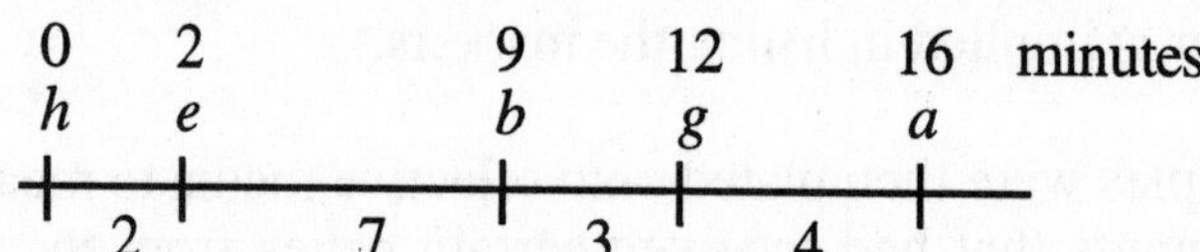

8.7 Distinguish between the lysogenic and lytic cycles.

Answer: See text, pp. 241-244, particularly Figure 8.15.

8.8 Distinguish between generalized and specialized transduction.

Answer See text, pp. 243-248, particularly Figures 8.16 and 8.17. The key difference between generalized and specialized transduction is that in generalized transduction, a phage can transduce *any* host chromosome gene, while in specialized transduction, a phage can transduce only those host chromosome genes near the specific site of prophage insertion.

8.9 Indicate whether each of the following occurs/or is a characteristic of

Generalized Transduction	(GT)
Specialized Transduction	(ST)
Occurs in Both	(B)
Occurs in Neither	(N)

a. Phage carries DNA of bacterial or viral DNA origin, never both.
b. Phage carries viral DNA covalently linked to bacterial DNA.
c. Phage integrates into a specific site on the host chromosome.
d. Phage integrates at a random site on the host chromosome.
e. "Headful" of bacterial DNA encoated.

f. Host lysogenized.
g. Prophage state exists.
h. Temperate phage involved.
i. Virulent phage involved.

Answer: a. GT
b. ST
c. ST
d. GT
e. GT
f. B
g. B
h. B
i. N

8.10 Consider the following transduction data:

DONOR	RECIPIENT	SELECTED MARKER	UNSELECTED MARKER	%
aceF$^+$ *dhl*	*aceF dhl*$^+$	*aceF*$^+$	*dhl*	88
aceF$^+$ *leu*	*aceF leu*$^+$	*aceF*$^+$	*leu*	34

Is *dhl* or *leu* closer to *aceF?*

Answer: The notation *aceF* represents a specific insertion site for an *F* factor. This table shows that cells selected for transduction of *F*$^+$ were isolated and then tested for the co-transduction of *dhl* or *leu.* If shows that in one experiment, of the *aceF*$^+$ transductants that were isolated, 88 percent were also transduced with the *dhl* marker. In another experiment, of the *aceF*$^+$ transductants that were isolated, only 34 percent were also transduced with the *leu* marker.

In order to obtain a transductant, a double crossover must occur between the transduced, donor DNA and the host chromosome. The closer two loci are together on the same chromosome, the greater the probability that they will both be included within the limits of a double crossover. Therefore, two loci that are closer together will show a greater frequency of co-transduction: *dhl* is closer to *aceF*$^+$ than *leu.*

8.11 Consider the following P1 transduction data:

DONOR	RECIPIENT	SELECTED MARKER	UNSELECTED MARKER	%
cysB$^+$ *trpE*	*cysB trpE*$^+$	*cysB*$^+$	*trpE*	37
cysB$^+$ *trpB*	*cysB trpB*$^+$	*cysB*$^+$	*trpB*	53

Is *trpE* or *trpB* closer to *cysB*?

Answer: The *trpB* marker is co-transduced with the *cysB*$^+$ selected marker more frequently than *trpE.* Hence, *trpB* is closer to *cysB* than *trpE.*

8.12 Consider the following data with P1 transduction:

DONOR	RECIPIENT	SELECTED MARKER	UNSELECTED MARKER	%
aroA pyrD+	*aroA*+ *pyrD*	*pryD*+	*aroA*	5
aroA+ *cmlB*	*aroA cmlB*+	*aroA*+	*cmlB*	26
cmlB pyrD+	*cmlB*+ *pyrD*	*pyrD*+	*cmlB*	54

Choose the correct order:

a. *aroA – cmlB – pyrD*
b. *aroA – pyrD – cmlB*
c. *cmlB – aroA – pyrD*

Answer: The frequency of cotransduction gives an indication of the closeness of each pair of genes: the higher the cotransduction frequency, the closer two genes. The *pryD* and *cmlB* genes show the highest cotransduction frequency, and are the closest together. This eliminates c as a possibility. The genes *aroA* and *pyrD* show the lowest co-transduction frequency, and so, they are the furthest apart. This eliminates b as a possibility. The *aroA* and *cmlB* genes show an intermediate cotransduction frequency, as would be expected if *cmlB* were between *aroA* and *pyrD*. Thus, the correct answer is a.

8.13 Order the mutants *trp, pyrF* and *qts* on the basis of the following three factor transduction cross:

Donor	*trp*+ *pyr*+ *qts*
Recipient	*trp pyr qts*+
Selected Marker	*trp*+

UNSELECTED MARKERS	NUMBER
pyr+ *qts*+	22
pyr+ *qts*	10
pyr qts+	68
pyr qts	0

Answer: Notice that in this cross, by selecting for *trp*+, one selects for recombinants. The frequency of the different classes of recombinants can be affected by two factors: the distance between genes and the number of crossovers that are needed to obtain a particular genotype. Considering the fact that the bacterial chromosome is circular, recombinants are only obtained by either two, or some multiple of two, crossovers. Obtaining four crossovers will be less common than obtaining two crossovers, and so a genotype that requires four crossovers to be produced is likely to be the least common. In this case, with the parents being *trp*+ *pyr*+ *qts* and *trp pyr qts*+, the gene order that requires a quadruple crossover to produce the least frequent *trp*+ *pyr qts* class is *trp – pyr – qts*. The other three transductant classes can be generated by double crossovers.

8.14 Order *cheA*, *cheB*, *eda* and *supD* from the following data:

MARKERS	% COTRANSDUCTION
cheA – eda	15
cheA – supD	5
cheB – eda	28
cheB – supD	2.7
eda – supD	0

Answer: The higher the frequency of cotransduction, the closer the loci. Thus, the relative proximity of the loci to each other is:

cheB – eda	closest together
cheA – eda	↓
cheA – supD	↓
cheB – supD	↓
eda – supD	furthest apart

A gene order that is consistent with these relationships is *eda – cheB – cheA – supD*.

8.15 A stock of T4 phage is diluted by a factor of 10^{-8} and 0.1 mL of it is mixed with 0.1 mL of 10^8 *E. coli B*/mL and 2.5 ml melted agar, and poured on the surface of an agar Petri dish. The next day 20 plaques are visible. What is the concentration of T4 phages in the original T4 stock?

Answer: When the phage dilution is mixed with the *E. coli*, any phage in the dilution will adsorb onto an *E. coli* cell and initiate a lytic cycle. Melted agar is used to limit the diffusion of the progeny phage that result from this infection. As the first, and later, subsequent generations of progeny phage infect nearby bacteria, a single plaque will develop. In this titering experiment, bacteria are present in large excess, so that a bacterial lawn will be formed on which plaques can be seen when the bacteria and phage exhaust the nutrient supply.

To calculate the titer of the phage stock, remember that only the volume and dilution of the phage suspension, and the number of plaques seen, are relevant. Since 0.1 mL of diluted phage gave 20 plaques, 1 mL would have 200 phage that would give 200 plaques. Since the dilution factor was 10^8, there would be 200 x 10^8 = 2 x 10^{10} phage/mL in the initial phage stock.

8.16 Wild-type phage T4 grows on both *E. coli B* and *E. coli K12(λ)*, producing turbid plaques. The *rII* mutants of T4 grow on *E. coli B*, producing clear plaques, but do not grow on *E. coli K12(λ)*. This host range property permits the detection of a very low number of r^+ phages among a large number of *rII* phages. With this sensitive system it is possible to determine the genetic distance between two mutations within the same gene, in this case the *rII* locus. Suppose *E. coli B* is mixedly infected with rII^x and rII^y, two separate mutants in the *rII* locus. Suitable dilutions of progeny phages are plated on *E. coli B* and *E. coli K12(λ)*. A 0.1-mL sample of a thousandfold dilution plated on *E. coli B* produced 672 plaques. A 0.2-mL sample of undiluted phage plated on *E. coli K12(λ)* produced 470 turbid plaques. What is the genetic distance between the two *rII* mutations?

Answer: The plaques produced on *E. coli K12(λ)* are r^+, while those on *E. coli B* may be either r^+ or r^-. Thus, the total number of progeny can be inferred from the number of plaques formed on *E. coli B*. Since *E. coli B* is co-infected with rII^x and rII^y, the only way to obtain an r^+ progeny phage is to have a crossover within the *rII* locus. The progeny resulting from a crossover would be 1/2 r^+ and 1/2 $r^{x,y}$ recombinants. The number of r^+ phage can be assayed for by growth on *E. coli K12(λ)*.

recombinant progeny in 1 mL = 2 x (number of r^+ phage/mL)
= 2 x (470/0.2)
= 4,700/mL.

total # of progeny in 1 mL = (dilution factor) x (# progeny phage/mL)
= 1,000 x (672/0.1)
= 6.72×10^6/mL.

RF = [4,700/(6.72×10^6)] x 100%
= 0.07%.

The map distance between r^x and r^y is 0.07 mu.

8.17 Construct a map from the following two factor phage cross data (show map distance):

CROSS	% RECOMBINATION
r1 x *r2*	0.10
r1 x *r3*	0.05
r1 x *r4*	0.19
r2 x *r3*	0.15
r2 x *r4*	0.10
r3 x *r4*	0.23

Answer: Recall that, as discussed in Chapter 5, the most accurate map distances are those obtained over the shortest intervals.

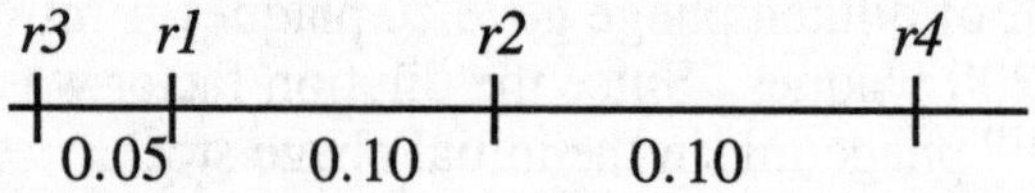

r3 — 0.05 — *r1* — 0.10 — *r2* — 0.10 — *r4*

8.18 The following two-factor crosses were made to analyze the genetic linkage between four genes in phage λ: *c, mi, s,* and *co.*

PARENTS	PROGENY
c + x + *mi*	1,213 *c* +, 1,205 + *mi*, 84 + +, 75 *c mi*
c + x + *s*	566 *c* +, 808 + *s*, 19 + +, 20 *c s*
co + x + *mi*	5,162 *co* +, 6,510 + *mi*, 311 + +, 341 *co mi*
mi + x + *s*	502 *mi* +, 647 + *s*, 65 + +, 56 *mi s*

Construct a genetic map of the four genes.

Answer: Analyze the data as you would a set of two-factor crosses:

CROSS	NUMBER OF PROGENY	# RECOMBINANTS	MU
c + x + *mi*	2,577	159	6.2
c + x + *s*	1,413	39	2.8
co + x + *mi*	12,324	652	5.3
mi + x + *s*	1,270	121	9.5

Two maps are compatible with these data:

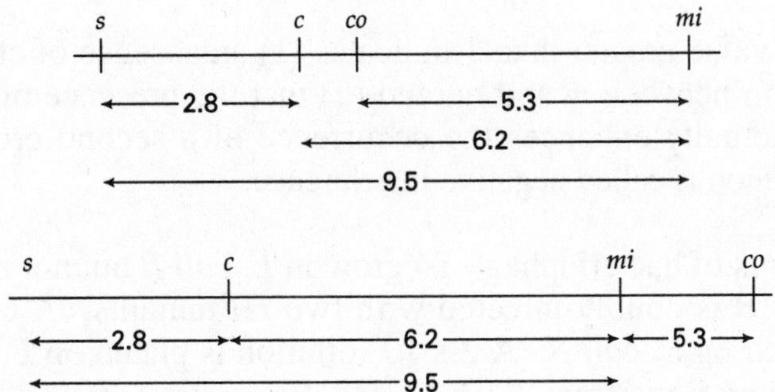

8.19 Three gene loci in T4 that affect plaque morphology in easily distinguishable ways are *r* (rapid lysis), *m* (minute), and *tu* (turbid). A culture of *E. coli* is mixedly infected with two types of phage *r m tu* and r^+ m^+ tu^+. Progeny phage are collected and the following genotype classes are found:

CLASS	NUMBER
r^+ m^+ tu^+	3,729
r^+ m^+ tu	965
r^+ m tu^+	520
r m^+ tu^+	172
r^+ m tu	162
r m^+ tu	474
r m tu^+	853
r m tu	3,467
	10,342

Construct a map of the three genes. What is the coefficient of coincidence, and what does the value suggest?

Answer: The phage progeny values are handled just as they are in any other three-point cross. First assign each progeny class to a parental, single or double crossover class, then determine the gene order by comparing double crossover classes to parental classes, then determine the distances between adjacent genes and the coefficient of coincidence.

The parental type chromosomes are *r m tu* and r^+ m^+ tu^+, and the gene order is *m* – *r* – *tu*.

RF($m - r$) = [(162 + 520 + 474 + 172)/10,342] x 100%
= 12.8%
There are 1.8 mu between m and r.

RF($r - tu$) = [(162 + 965 + 853 + 172)/10,342] x 100%
= 20.8%
There are 20.8 mu between r and tu.

c.o.c. = (162 + 172 /(0.208 x 0.128 x 10,342) = 334/275 = 1.21

A c.o.c. value greater than 1 indicates (1) an absence of chiasma interference in the region between m and tu, and (2) that the presence of one crossover in this region actually enhances the occurrence of a second crossover nearby. This phenomenon is called negative interference.

8.20 The *rII* mutants of bacteriophage T4 grow in *E. coli B* but not in *E. coli K12(λ)*. The *E. coli* strain *B* is doubly infected with two *rII* mutants. A 6 x 10^7 dilution of the lysate is plated on *E. coli B*. A 2 x 10^5 dilution is plated on *E. coli K12(λ)*. Twelve plaques appeared on strain *K12(λ)*, and 16 on strain *B*. Calculate the amount of recombination between these two mutants.

Answer: The plaques that appear on strain *B* can be used to determine the total number of phage progeny in the lysate. The plaques that appear on strain *K12(λ)* can be used to determine the number of recombinant progeny. They are r^+, and are half of the recombinants in the tested sample. See the solution to problem 8.16 and text pp. 252-254.

RF = [2 x 12 x (2 x 10^5)]/[16 x (6 x 10^7)] x 100% = 0.5%

8.21 Wild-type (r^+) strains of T4 produce turbid plaques, whereas *rII* mutant strains produce larger, clearer plaques. Five *rII* mutations ($a - e$) in the *A* cistron of the *rII* region of T4 give the following percentages of wild-type recombinants in two-point crosses:

CROSS	% OF WILD-TYPE RECOMBINANTS
$a \times b$	0.2 percent
$a \times c$	0.9 percent
$a \times d$	0.4 percent
$b \times c$	0.7 percent
$e \times a$	0.3 percent
$e \times d$	0.7 percent
$e \times c$	1.2 percent
$e \times b$	0.5 percent
$b \times d$	0.2 percent
$d \times c$	0.5 percent

What is the order of the mutational sites and what are the map distances between the sites?

Answer: There are two products of intragenic recombination between two *rII* mutants rII^x, rII^y:rII^{++} and rII^{xy}. Thus, the frequency of wild-type recombinants is half the total recombinant frequency. These data give the following map:

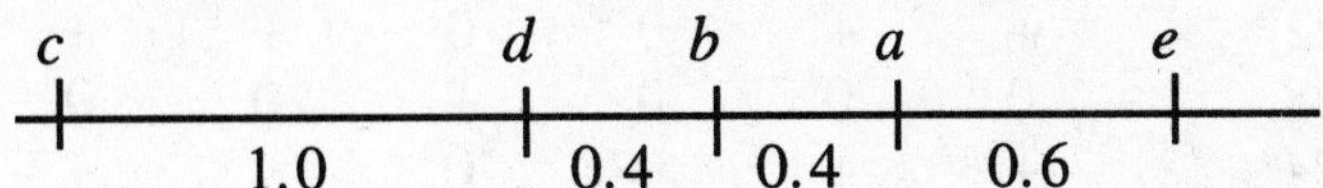

8.22 Choose the correct answer in each case:

a. If one wants to know if two different *rII* point mutants lie at exactly the same site (nucleotide pair), one should:
i. Coinfect *E. coli K12(λ)* with both mutants. If phage are produced, they lie at the same site.
ii. Coinfect *E. coli K12(λ)* with both mutants. If phage are not produced, they lie at the same site.
iii. Coinfect *E. coli B* with both mutants and plate the progeny phage on both *E. coli B* and *E. coli K12(λ)*. If plaques appear on *B* but not *K12(λ)*, they lie at the same site.
iv. Coinfect *E. coli K12(λ)* with both mutants and plate the progeny phage on both *E. coli B* and *E. coli K12(λ)*. If plaques appear on *K12(λ)* but not *B* they lie at the same site.

b. If one wants to know if two different *rII* point mutants lie in the same cistron, one should:
i. Coinfect *E. coli K12(λ)* with both mutants. If phage are produced, they lie at the same cistron.
ii. Coinfect *E. coli K12(λ)* with both mutants. If phage are not produced, they lie in the same cistron.
iii. Coinfect *E. coli B* with both mutants and plate the progeny phage on both *E. coli B* and *E. coli K12(λ)*. If plaques appear on *B* but not *K12(λ)*, they lie in the same cistron.
iv. Coinfect *E. coli K12(λ)* with both mutants and plate the progeny phage on both *E. coli B* and *E. coli K12(λ)*. If plaques appear on *Kl2(λ)* but not *B* they lie in the same cistron.

Answer: a. iii.
b. ii.

8.23 Given the following map with point mutants, and given the data in the table below, draw a topological representation of deletion mutants *r*21, *r*22, *r*23, *r*24 and *r*25. (Be sure to indicate clearly the endpoints of the deletions.)

(+ = r^+ recombinants are obtained. 0 = r^+ recombinants are not obtained.)

Map:

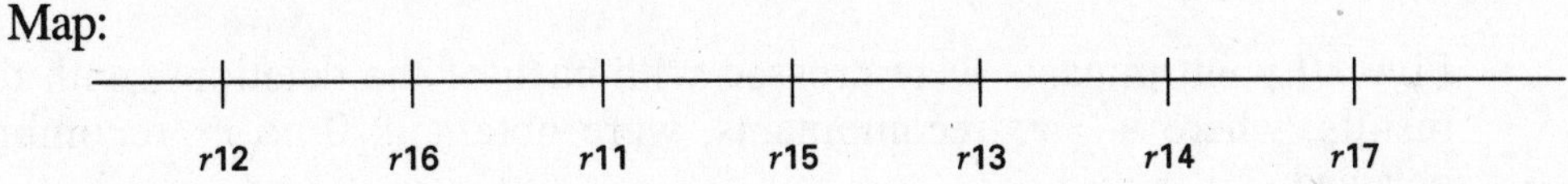

DELETION MUTANTS	POINT MUTANTS r11	r12	r13	r14	r15	r16	r17
r21	0	+	0	+	0	+	+
r22	+	+	0	0	+	+	0
r23	0	0	0	+	0	0	+
r24	+	+	0	0	+	+	+
r25	+	+	0	0	0	+	+

Answer: If no r^+ recombinants are obtained, the deletion removes the site of the point mutant. If r^+ recombinants are obtained, the site of the point mutation is not within the boundaries of the deletion. To determine the deleted region then, define the region that includes all of the point mutations unable to recombine with the deletion. Check your answer by verifying that all of the point mutations outside of this region do recombine with the deletion.

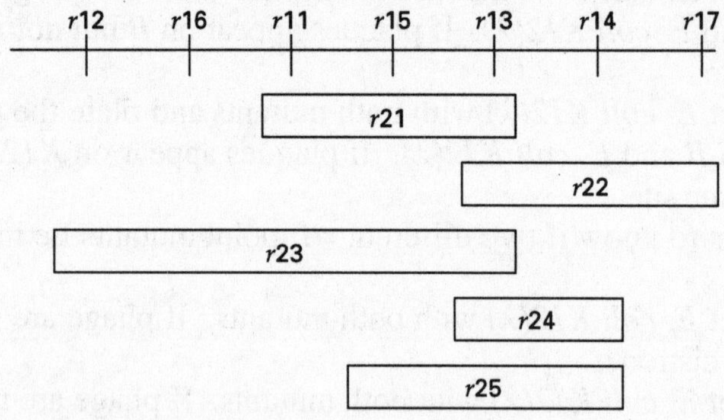

8.24 A set of seven different *rII* deletion mutants of bacteriophage T4, *1* through *7*, were mapped, with the following result:

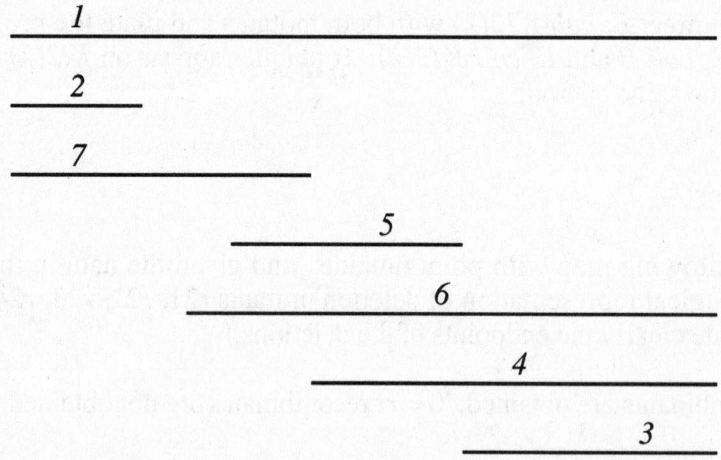

Five *rII* point mutants were crossed with each of the deletions, with the following results, where + = r^+ recombinants, were obtained, 0 no r^+ recombinants were obtained:

POINT MUTANTS	DELETION MUTANTS *1*	*2*	*3*	*4*	*5*	*6*	*7*
a	0	+	+	+	0	0	0
b	0	0	+	+	+	+	0
c	0	+	+	0	0	0	+
d	0	+	0	0	+	0	+
e	0	+	+	+	+	0	0

Map the locations of the point mutants.

Answer: Approach this problem systematically, utilizing two facts: (1) if a point mutant is *able* to recombine with a deletion mutant, it must lie *outside* of the deleted region and (2) if a point mutant is *unable* to recombine with a deletion mutant, it must lie *within* the deleted region. This means that a point mutant must lie within a region remaining intact in each of the deletions with which it does recombine, and conversely, it must lie in a deleted region that is shared by each of the deletions with which it does not recombine. Employ both of these inferences to define the region in which the point mutant lies.

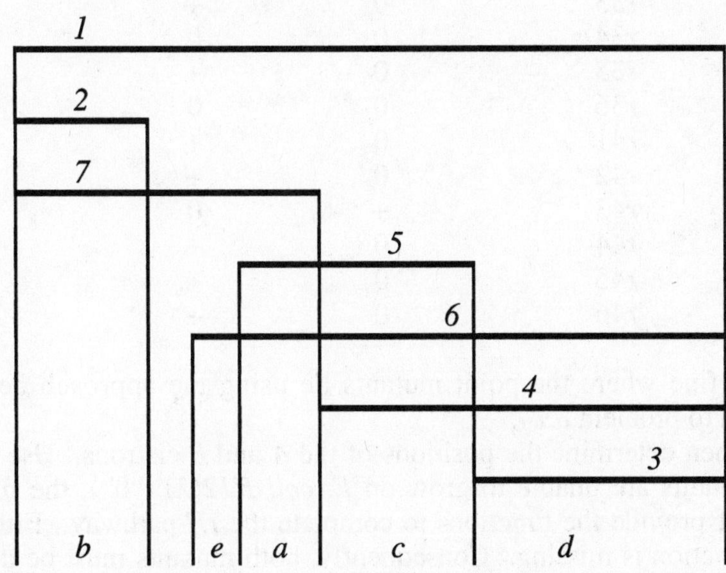

8.25 Given the following deletion map with deletions *r*31, *r*32, *r*33, *r*34, *r*35 and *r*36, place the point mutants *r*41, *r*42 etc., on the map. Be sure you show where they lie with respect to end points of the deletions.

*r*31

*r*32

*r*33

*r*34

*r*35 *r*36

POINT MUTANTS	DELETION MUTANTS (+ = RECOMBINANTS PRODUCED) 0 = NO r^+ RECOMBINANTS PRODUCED)					
	*r*31	*r*32	*r*33	*r*34	*r*35	*r*36
*r*41	0	0	0	0	+	0
*r*42	0	0	0	+	0	+
*r*43	0	0	+	+	+	0
*r*44	0	0	0	0	+	+
*r*45	0	+	0	+	+	+
*r*46	0	0	+	0	+	0

Show the dividing line between the *A* cistron and the *B* cistron on your map above from the following data: [+ = growth on strain *K12(λ)*, 0 = no growth on strain *K12(λ)*]:

MUTANT	COMPLEMENTATION WITH *rIIA*	*rIIB*
*r*31	0	0
*r*32	0	0
*r*33	0	+
*r*34	0	0
*r*35	0	+
*r*36	0	0
*r*41	0	+
*r*42	0	+
*r*43	+	0
*r*44	0	+
*r*45	0	+
*r*46	0	+

Answer: First define where the point mutants lie using the approach described in the solution to problem 8.24.

Then determine the positions of the *A* and *B* cistrons. Use the fact that if two mutants are unable to grow on *E. coli K12(λ)* ("0"), the mutants cannot together provide the functions to complete the *rII* pathway. Either the *rIIA* or *rIIB* function is missing. Consequently, both mutants must be defective in the same function. In other words, the mutants do not complement each other. For example, if a mutant does not complement a known *rIIA* mutant (i.e., there is no growth on *E. coli K12(λ)*, "0") then the mutation is in the *A* cistron. Note that deletions can affect both the *A* and *B* cistrons, while point mutations can only affect one or the other cistron.

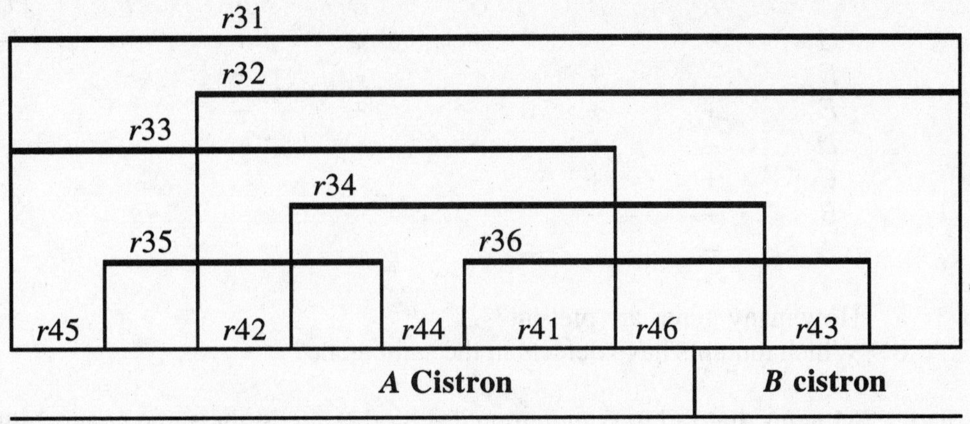

8.26 Mutants in the *ade2* gene of yeast require adenine and are pink because of the intracellular accumulation of a red pigment. Diploid strains were made by mating haploid mutant strains. The diploids exhibited the following phenotypes:

CROSS	DIPLOID PHENOTYPE
1 x 2	pink, adenine-requiring
1 x 3	white, prototrophic
1 x 4	white, prototrophic
3 x 4	pink, adenine-requiring

How many genes are defined by the four different mutants? Explain?

Answer: When two haploid strains mate to form a diploid cell, each contributes its genome. Consequently, there are two sets of genes in a diploid cell, and complementation between two mutants can be observed. If two mutations are in the same unit of function (i.e., the same gene), then they will not complement when together in a diploid cell and the mutant phenotype will be exhibited. If two mutations are in different genes, they will complement each other in a diploid cell and the wild-type phenotype will result.

In the data presented here, mutations 1 and 3, and mutations 1 and 4 complement each other, indicating that mutations 1 and 3 are in different genes, and mutations 1 and 4 are in different genes. On the other hand, mutations 1 and 2, and mutations 3 and 4 fail to complement each other. Thus, mutations 1 and 2 are in the one gene, while mutations 3 and 4 are in a different gene. There are two genes.

8.27 In *Drosophila* mutants *A, B, C, D, E, F*, and *G* all have the same phenotype: the absence of red pigment in the eyes. In pairwise combinations in complementation tests the following results were produced, where + = complementation and – = no complementation.

	A	*B*	*C*	*D*	*E*	*F*	*G*
G	+	−	+	+	+	+	−
F	−	+	+	−	+	−	
E	+	+	−	+	−		
D	−	+	+	−			
C	+	+	−				
B	+	−					
A	−						

a. How many genes are present?
b. Which mutants have defects in the same gene?

Answer: Mutants that fail to complement each other are in the same gene. Mutants *G* and *B* fail to complement each other, mutants *C* and *E* fail to complement each other, and mutants *A, F* and *D* fail to complement each other. Therefore there are three genes, with mutants *G* and *B* in one, mutants *A, F* and *D* in a second, and mutants *C* and *D* in a third.

8.28 A homozygous white-eyed Martian fly (w_1/w_1) is crossed to a homozygous white-eyed fly from a different stock (w_2/w_2). It is well known that wild-type Martian flies have red eyes. This cross produces all white-eyed progeny. State whether the following is true or false. Explain your answer.

i) w_1 and $w_2$are allelic genes.
ii) w_1 and w_2 are non-allelic.
iii) w_1 and w_2 affect the same function.
iv) The cross was a complementation test.
v) The cross was a cis-trans test.
vi) w_1 and w_2, are allelic by terms of the functional test.

The F_1 white-eyed flies were allowed to interbreed, and when you classified the F_1 you found 20,000 white-eyed flies and ten red-eyed progeny. Concerned about contamination, you repeat the experiment and get exactly the same results. How can you best account for the presence of the red-eyed progeny? As part of your explanation give the genotypes of the F_1 and F_2 generation flies.

Answer: a.

i) True. In the cross w_1/w_1 x w_2/w_2, the offspring are w_1w_2. That they are white-eyed indicates that w_1 and w_2 are mutations in the same function, i.e., the same gene. Thus, w_1 and w_2 are allelic genes.
ii) False. If the genes were non-allelic, each mutation would be in a different function. The two mutations would each be in different steps of a pathway, with each mutant retaining the function of a different step. Since the progeny obtain a set of genes from each mutant parent, they would have the function of both steps, and have red eyes. Since they have white eyes, this statement is false.
iii) True.
iv) True. A complementation test examines whether two mutations affect the same function.
v) True. The F_1 is a trans-heterozygote.
vi) True. If two genes affect the same function, they are allelic.

b. The complementation test indicates that the two mutations affect the same function, and are therefore alleles at the same gene. However, it does not indicate whether they are *identical* alleles or not. To evaluate if two allelic mutations lie in exactly the same position in a gene, one needs to assess whether recombination can occur between them. If two alleles are identical, and lie in the same site, *intragenic* recombination between them cannot occur. If they lie in different sites, *intragenic* recombination can occur, giving rise to wild-type recombinant chromosomes. The appearance of wild-type F_2 progeny results from intragenic recombination, as is discussed below.

If w_1 and w_2 lie in different sites in the same gene, the cross and the F_1 progeny can be written as

$$\text{P:} \quad \frac{w_1 \; +}{w_1 \; +} \times \frac{+ \; w_2}{+ \; w_2} \qquad\qquad \text{F}_1\text{:} \quad \frac{w_1 \; +}{+ \; w_2}$$

A rare *intragenic* crossover between the two mutant sites can be diagrammed as follows:

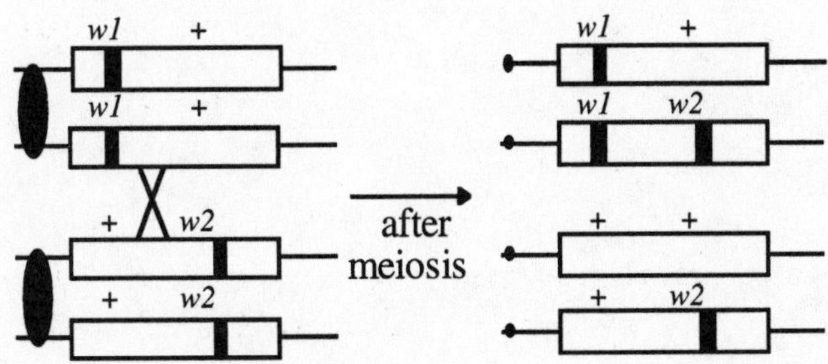

This crossover gives rise to gametes that are w_1 w_2 and ++. The w_1 w_2 gamete would not be detected (w_1 w_2/w_1 + and w_1 $w_2/+$ w_2 flies are white-eyed, as are their non-crossover siblings), but a ++ gamete would give rise to a red-eyed offspring that has a genotype of + +/w_1 + or + +/+ w_2. It is interesting to note that from the frequency of the white-eyed progeny, one can determine the recombination frequency (RF) between the two point mutants, just as in Benzer's *cis-trans* tests in the *rII* region of T4.

$$\text{RF } (w_1, w_2) = [(2 \text{ x } 10)/20{,}000] \text{ x } 100\% = 0.1\%.$$

CHAPTER 9

THE BEGINNINGS OF MOLECULAR GENETICS: GENE FUNCTION

I. CHAPTER OUTLINE

GENE CONTROL OF ENZYME STRUCTURE
- Garrod's Hypothesis of Inborn Errors of Metabolism
- One Gene–One Enzyme Hypothesis

GENETICALLY BASED ENZYME DEFICIENCIES IN HUMANS
- Phenylketonuria
- Albinism
- Lesch-Nyhan Syndrome
- Tay-Sachs Disease

GENE CONTROL OF PROTEIN STRUCTURE
- Sickle-Cell Anemia: Symptoms and Causes
- Other Hemoglobin Mutants
- Biochemical Genetics of the Human ABO Blood Groups
- Cystic Fibrosis

GENETIC COUNSELING
- Carrier Detection
- Fetal Analysis

II. REVIEW OF KEY TERMS, SYMBOLS AND CONCEPTS

Without consulting the text and in your own words, write a brief definition of each term in the groups below. Then, either using a short phrase or a simple diagram, identify the relationship(s) between specific pairs of terms within a set. Finally, consult the text (and perhaps a friend who has also done the exercise) to check your answers.

1	2	3	4
inborn error of metabolism	alkaptonuria (AKU)	amino acid	sickle-cell anemia
metabolic pathway	phenylketonuria (PKU)	polypeptide	hemoglobin
one gene–one enzyme hypothesis	albinism	protein	α-polypeptide
biochemical pathway	Lesch-Nyhan Syndrome	N-terminus	β-polypeptide
one gene–one polypeptide hypothesis	Tay-Sachs disease	C-terminus	electrophoresis
	recessive trait		amino-acid substitution
			multiple alleles

5	6
enzymatic vs. non-enzymatic proteins	genetic counseling
ABO blood type	carrier detection
antigen	fetal analysis
antibody	amniocentesis
polysaccharide	chorionic villus sampling
glycosyltransferase	
cystic fibrosis (CF)	
transmembrane conductance regulator	
active transport	

III. THINKING ANALYTICALLY

The problems in this chapter require you to:

(1) relate the genetic properties of a mutation to its visible phenotype,
(2) relate the visible phenotype to a biochemical deficit (a 'biochemical phenotype'),
(3) relate the biochemical deficit associated with a mutation to its genetic properties.

The layers of complexity that unfold in a problem as you make these connections are somewhat like those found in a good detective novel. Clues at one level (the genetic properties of the mutations, the mutant phenotypes, the biochemical abnormalities found in the mutant) can be used to infer what might be happening at another level. Approach the problems systematically, attempting to connect these levels wherever possible.

To relate biochemical pathways to mutant phenotypes, diagram as much of the biochemical pathway as you can, and then try to understand how specific phenotypic consequences arise when a particular step is blocked. Go slowly and be thorough. It is especially important to realize that mutations in different steps of a biochemical pathway do not always cause identical mutant phenotypes. In addition, mutations affecting a particular step can have multiple consequences. They may lead to the absence of the final product and cause one phenotype. They may also lead to the accumulation of intermediate metabolites of the pathway and cause a different phenotype. Furthermore, mutations at different steps of a pathway may be able to be "rescued" if different biochemical intermediates are provided to the mutant.

By doing the problems in this chapter, you will be able to enhance your understanding of one of the most challenging, but also one of the most valuable, concepts in genetics, that of complementation. Two identically appearing mutants may have the same visible phenotype because each lacks the final product of one biochemical pathway. Still, if these mutants are blocked in different steps of that pathway, two genes could be involved. Each gene would code for an enzyme carrying out a different step. If this is the case, each of the mutants has a functional enzyme for the step that is missing in the other. Consider what happens when a haploid complement of genes from each mutant is placed together in one diploid cell, as in a complementation test. The cell will have all the enzymes necessary to complete the pathway and be normal. Thus, complementation tests can help you to infer the details of biochemical pathways. If you already know what the biochemical pathway is, you should be able to predict that mutations in different steps complement each other.

IV. QUESTIONS FOR PRACTICE

A. Multiple Choice Questions

Choose the correct answer or answers for the following questions.

1. An inborn error of metabolism is
   a. any biochemical abnormality that results from taking a drug.
   b. any heritable biochemical abnormality.
   c. any recessive mutation.
   d. any dominant mutation.

2. Why was the one-gene one-enzyme hypothesis recast as the one-gene one-polypeptide hypothesis?
   a. Genes can encode proteins that are not enzymes.
   b. Some enzymes are not polypeptides.
   c. Some enzymes have more than one polypeptide subunit.
   d. Some polypeptides are not enzymes.

3. Suppose an individual was diagnosed with PKU as a child and was successfully treated. If that individual and a normal, non-carrier (pku^+/pku^+) partner have children, what is the probability that they will have offspring that are carriers for PKU?
   a. 0.00
   b. 0.25
   c. 0.50
   d. 1.00

4. In individuals affected with AKU,
   a. a block in the pathway leads to the accumulation of homogentisic acid.
   b. if homogentisic acid is provided, the pathway can be completed.
   c. a block in the pathway leads to the accumulation of phenylalanine.
   d. if phenylalanine is provided, the pathway can be completed.

5. Two *Neurospora* auxotrophs are found that are unable to grow on minimal medium but that are able to grow on minimal medium supplemented with arginine. When each is tested, only one can grow on minimal medium supplemented with ornithine, a biochemical precursor to arginine. Which statement(s) below are supported by these findings?
   a. The auxotrophs are blocked in different biochemical steps.
   b. At least one auxotroph is blocked in a step before ornithine is made.
   c. Both of the auxotrophs are blocked in a step before ornithine is made.
   d. Both of the auxotrophs accumulate ornithine.

6. In complex metabolic pathways such as the phenylalanine-tyrosine metabolic pathway,
   a. blocks at different steps always result in the same phenotype.
   b. a block at one step results solely in the accumulation of the biochemical made just before that step.
   c. blocks at different steps can lead to very different phenotypes.

d. a block at one step can lead to the accumulation of potentially toxic derivatives of biochemicals synthesized before that step.
e. a and b.
f. c and d.

7. If two albinos marry and have normal children,
   a. complementation occurred, the adults are blocked in different enzymatic steps.
   b. complementation occurred, the adults are blocked in the same enzymatic step.
   c. no complementation occurred, the adults are blocked in different enzymatic steps.
   d. no complementation occurred, the adults are blocked in the same enzymatic step.

8. Both Tay-Sachs disease and Lesch-Nyhan syndrome are examples of human diseases that are
   a. pleiotropic.
   b. caused by a deficiency in one enzyme activity.
   c. recessive.
   d. a, b and c.

9. An example of a heritable disease that results from biochemical defect in a non-enzymatic protein is
   a. PKU.
   b. AKU.
   c. sickle-cell anemia.
   d. Tay-Sachs disease.

10. Which of the following is *not true* about both amniocentesis and chorionic villus sampling?
   a. Both can be performed with an acceptably low risk as early as the 8th week of pregnancy.
   b. Each procedure has a risk of fetal loss.
   c. One is more likely that the other to provide an accurate diagnosis.
   d. Both are valid, useful procedures in genetic counseling.

Answers: 1b; 2c; 3d; 4a; 5a & b; 6f; 7a; 8d; 9c; 10a.

B. Thought Questions

1. If you were a genetics counselor, how would you counsel the following couple that is contemplating having children? The prospective father is apparently normal, except for a nervous twitch of his nose, but his maternal uncle had PKU as a child and his brother, who plays in a heavy metal band, is hard-of-hearing. The prospective mother also appears normal, but her father has hemophilia and her brother, who plays in the same band as her brother-in-law, is also hard-of-hearing. You know that hemophilia is often X-linked, and that PKU and sometimes deafness are autosomal recessive.

2. Distinguish between the procedures of amniocentesis and chorionic villus sampling. How is each useful in genetic counseling? What reasons are there for or against choosing each procedure?

3. All states currently require testing for PKU in infants. Do you expect the frequency of PKU to decline, or increase, over time? Justify your answer.

4. Over two hundred hemoglobin mutants have been detected. Explain why some hemoglobin mutations do not lead to as severe sickle-cell anemia phenotypes as others.

5. Some alleles of sickle-cell anemia, when heterozygous with a normal allele, confer some resistance to malarial infection. What consequences might this have on selection for this allele in different human populations?

6. Keep in mind your answer to questions 4 and 5 as you respond to the following question. Suppose you have the ability to make point mutations in a gene that encodes an enzyme, and are able to determine the percentage of normal enzyme activity that remains in the mutant enzyme. What effects would you expect to see if you systematically mutate the gene for the enzyme, changing only one amino acid at a time?

7. As you have seen in the study of PKU, Tay-Sachs disease, Lesch-Nyhan syndrome and sickle-cell anemia, mutations that result in the blockage of a single enzymatic step, or the function of a single protein, can have pleiotropic phenotypes. Generate an hypothesis about a potentially general principle from these data.

8. What functions do (should) genetic counseling services provide? What do they look for, how do they proceed, what do they recommend, and what, if any, advice, do (should) they give?

V. SOLUTIONS TO TEXT PROBLEMS

9.1 Phenylketonuria (PKU) is an inheritable metabolic disease of humans; its symptoms include mental deficiency. This phenotypic effect is due to:

a. accumulation of phenylketones in the blood.
b. accumulation of maple sugar in the blood.
c. deficiency of phenylketones in the blood.
d. deficiency of phenylketones in the diet.

Answer: a.

9.2 If a person were homozygous for both PKU (phenylketonuria) and AKU (alkaptonuria), would you expect him or her to exhibit the symptoms of PKU or AKU or both? Refer to the pathway below.

Phenylalanine
↓ (blocked in PKU)
tyrosine → DOPA → melanin
↓
ρ-Hydroxyphenylpyruvic acid
↓
Homogentisic acid
↓ (blocked in AKU)
Maleylacetoacetic acid

Answer: Given the pathway that is presented, a double homozygote will have PKU, but not AKU. The PKU block will prevent homogentisic acid from being formed, so that it could not accumulate to high levels.

9.3 Refer to the pathway shown in Question 9.2. What effect, if any, would you expect PKU (phenylketonuria) and AKU (alkaptonuria) to have on pigment formation?

Answer: The block in PKU leads to decreased tyrosine levels and so should lead to a decrease in melanin formation and pigmentation. However, some tyrosine will be obtained from food, and so the block in the pathway can be partially circumvented in this way. It has been reported that PKU patients are sometimes lighter in pigmentation than their normal relatives.

Since the block in AKU patients lies after (downstream of) the formation of melanin, pigmentation should not be affected in AKU patients. However, the levels of the products of an enzymatic reaction can affect the efficiency with which that reaction proceeds. If the products of an enzymatic reaction accumulate, the reaction may be inhibited. This in turn can lead to the accumulation of the substrates of the reaction. In this case, the accumulation of homogentisic acid in AKU individuals may ultimately lead to elevation of tyrosine levels. This tyrosine might be available for conversion to melanin. Hence, if AKU has any affect, it may be to increase pigmentation levels.

9.4 a^+, b^+, c^+, and d^+ are independently assorting Mendelian genes controlling the production of a black pigment. The alternate alleles that give abnormal functioning of these genes are a, b, c, and d. A black individual of genotype $a^+/a^+\ b^+/b^+\ c^+/c^+\ d^+/d^+$ is crossed with a colorless individual of genotype $a/a\ b/b\ c/c\ d/d$ to produce a black F_1. $F_1 \times F_1$ crosses are then done. Assume that a^+, b^+, c^+, and d^+ act in a pathway as follows:

$$\text{colorless} \xrightarrow{a^+} \text{colorless} \xrightarrow{b^+} \text{colorless} \xrightarrow{c^+} \text{brown} \xrightarrow{d^+} \text{black}$$

a. What proportion of the F_2 are colorless?
b. What proportion of the F_2 are brown?

Answer: The F_1 cross is $a^+/a\ b^+/b\ c^+/c\ d^+/d \times a^+/a\ b^+/b\ c^+/c\ d^+/d$.

a. A colorless F_2 individual would result if an individual has an a/a, b/b, and/or c/c genotype. This would consist of many possible genotypes. Rather than identify all of these combinations, it is possible to use the fact that [the proportion of colorless individuals] = 1 - [the proportion of pigmented individuals]. The proportion of pigmented individuals ($a^+/–\ b^+/–\ c^+/–$) is 3/4 x 3/4 x 3/4 = 27/64. The chance of not obtaining this genotype is 1 - 27/64 = 37/64.
b. A brown individual is $a^+/–\ b^+/–\ c^+/–\ d/d$. The proportion of brown individuals is 3/4 x 3/4 x 3/4 x 1/4 = 27/256.

9.5 Using the genetic information given in Problem 9.4, now assume that a^+, b^+, and c^+ act in a pathway as follows:

$$\left.\begin{array}{l}\text{colorless} \xrightarrow{a^+} \text{red} \\ \text{colorless} \xrightarrow{b^+} \text{red}\end{array}\right\rangle \xrightarrow{c^+} \text{black}$$

Black can be produced only if both red pigments are present, i.e., *C* converts the two red pigments together into a black pigment.

a. What proportion of the F_2 are colorless?
b. What proportion of the F_2 are red?
c. What proportion of the F_2 are black?

Answer: a. Colorless individuals are *a/a b/b –/–*. The chance of obtaining such an individual from a cross of a^+/a b^+/b c^+/c x a^+/a b^+/b c^+/c is 1/4 x 1/4 = 1/16.

b. There are two ways to determine the proportion of red individuals. First, notice that red individuals are obtained either when only one of the a^+ or b^+ functions is present (no matter what genotype is at *c*), or when both a^+ and b^+ functions are present but c^+ is not. Thus, these phenotypes are obtained from the following genotypes: $a^+/–$ *b/b –/–*, *a/a* $b^+/–$ *–/–*, or $a^+/–$ $b^+/–$ *c/c*. The probability of obtaining one of these genotypes is

(3/4 x 1/4 x 1) + (1/4 x 3/4 x 1) + (3/4 x 3/4 x 1/4) = 33/64.

A second way is to use the information from parts a and c. From part c, one has that 27/64 individuals are black, so the remaining 37/64 are either red or colorless. From part a, 1/16 = 4/64 are colorless. Thus there are (37 - 4)/64 = 33/64 that are red.

c. Black individuals are formed only if both red pigments are available and are then converted to black pigment. Thus, an individual must be $a^+/–$ $b^+/–$ $c^+/–$. The chance of obtaining such an individual is 3/4 x 3/4 x 3/4 = 27/64.

9.6 a. Three genes on different chromosomes are responsible for three enzymes that catalyze the same reaction in corn:

$$\text{colorless compound} \xrightarrow{a^+,\ b^+,\ c^+} \text{red compound}$$

The normal functioning of any one of these genes is sufficient to convert the colorless compound to the red compound. The abnormal functioning of these genes is designated by *a*, *b*, and *c*, respectively. A red a^+/a^+ b^+/b^+ c^+/c^+ is crossed by a colorless *a/a b/b c/c* to give a red F_1, a^+/a b^+/b c^+/c. The F_1 is selfed. What proportion of the F_2 are colorless?

b. It turns out that another step is involved in the pathway. It is controlled by gene d^+, which assorts independently of a^+, b^+, and c^+:

$$\text{colorless compound 1} \xrightarrow{d^+} \text{colorless compound 2} \xrightarrow{a^+,\ b^+,\ c^+} \text{red compound}$$

The inability to convert colorless 1 to colorless 2 is designated *d*. A red a^+/a^+ b^+/b^+ c^+/c^+ d^+/d^+ is crossed by a colorless *a/a b/b c/c d/d*. The F_1 are all red. The red F_1's are now selfed. What proportion of the F_2 are colorless?

Answer: a. Since any of the normal alleles a^+, b^+, or c^+ is sufficient to catalyze the reaction leading to color, in order for color to fail to develop, all three normal alleles must be missing. That is, the colorless F_2 must be *a/a b/b c/c*. The chance of obtaining such an individual is 1/4 x 1/4 x 1/4 = 1/64.

b. Now, colorless F_2 are obtained if *either* one or both steps of the pathway are blocked. That is, colorless F_2 are obtained in either of the following genotypes: *d/d –/– –/– –/–* (the first or both steps blocked) or *d⁺/– a/a b/b c/c* (second step blocked). The chance of obtaining such individuals is

$$(1/4 \times 1 \times 1 \times 1) + (3/4 \times 1/4 \times 1/4 \times 1/4) = (64 + 3)/256 = 67/256.$$

9.7 In hypothetical diploid organisms called mongs, the recessive allele *bw* causes a brown eye, and the (unlinked) recessive allele *st* causes a scarlet eye. Organisms homozygous for both recessives have white eyes. The genotypes and corresponding phenotypes, then, are as follows:

$bw^+/–$	$st^+/–$	red eye
bw/bw	$st^+/–$	brown
$bw^+/–$	st/st	scarlet
bw/bw	st/st	white

Outline a hypothetical biochemical pathway that would produce this type of gene interaction. Demonstrate why each genotype shows its specific phenotype.

Answer: Since the $bw/bw\ st^+/–$ individuals have brown pigmentation, but only have st^+ function, the st^+ gene must make brown pigment. Similarly, the scarlet pigmentation of $bw^+/–\ st/st$ individuals indicates that the bw^+ gene must make scarlet pigment. The combination of the brown and scarlet pigments in $bw^+/–\ st^+/–$ individuals leads to red eyes. The absence of either pigment in $bw/bw\ st/st$ individuals leads to white (colorless) eyes.

The production of two different colored pigments, with each affected by a separate gene, suggests that there are two distinct biochemical pathways. Each of these pathways might have a number of steps. One pathway produces a brown pigment and the other produces a scarlet pigment. This can be illustrated in the following diagram:

Two pathways, each with an unknown number of steps:

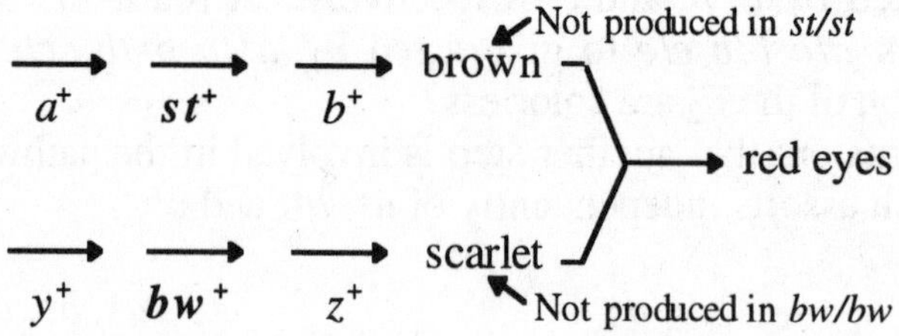

Not produced in *st/st*

→ → → brown

a^+ st^+ b^+

→ red eyes

→ → → scarlet

y^+ bw^+ z^+

Not produced in *bw/bw*

9.8 In J. R. R. Tolkien's *The Lord of the Rings*, the Black Riders of Mordor ride steeds with eyes of fire. As a geneticist, you are very interested in the inheritance of the fire-red eye color. You discover that the eyes contain two types of pigments, brown and red, that are usually bound to core granules in the eye. In wild-type steeds precursors are converted by these granules to the above pigments, but in steeds homozygous for the recessive X-linked gene *w* (white eye), the granules remain unconverted and a white eye results. The metabolic pathways for the synthesis of the two pigments are shown in the following figure. Each step of the pathway is controlled by a gene: A mutation *v* results in vermilion eyes; *cn* results in cinnabar eyes; *st* results in scarlet eyes; *bw* results in brown eyes; and *se* results in black eyes. All these mutations are recessive to their wild-type alleles and all are unlinked. For the following genotypes, show the phenotypes and proportions of steeds that would be obtained in the F_1 of the given matings.

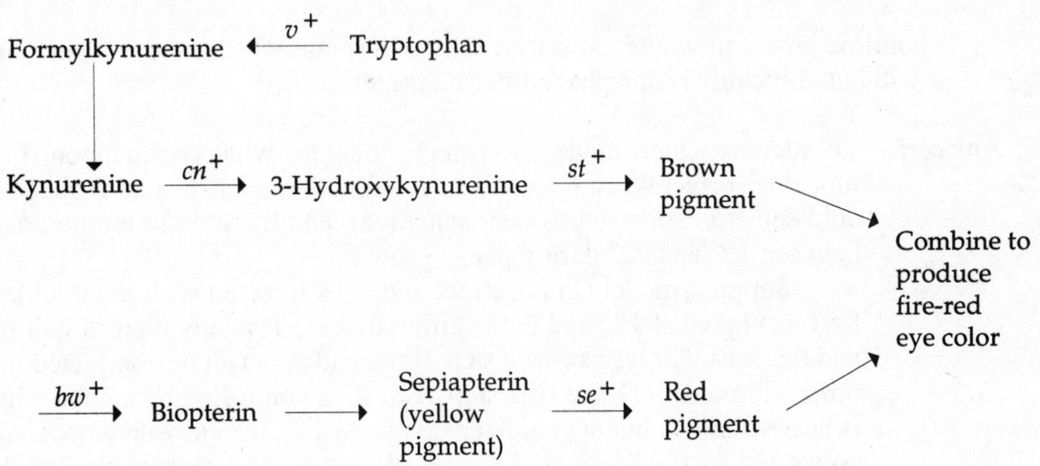

a. *w/w bw⁺/bw⁺ st/st* x *w⁺/Y bw/bw st/st*
b. *w⁺/w⁺ se/se bw/bw* x *w/Y se⁺/se⁺ bw⁺/bw⁺*
c *w⁺/w⁺ v⁺/v⁺ bw/bw* x *w/Y v/v bw/bw*
d. *w⁺/w⁺ bw⁺/bw st⁺/st* x *w/Y bw/bw st/st*

Answer: a. 1/2 *w/w⁺ bw/bw⁺ st/st,* scarlet-eyed daughters;
1/2 *w/Y bw/bw⁺ st/st⁺*, white-eyed sons.
b. *w/w⁺ se/se⁺ bw/bw⁺* and *w⁺/Y se/se⁺ bw/bw⁺*, all fire-red eyes.
c. *w/w⁺ v/v⁺ bw/bw* and *w⁺/Y v/v⁺ bw/bw*, all brown eyes.
d. 1/4 *w⁺/w* or *w⁺/Y, bw/bw⁺ st/st⁺*, fire-red eyes;
1/4 *w⁺/w* or *w⁺/Y, bw/bw st/st⁺*, brown eyes;
1/4 *w⁺/w* or *w⁺/Y, bw/bw⁺ st/st* , scarlet-eyes;
1/4 *w⁺/w* or *w⁺/Y, bw/bw st/st* (the color of 3-hydroxykyrurenine plus the color of the precursor to biopterin, or colorless = white).

9.9 Upon infection of *E. coli* with bacteriophage T4, a series of biochemical pathways result in the formation of mature progeny phages. The phages are released following lysis of the bacterial host cells. Let us suppose that the following pathway exists:

$$A \xrightarrow{\text{enzyme}} B \xrightarrow{\text{enzyme}} \text{mature phage}$$

Let us also suppose that we have two temperature-sensitive mutants that involve the two enzymes catalyzing these sequential steps. One of the mutations is cold-sensitive (*cs*) in that no mature phages are produced at 17°C. The other is heat-sensitive (*hs*) in that no mature phages are produced at 42°C. Normal progeny phages are produced when phages carrying either of the mutations infect bacteria at 30°C. However, let us assume that we do not know the sequence of the two mutations. Two models are therefore possible:

$$\text{(1)}\quad A \xrightarrow{hs} B \xrightarrow{cs} \text{phage}$$

$$\text{(2)}\quad A \xrightarrow{cs} B \xrightarrow{hs} \text{phage}$$

Outline how you would experimentally determine which model is the correct model without artificially lysing phage-infected bacteria.

Answer: To address which model is correct, consider what will happen if a particular model is correct when *E. coli* is infected with a doubly mutant phage (one step is cold-sensitive, one step is heat-sensitive), and the growth temperature is shifted between 17° and 42° during phage growth.

Suppose model (1) is correct and cells infected with the double mutant are first incubated at 17° and then shifted to 42°. Progeny phages will be produced and the cells will lyse as each step of the pathway can be completed in the correct order. In model (1), the first step, *A* to *B*, is controlled by a gene whose product is heat-sensitive, but not cold-sensitive. At 17°, the enzyme works, and *A* will be converted to *B*. While at 17°, the second, cold-sensitive step of the pathway prevents the production of mature phage. However, when the temperature is shifted to 42°, the accumulated *B* product can be used to make mature phage, so that lysis will occur.

Under model (1), a temperature shift performed in the reverse direction does not allow for growth. When *E. coli* cells are infected with a doubly mutant phage, and placed at 42°, the heat-sensitive first step precludes the accumulation of *B*. When the culture is shifted to 17°, *B* can accumulate, but now, the second step cannot occur, so that no progeny phage can be produced. Therefore, if model (1) is correct, lysis will be seen only in a temperature shift from 17° to 42°.

If model (2) is correct, growth will be seen only in a temperature shift from 42° to 17°, and not in a temperature shift from 17° to 42°. Hence, the correct model can be deduced by performing the temperature-shift experiment in each direction, and observing which direction allows progeny phage to be produced.

9.10 Four mutant strains of *E. coli* (*a, b, c,* and *d*) all require substance X in order to grow. Four plates were prepared, as shown in the following figure. In each case the medium was minimal, with just a trace amount of substance X, to allow a small amount of growth of the mutant cells. On plate *a*, cells of mutant strain *a* were spread over the agar, and grew to form a thin lawn. On plate *b* the lawn is composed of mutant *b* cells, and so on. On each plate, cells of the four mutant types were inoculated over the lawn, as indicated by the circles. Dark circles indicate luxuriant growth occurred. That is, this experiment tests whether the bacterial strain spread on the plate can "feed" the four strains inoculated on the plate, allowing them to grow. What do these results show about the relationship of the four mutants to the metabolic pathway leading to substance X?

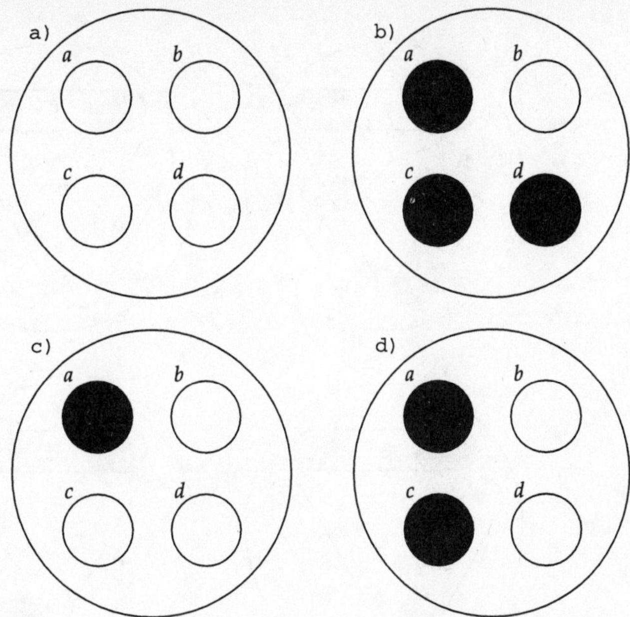

Answer: A strain that is blocked at a later step in a pathway can *"feed"* a strain that is blocked in an earlier step. It can secrete into the medium a metabolic intermediate that allows a feeding strain's block to be bypassed. Hence, a strain that cannot feed any other is blocked in the earliest step, and a strain that feeds all others (but itself) is blocked in the last step. Mutant *a* is blocked in the earliest step in the pathway, as it cannot feed any of the others. Mutant *c* is next, as it can supply the substance *a* needs, but cannot feed *b* or *d*. Mutant *d* is next, and mutant *b* is last in the pathway, as it can feed all of the others. The pathway is $a \rightarrow c \rightarrow d \rightarrow b$.

9.11 The following table indicates what enzyme is deficient in six different complementing mutants of *E. coli*, none of which can grow on minimal medium. All of them will grow if tryptophan (Trp) is added to the medium.

MUTANT	ENZYME MISSING
trpE	anthranilate synthetase
trpA	tryptophan synthetase
trpF	IGP synthetase
trpB	tryptophan synthetase
trpD	PRA transferase
trpC	PRA isomerase

Each of the plates shown in the following figure the results of streaking three of the mutants on minimal medium with just a trace of added tryptophan. Heavy shading indicates regions of heavy growth, indicating that in order to permit a few cycles of replication a strain can be fed by the strain streaked next to it on the plate. In what order do the enzymes listed above act in the tryptophan synthetic pathway?

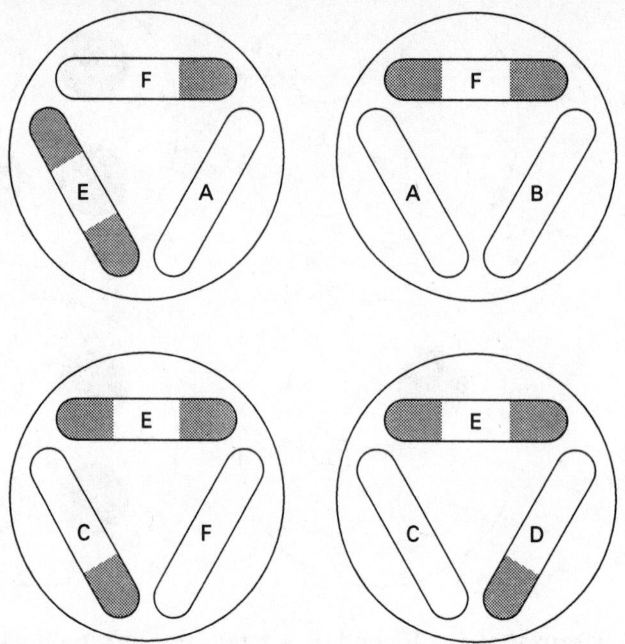

Answer: In this problem, cells accumulate different metabolic intermediates, depending on the step in which they are blocked. These intermediates are secreted into the media, where they diffuse locally, and are taken up by neighboring, mutant cells that are blocked in different steps. A mutant blocked in a particular metabolic step can use a substance to grow if the substance circumvents its block. Put another way, if the mutant is blocked in a step prior to the formation of the substance, its growth can be facilitated by feeding it that substance. A mutant will be able to grow by cross-feeding off a second mutant if it is blocked at an earlier step than the second mutant.

Since *trpE* can be fed by *F, A, C,* and *D*, it is blocked earlier in the pathway than they are. *E* is also earlier than *B*, since *B* can feed *F*. Thus, anthranilate synthetase is the first enzyme in the pathway. *F* can feed *C*, indicating *C* is blocked earlier than *F*. *C* can feed *D*, indicating *D* is blocked earlier than *C*. *A* and *B* can feed *F*, indicating *F* is blocked earlier than either *A* or *B*. Thus, the order is $E \rightarrow D \rightarrow C \rightarrow F \rightarrow [A\ B]$. After anthranilate synthetase, the enzymes are, in order, PPA transferase, PRA isomerase, IGP synthetase, and tryptophan synthetase.

9.12 Refer to the list of mutants and enzymes given in Problem 9.11, and explain how it can be that two different complementing mutants (*trpA* and *trpB*) can affect the activity of the same enzyme. How will two such mutants be related in terms of their position within the metabolic pathway?

Answer: Since *trpA* and *trpB* complement each other, one can hypothesize that they affect different polypeptides. Since they both affect the activity of one enzyme, tryptophan synthetase, this enzyme must contain (at least) two different polypeptide chains. The position in the metabolic pathway is purely a function of which enzyme is affected (which catalytic step). Thus *trpA* and *trpB* affect the same step of the pathway.

9.13 Two mutant strains of *Neurospora* lack the ability to make compound Z. When crossed, the strains usually yield asci of two types: (1) those with spores that are all mutant and (2) those with four wild-type and four mutant spores. The two types occur in a 1:1 ratio.

a. Let *c* represent one mutant, and let *d* represent the other. What are the genotypes of the two mutant strains?
b. Are *c* and *d* linked?
c. Wild-type strains can make compound Z from the constituents of the minimal medium. Mutant *c* can make Z if supplied with X but not if supplied with Y, while mutant *d* can make Z from either X or Y. Construct the simplest linear pathway of the synthesis of Z from the precursors X and Y, and show where the pathway is blocked by mutations *c* and *d*.

Answer:
a. $c^+ d$ and $c d^+$.
b. The genes are not linked, since parental ditype (PD) and non-parental ditype (NPD) tetrads occur in equal frequencies.
c. The simplest linear pathway for the synthesis of Z is Y → X → Z, where *d* is blocked in the synthesis of Y and *c* is blocked in the synthesis of X from Y.

9.14 The following growth responses (where + = growth and 0 = no growth) of mutants *1 - 4* were seen on the related biosynthetic intermediates A, B, C, D, and E. Assume all intermediates are able to enter the cell, that each mutant carries only one mutation, and that all mutants affect steps after B in the pathway.

	Growth On				
Mutant	A	B	C	D	E
1	+	0	0	0	0
2	0	0	0	+	0
3	0	0	+	0	0
4	0	0	0	+	+

Which of the schemes in the figure fits best with the data with regard to the biosynthetic pathway?

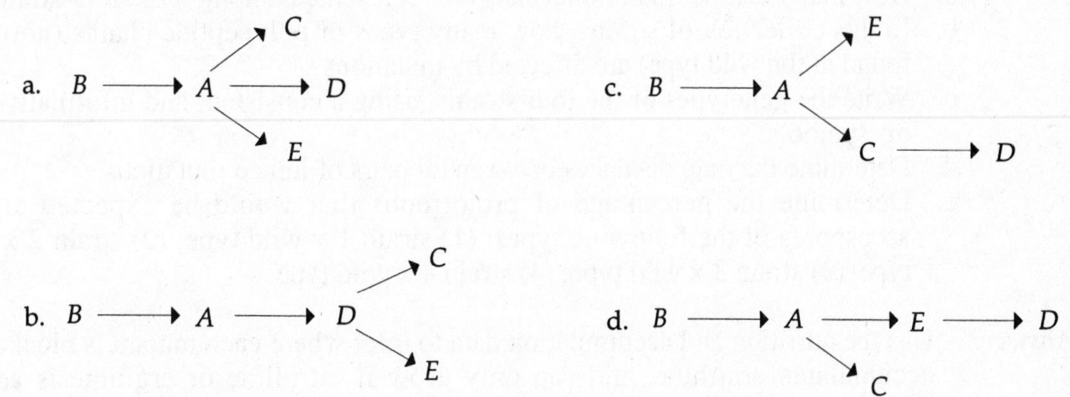

Answer: Approach this problem by trying to fit the data to each pathway, as if it were correct. Check where each mutant would be blocked, and whether the mutant would be able to grow if supplemented with the single nutrient listed. Notice that it is possible to fit the data for mutants *1*, *2* and *3* to each pathway shown, but the data for mutant 4 limit the choice to the pathway shown in *d*.

9.15 Four strains of *Neurospora*, all of which require arginine but have unknown genetic constitution, have the following characteristics. The nutrition and accumulation characteristics are as follows:

	GROWTH ON				
STRAIN	MINIMAL MEDIUM	ORNITHINE	CITRULLINE	ARGININE	ACCUMULATES
1	–	–	+	+	Ornithine
2	–	–	–	+	Citrulline
3	–	–	–	+	Citrulline
4	–	–	–	+	Ornithine

The pairwise complementation tests of the four strains gave the following results (+ = growth on minimal medium and 0 = no growth on minimal medium):

	4	3	2	1
1	0	+	+	0
2	0	0	0	
3	0	0		
4	0			

Crosses among mutants yielded prototrophs in the following percentages:

1 x 2: 25 percent
1 x 3: 25 percent
1 x 4: none detected among 1 million ascospores
2 x 3: 0.002 percent
2 x 4: 0.001 percent
3 x 4: none detected among 1 million ascospores

Analyze the data and answer the following questions.

a. How many distinct mutational sites are represented among these four strains?
b. In this collection of strains, how many types of polypeptide chains (normally found in the wild type) are affected by mutations?
c. Write the genotypes of the four strains, using a consistent and informative set of symbols.
d. Determine the map distances between all pairs of linked mutations.
e. Determine the percentage of prototrophs that would be expected among ascospores of the following types: (1) strain 1 x wild type; (2) strain 2 x wild type; (3) strain 3 x wild type; (4) strain 4 x wild type.

Answer: Use the nutrition and accumulation data to infer where each mutant is blocked. 1 accumulates ornithine, and can only grow if citrulline or arginine is added. Hence, 1 is blocked in the conversion of ornithine to citrulline. 2 and 3 are blocked in the conversion of citrulline to arginine. Mutant 4 is more complex. It can only grow if supplemented with arginine, but accumulates ornithine. 4 must be a double mutant, and is blocked in the conversion of ornithine to citrulline as well as from citrulline to arginine. One has:

$$\text{precursor} \xrightarrow{} \text{ornithine} \xrightarrow{1,4} \text{citrulline} \xrightarrow{2,3,4} \text{arginine}$$

Enzyme: A B C

Use the complementation-test data to determine whether two mutants affect the same function. Two mutants able to grow on minimal media complement each other: one provides a function missing in the other. Failure to grow indicates no complementation: both are missing the same function. 2 and 3 complement 1, so 2 and 3 affect a different step than 1. 2 and 3 fail to complement each other, so affect the same step. 4 fails to complement 1, 2, or 3, and affects both steps. This is consistent with the above diagram: 1 and 4 affect polypeptide B, and 2, 3, and 4 affect polypeptide C.

Start to analyze the recombination data by assigning symbols to the genes and writing out the genotypes and crosses. Let b^+ encode enzyme B and c^+ encode enzyme C. Then 1 = $b^1\ c^+$, 2 = $b^+\ c^2$, 3 = $b^+\ c^3$, and 4 = $b^4\ c^4$. The prototrophic ascospores produced in the crosses 1 x 2, 1 x 3, and 2 x 3 are $b^+\ c^+$ and are half of the recombinant progeny (e.g., in 1 x 2, parental-types are $b^1\ c^+$ and $b^+\ c^2$, recombinant-types are $b^+\ c^+$ and $b^1\ c^2$). The results can be tabulated as:

Cross	Genotypes	$b^+\ c^+$ spores
1 x 2	$b^1\ c^+ \times b^+\ c^2$	25.0%
1 x 3	$b^1\ c^+ \times b^+\ c^3$	25.0%
1 x 4	$b^1\ c^+ \times b^4\ c^4$	$< 1\text{x}10^{-6}$
2 x 3	$b^+\ c^2 \times b^4\ c^4$	0.002%
2 x 4	$b^+\ c^2 \times b^4\ c^4$	0.001%
3 x 4	$b^+\ c^3 \times b^4\ c^4$	$< 1\text{x}10^{-6}$

Use the recombination frequencies to infer whether the *b* and *c* genes are linked, and whether two alleles at one gene are identical or can be separated by recombination. 1 x 2 and 1 x 3 give 25 percent prototrophs (show 50 percent recombinants), so *b* and *c* are unlinked. 2 x 3 produces very few prototrophs, as they can only arise from rare, intragenic recombination. The 0.002% prototrophs (= half the recombinants) indicate that c^2 and c^3 are 0.004 mu apart. 4 fails to recombine with 1 or 3, indicating that $b^4 = b^1$ and $c^4 = c^3$ (they occupy the same site). The 0.001% prototrophs from 2 x 4 arise when a c^+ allele is obtained after intragenic recombination between c^2 and c^4, <u>and</u> the b^+ allele (in mutant 2) assorts with it into the ascospore. When c^+ assorts with b^2, (half of the time) a prototrophic ascospore is not obtained. This is why only 0.001% prototrophs are recovered from 2 x 4, instead of the 0.002% prototrophs recovered from 2 x 3. This results in the following map:

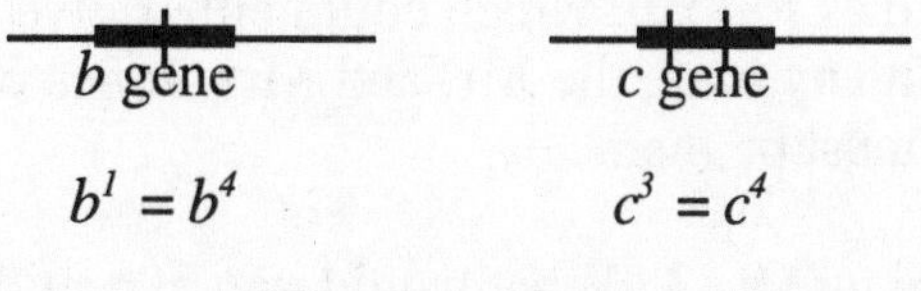

b^1 c^2 c^3

b gene *c* gene

$b^1 = b^4$ $c^3 = c^4$

a. Three distinct mutational sites exist based on the recombination analysis: one in the *b* gene, and two in the *c* gene.

b. Two polypeptide chains are affected based on the complementation analysis: one encoded by the *b* gene, and one encoded by the *c* gene.
c. The genotypes are as given in the table above.
d. There are 0.004 mu between the two mutant sites in the *c* gene identified by c^2 and c^3. The *b* and *c* genes are unlinked.
e. The *b* and *c* genes assort independently of one another. Thus, strains 1, 2, or 3, when mated with wild-type, will give 50 percent prototrophs and 50 percent auxotrophs. As strain 4 is a double mutant, it will only give 25 percent prototrophs. Four equally frequent genotypes would be expected from the cross b^4 c^4 x + +. Only one of the four will be + +.

9.16 A breeder of Irish setters has a particularly valuable show dog that he knows is descended from the famous bitch Rheona Didona, who carried a recessive gene for atrophy of the retina. Before he puts the dog to stud, he must ensure that it is not a carrier for this allele. How should he proceed?

Answer: He should testcross the male to a retinal atrophic female. Retinal atrophic pups (expected half of the time) would indicate that the male is heterozygous. If the retinal atrophy is associated with a known biochemical phenotype (e.g., altered enzyme activity), an alternative method might be to take a tissue biopsy or blood sample from the male, and assess whether normal levels of enzyme activity are present in the male. Recessive mutations in many enzymes, when heterozygous with a normal allele, lead to reduced levels of enzyme activity even though no visible phenotype is observed.

9.17 Suppose you were on a jury to decide the following case:

The Jones family claims that Baby Jane, given to them at the hospital, does not belong to them but to the Smith family, and that the Smith's baby Joan really belongs to the Jones family. It is alleged that the two babies were accidentally exchanged soon after birth. The Smiths deny that such an exchange has been made. Blood group determinations show the following results:

Mrs. Jones, AB
Mr. Jones, O
Mrs. Smith, A
Mr. Smith, O
Baby Jane, A
Baby Joan, O

Which baby belongs to which family?

Answer: Since Mrs. Jones is type AB and Mr. Jones is type O, the cross can be diagrammed as I^AI^B x *ii*, which would give I^Ai (type A) or I^Bi (type B) children. Since Mrs. Smith is type A and Mr. Smith is type O, the cross can be diagrammed as I^A– x *ii*, which could either give I^ai (type A) or potentially type O children. In any case, the Mr. and Mrs. Jones could not have a type O child, so Baby Jane must be theirs.

9.18 Refer back to Figure 9.13. What would you expect the phenotype to be in individuals heterozygous for the following two hemoglobin mutations?
a. Hb–Norfolk and Hb-S
b. Hb-C and Hb-S

Answer: a. From Figure 9.13, Hb–Norfolk affects the α-chain, while Hb–S affects the β-chain, of hemoglobin. Since each chain is encoded by a separate gene, and in a heterozygote, there remains one normal gene, a normal copy of the gene for each of the α-chain and the β-chain will be present. Thus, the mutants will complement each other, as normal hemoglobin molecules can be formed. The phenotype will be normal (or near normal, since some abnormal hemoglobin molecules may form).

b. Both Hb-C and Hb-S affect the β-chain. No normal β-chains will be present, and so all hemoglobin molecules will be non-functional. The individual will be anemic.

9.19 In evaluating my teacher, my sincere opinion is that:

a. He/she is a swell person whom I would be glad to have as a brother-in-law/sister-in-law.
b. He/she is an excellent example of how tough it is when you do not have either genetics or environment going for you.
c. He/she may have okay DNA to start with, but somehow all the important genes got turned off.
d. He/she ought to be preserved in tissue culture for the benefit of other generations.

Answer: Your choice!

CHAPTER 10

THE STRUCTURE OF GENETIC MATERIAL

I. CHAPTER OUTLINE

THE NATURE OF GENETIC MATERIAL: DNA AND RNA
- The Discovery of DNA as Genetic Material
- The Discovery of RNA as Genetic Material

THE CHEMICAL COMPOSITION OF DNA AND RNA
- The Physical Structure of DNA: The Double Helix
- Other DNA Structures
- DNA in the Cell
- Bends in DNA

II. REVIEW OF KEY TERMS, SYMBOLS AND CONCEPTS

Without consulting the text and in your own words, write a brief definition of each term in the groups below. Then, either using a short phrase or a simple diagram, identify the relationship(s) between specific pairs of terms within a set. Finally, consult the text (and perhaps a friend who has also done the exercise) to check your answers.

1	2	3	4
transformation	^{35}S and ^{32}P	macromolecule	adenine
transforming principle	Hershey-Chase experiment	nucleotide	guanine
nuclease	tobacco mosaic virus (TMV)	pentose sugar	thymine
ribonuclease (RNase)	DNA as genetic material	nitrogenous base	cytosine
deoxyribonuclease (DNase)	RNA as genetic material	ribose	uracil
Griffith's experiment		deoxyribose	RNA
Avery, MacLeod, & McCarty's experiment		purine	DNA
		pyrimidine	

5	6	7
deoxyribonucleotide	Chargaff's rules	oligomer
ribonucleotide	X-ray diffraction	A-DNA
nucleotide	3' OH, 5' P	B-DNA
nucleoside	polarity	Z-DNA
phosphodiester bond	antiparallel strands	double helix
3' OH, 5' P	complementary base pairing	major groove
nucleoside monophosphate	double helix	minor groove
nucleoside triphosphate	major groove	
polynucleotide	minor groove	

III. THINKING ANALYTICALLY

The material that is presented in this chapter, and the problems that accompany it, require you to pay attention to detail, and retain a considerable amount of information. The material may seem somewhat more "factual," and less "conceptual" than that covered in previous chapters. As you probably have developed good skills to master this type of material from other areas of study, the material in this chapter may seem easier to learn.

As you approach the material however, do not view it as "just a set of facts." This chapter presents key experimental evidence for several areas of genetics, including evidence that nucleic acids comprise the genetic material, and evidence for the structure of DNA. Pay close attention to the subtleties of the experiments that are presented, and *how* they are used to gather evidence for or against a particular view. What we now take for granted as fact was previously hotly debated. Look to the experiments presented in this chapter to discover how insightful experimentation can lead to resolution of fundamentally important issues.

As you approach the material in this chapter, be sure to learn the chemical structures of nucleic acid components. Learning them now will allow you to develop a substantially better understanding of the material that is coming in future chapters. Start by taking a paper and pencil, and copying the component parts of a nucleotide (the sugar, the bases, the phosphodiester bond) from the text figures. Pay particular attention to the polarity of the molecules. Practice these awhile, and then try to draw them from memory. Practice your drawings several days in a row, until you can draw them without text assistance.

The material in this chapter requires you to learn (or to review) units of measurement. It is important to be able to use them fluently, and have a general sense of dimension. If you are working on a problem that presents values in a variety of units, it is often useful to convert the units to as small an assortment as possible. Several measurements are particularly important to review:

UNIT/CONVERSION	EXAMPLE
1 Å = 1 Ångstrom = 1 x 10^{-10} meter	distance between 2 protons in H_2 = 0.74Å
1 nm = 1 nanometer = 1 x 10^{-9} meter	width of a DNA double helix = 2 nm
1 μm = 1 micrometer = 1 x 10^{-6} meter	typical red blood cell diameter = 7 μm
1 dalton (unit of mass) = 1.66 x 10^{-24} gram	1 H atom = 1 dalton

IV. QUESTIONS FOR PRACTICE

A. Multiple Choice Questions

Choose the correct answer or answers for the following questions.

1. DNA and RNA are polymers of
   a. nucleosides.
   b. nucleotides.
   c. pentose sugars connected by phosphodiester bonds.
   d. ribonucleotides.

2. A molecule consisting of ribose covalently bonded to a purine or pyrimidine base is a
   a. ribonucleoside.
   b. ribonucleotide.
   c. nuclease.
   d. deoxyribonucleotide.

3. The transforming principle was found to be
   a. a cellular material that could alter a cell's heritable characteristic.
   b. a substance derived from killed viruses.
   c. modified RNA that could change a living cell.
   d. a transmissible substance that revives dead cells.

4. When Griffith injected mice with a mixture of live *R* pneumococcus that had been derived from a *IIS* strain and heat killed *IIIS* bacteria,
   a. the mice survived and he recovered live type *IIIR* organisms.
   b. the mice died, but he recovered live type *IIIR* cells.
   c. the mice died, but he recovered live type *IIS* cells.
   d. the mice died, but he recovered live type *IIIS* cells.

5. In the Hershey-Chase experiment, T2 phage were radioactively labeled with either ^{35}S or ^{32}P. After radioactively labeled phage were allowed to infect *E. coli,*
   a. the ^{35}S was found in progeny phage, and the ^{32}P was found in phage ghosts, indicating that DNA was the genetic material.
   b. the ^{35}S was found in phage ghosts, and the ^{32}P was found in progeny phage, indicating that DNA was the genetic material.
   c. the ^{35}S was found in both progeny phage and in phage ghosts, indicating that DNA was the genetic material.
   d. the ^{32}P was found in both progeny phage and in phage ghosts, indicating that DNA was the genetic material.

6. Analysis of the bases of a sample of nucleic acid yielded these percentages: A–20 percent, G–30 percent, C–20 percent, T–30 percent. The sample must be
   a. double-stranded RNA.
   b. single-stranded RNA.
   c. double-stranded DNA.
   d. single-stranded DNA.

7. Which of the following is *not* true about a linear molecule of double-stranded DNA?
   a. It is a double helix composed of antiparallel strands.

b. Bases are paired via hydrogen bonds.
c. At one end, two 5' phosphate groups can be found.
d. Pentose sugars are linked via covalent phosphodiester bonds.

8. Which kind of DNA is *not* likely to be found in cells?
   a. A-form DNA
   b. B-form DNA
   c. Z-form DNA
   d. single-stranded DNA
   e. none of the above

9. Physical evidence indicates that DNA is not a rigid, rod-like molecule. How might bending DNA play an important role?
   a. It might facilitate the binding of DNA by regulatory proteins.
   b. Certain enzymes that cleave DNA may bend it in the cleavage process.
   c. It plays an important role in enzyme-mediated recombination events.
   d. All of the above.

10. Two double-stranded 25 base-pair DNA fragments are heated in solution. Fragment A has 60 percent GC, and fragment B has 40 percent GC. Which of the following observations might one make as the solution temperature increases?
    a. At a low enough temperature, both fragments will remain double-stranded.
    b. A will separate into single strands at a lower temperature than B.
    c. B will separate into single strands at a lower temperature than A.
    d. At a high enough temperature, both A and B will separate into single strands.

Answers: 1b; 2a; 3a; 4c; 5b; 6d; 7c; 8a; 9d; 10a, c & d.

B. Thought Questions

1. Describe the logic behind the series of experiments that conclusively demonstrated that DNA was the genetic material in some cells. Why was this hotly debated for so long? When RNA was found to be the genetic material for some viruses, was the debate re-kindled, or was this "the exception that proved the rule"?

2. Consider a double-stranded DNA molecule. How many strands of DNA exist in a single chromosome during the following stages of the cell cycle, mitosis and meiosis: G_1, G_2, prophase, metaphase, anaphase, pachynema, diplonema, anaphase I, and anaphase II.

3. Describe the various forms of DNA (A, B, Z) that have been identified. What are their most significant differences? Are any of these differences likely to have functional significance? If so, what are they?

4. Watson and Crick reputedly deduced the structure of DNA by applying observations made by others to molecular models, without any direct experimental observations of their own. Is this legitimate scientific procedure? If so, why? If not, why not?

5. Draw out the chemical structure of a double-stranded DNA molecule so that one strand has the sequence 5'-ATG-3'. Indicate the polarity of each strand, the location and the kinds of bonds that exist, and the approximate dimensions of the molecule.

6. Double-stranded DNA molecules have negatively charged sugar-phosphate backbones, and a major and minor groove in which the chemical groups of the bases project. Some proteins that interact with DNA do so in a highly sequence-specific manner, while others interact in a

largely sequence non-specific manner. What different features of the DNA molecule might these classes of protein be recognizing?

V. SOLUTIONS TO TEXT PROBLEMS

10.1 The experiment by Griffith by which a mixture of dead and live bacteria were injected into mice demonstrated (choose the right answer):

a. DNA is double stranded.
b. mRNA of eukaryotes differs from mRNA of prokaryotes.
c. bacterial transformation.
d. bacteria can recover from heat treatment if live helper cells are present.

Answer: c.

10.2 In the 1920s while working with *Diplococcus pneumoniae*, the agent that causes pneumonia, Griffith discovered an interesting phenomenon. In the experiments mice were injected with different types of bacteria. For each of the following bacteria type(s) injected, indicate whether the mice lived or died:

a. type *IIR*
b. type *IIIS*
c. heat-killed *IIIS*
d. type *IIR* + heat-killed *IIIS*

Answer: a. lived
b. died
c. lived
d. died (DNA from the *IIIS* bacteria transformed the *IIR* bacteria to a virulent form)

10.3 Several years after Griffith described the transforming principle, Avery, MacLeod, and McCarty investigated the same phenomenon.

a. Describe their experiments.
b. What did their experiments demonstrate beyond Griffith's?
c. How were enzymes used as a control in their experiments?

Answer: a. They extracted *S* cells, and showed that they could transform *R* cells with these extracts *in vitro*. They showed that the transforming principle co-purified with DNA. This indicated that the genetic material was DNA.
b. They identified the transforming principle to be DNA.
c. Ribonuclease did not destroy the transforming principle, but deoxyribonuclease did abolish transforming activity. This result substantiated the fact that DNA was the genetic material.

10.4 The Hershey-Chase experiment in which T2 phages grown in bacteria on media containing either radioactively-labeled sulfur or radioactively-labeled phosphorus demonstrated that (choose the correct answer):

a. The coat material of the phage controls the kind of DNA replicated in the host cell.
b. The nucleic acid of the phage contains the genetic information.

c. The prophage state is necessary for generalized transduction.
d. Conjugation is dependent upon some of the cells being *Hfr*.
e. A metaphase chromosome is composed of two chromatids, each containing a single DNA molecule.

Answer: b.

10.5 Hershey and Chase showed that when phages were labeled with ^{32}P and ^{35}S, the ^{35}S remained outside the cell and could be removed without affecting the course of infection, whereas the ^{32}P entered the cell and could be recovered in progeny phages. What distribution of isotope would you expect to see if parental phages were labeled with isotopes of
a. C?
b. N?
c. H?
Explain your answer.

Answer: a. Phage ghosts (supernatant) and progeny would have isotope.
b. Phage ghosts (supernatant) and progeny would have isotope.
c. Phage ghosts (supernatant) and progeny would have isotope.

Both amino acids and nucleic acids have carbon, nitrogen and hydrogen. If a parental phage was labeled with isotopes of C, N, or H, one would expect the phage to have a labeled protein coat as well as labeled DNA. As a consequence, these isotopes would be recovered in the DNA of the progeny phage, as well as in the phage ghosts left behind in the supernatant after phage infection. This experiment would not be very helpful to determine whether DNA or protein was the genetic material, as neither material would be selectively labeled. The elegance of the Hershey-Chase experiment was to use isotopes that *selectively* labeled only one component of the phage, so that it could be followed from one generation to the next.

10.6 What is the evidence that the genetic material of TMV (tobacco mosaic virus) is RNA?

Answer: See text, pp. 302-303.

10.7 The X-ray diffraction data obtained by Rosalind Franklin suggested (choose the correct answer):
a. DNA is a helix with a pattern which repeats every 3.4 nanometers.
b. purines are hydrogen bonded to pyrimidines.
c. replication of DNA is semiconservative.
d. mRNA of eukaryotes differs from mRNA of prokaryotes.

Answer: a.

10.8 In DNA and RNA, which carbon atoms of the sugar molecule are connected by a phosphodiester bond?

Answer: The 5' carbon of one sugar is connected by a phosphodiester (O–P–O) bond to the 3' carbon of another. See text Fig. 10.11, p. 305.

10.9 Which base is unique to DNA, and which base is unique to RNA?

Answer: Thymine (T) is a pyrimidine base unique to DNA, while uracil (U) is a pyrimidine base unique to RNA. See text pp. 304-306.

10.10 How do nucleosides and nucleotides differ?

Answer: A nucleoside has a sugar (ribose in RNA, 2' deoxyribose in DNA) plus a base (a pyrimidine or purine base attached to the sugar at its 1' carbon). A nucleotide has the sugar and the base, as well as a phosphate group attached to its 5' carbon via an ester (O) bond. See text Fig. 10.11, p. 305.

10.11 What chemical group is found at the 5' end of a DNA chain? At the 3' end of a DNA chain?

Answer: A phosphate group is found at the 5' end of a DNA polynucleotide chain, and a hydroxyl (OH) group is found at the 3' end of a DNA polynucleotide chain.

10.12 What evidence do we have that in the helical form of the DNA molecule that the base pairs are composed of one purine and one pyrimidine?

Answer: Two different lines of evidence support the view that a base pair is composed of one purine and one pyrimidine.

(1) When the chemical components of double-stranded DNA from a wide variety of organisms were analyzed quantitatively by Chargaff, it was found that the amount of purines equaled the amount of pyrimidines. More specifically, it was found that the amount of adenine equaled the amount of thymine, and that the amount of cytosine equaled the amount of guanine. The simplest hypothesis to explain these observations was the existence of complementary base pairing, A on one strand paired with T on the other strand, and G paired with C.

(2) More direct physical evidence was provided by X-ray diffraction studies. These established the dimensions of the DNA double helix and allowed for comparison with the known sizes of the bases. The diameter of the double helix is constant throughout its length at 2 nm. This is the right size to accommodate a purine paired with a pyrimidine, but too small for a purine-purine pair, and too big for a pyrimidine-pyrimidine pair.

10.13 What evidence is there to substantiate the statement: "There are only two base pair combinations in DNA, A-T and C-G"?

Answer: The evidence stems from the experiments by Chargaff. Quantitative measurements of the four bases in double-stranded DNA isolated from a wide variety of organisms indicated that, in all cases, the amount of A equaled the amount of T, and the amount of G equaled the amount of C. Moreover, different DNAs exhibited different base ratios so that while (A) = (T) and (G) = (C), in most organisms (A + T) ≠ (G + C). Put another way, the %GC content of different DNA samples varies. The simplest hypothesis is that there are two base pairs in DNA, A–T and G–C, and the proportion of the two base pairs varies from organism to organism.

10.14 How many different kinds of nucleotides are there in DNA molecules?

Answer: There are four different kinds of nucleotides in DNA. A DNA nucleotide consists of 2'-deoxyribose plus a phosphate plus a base. Since there are four different nitrogenous bases in DNA, adenine (A), thymine (T), guanine (C) and cytosine (C), there are four different nucleotides. In a DNA molecule, one finds monophosphate nucleotides, so that there is deoxyadenosine monophosphate (dAMP), thymidine monophosphate (TMP), deoxyguanosine monophosphate (dGMP) and deoxycytidine monophosphate (dCMP).

10.15 What is the base sequence of the DNA strand, that would be complementary to the following single-stranded DNA molecules:

a. `5' AGTTACCTGATCGTA 3'`
b. `5' TTCTCAAGAATTCCA 3'`

Answer: a. `3' TCAATGGACTAGCAT 5'` (or `5' TACGATCAGGTAACT 3'`).
b. `3' AAGAGTTCTTAAGGT 5'` (or `5' TGGAATTCTTGAGAA 3'`).

10.16 Is an adenine-thymine or guanine-cytosine base pair harder to break apart? Explain your answer.

Answer: The adenine-thymine base pair is held together by two hydrogen bonds, while the guanine-cytosine base pair is held together by three hydrogen bonds. Thus, the guanine-cytosine base pair requires more energy to break apart and is the harder to break apart.

10.17 The double-helix model of DNA, as suggested by Watson and Crick, was based on a variety of lines of evidence gathered on DNA by other researchers. The facts fell into the following two general categories; give three examples of each:

a. chemical composition
b. physical structure

Answer: See text, pp. 304-309.

10.18 For double-stranded DNA, which of the following base ratios always equals 1?

a. (A + T)/(G + C)
b. (A + G)/(C + T)
c. C/G
d. (G + T)/(A + C)
e. A/G

Answer: Since (G) = (C), and (A) = (T), it follows that (G + A) = (C + T), and (G + T) = (A + C). Thus, b, c, and d are all equal to 1.

10.19 If the ratio of (A + T) to (G + C) in a particular DNA is 1.00, does this result indicate that the DNA is most likely constituted of two complementary strands of DNA or a single strand of DNA, or is more information necessary?

Answer: More information is needed. That (A + T)/(G + C) = 1 indicates only that (A + T) = (G + C). If the DNA was double-stranded, (G) = (C), and (A) = (T). For the observed ratio, there would need to be 25 percent A, 25 percent T, 25 percent C and 25 percent G. However, if the DNA were single-stranded, one could still observe this ratio. In single-stranded DNA, there are no restrictions on the relative amounts of the different bases. There are many ways in which one could observe (A + T)/(G + C) = 1. For example, suppose there were 35 percent A, 15 percent T, 20 percent G and 30 percent C.

10.20 Explain whether the (A + T)/(G + C) ratio in double-stranded DNA is expected to be the same as the (A + C)/(G + T) ratio.

Answer: In double-stranded DNA, the (A + C)/(G + T) ratio is expected to equal 1. One has A pairing with T, and C pairing with G. This means that (A) = (T) and (G) = (C), which in turn means that (A + C) = (G + T). This leads to (A + C)/(G + T) = 1. The (A + T)/(G + C) ratio need not equal 1. Unless the amount of A–T base pairs equals the amount of G–C base pairs, this ratio will not equal 1. Organisms have varying amounts of %GC, so that the (A + T)/(G + C) ratio need not equal 1.

10.21 The percent cytosine in a double-stranded DNA is 17. What is the percent of adenine in that DNA?

Answer: In double-stranded DNA, if (C) = 17 percent, then (G) = 17 percent. This means that the DNA has 34 percent G–C, and 66 percent A–T base pairs. Hence, the DNA will have (66/2)(100%) = 33% A.

10.22 Upon analysis, a double-stranded DNA molecule was found to contain 32 percent thymine. What percent of this same molecule would be made up of cytosine?

Answer: In double-stranded DNA, if (A) = 32 percent, then (T) = 32 percent. This means that the DNA will have 64 percent A–T, and 36 percent G–C base pairs. The DNA will have (36/2)(100%) = 18% C.

10.23 A sample of double-stranded DNA has 62 percent GC. What is the percentage of A in the DNA?

Answer: If double-stranded DNA has 62 percent GC, it has (100% - 62%) = 38% AT. The percentage of A will be (38/2)(100%) = 19%.

10.24 A double-stranded DNA polynucleotide contains 80 thymidylic acid and 110 deoxyguanylic acid residues. What is the total nucleotide number in this DNA fragment?

Answer: Since the DNA molecule is double-stranded, (A) = (T), and (G) = (C). If there are 80 T residues, there must be 80 A residues. If there are 110 G residues, there must be 110 C residues. The molecule has (110 + 110 + 80 + 80) = 380 nucleotides, or 190 base pairs.

10.25 Analysis of DNA from a bacterial virus indicates that it contains 33 percent A, 26 percent T, 18 percent G, and 23 percent C. Interpret these data.

Answer: First, notice that (A) ≠ (T), and (G) ≠ (C). Thus, the DNA is not double-stranded. The bacterial virus appears to have a single-stranded DNA genome.

10.26 The following are melting temperatures for different double-stranded DNA molecules. Arrange these molecules from lower to higher content of GC pairs.

a. 73°C
b. 69°C
c. 84°C
d. 78°C
e. 82°C

Answer: G–C base pairs have three hydrogen bonds, while A–T base pairs have two. As a consequence, G–C base pairs are stronger than A–T base pairs. If a double-stranded molecule in solution is heated, the thermal energy will "melt" the hydrogen bonds, denaturing the double-stranded molecule into single-strands. Double-stranded molecules with more G–C base pairs require more thermal energy to break the hydrogen bonds, and so dissociate into single strands at higher temperatures. Put another way, the higher the GC content of a double-stranded DNA molecule, the higher its melting temperature. Re-ordering the molecules from lowest to highest %GC, one has **b** (69°), then **a** (73°), **d** (78°), **e** (82°), and **c** (84°).

10.27 What is a DNA oligomer?

Answer: A DNA oligomer is a short stretch of DNA containing a small number of bases. It is often referred to as an oligonucleotide, or an oligo, for short. Sometimes, when the exact size of the oligomer can be specified, this information might be included. For example, a single-stranded oligomer having 20 bases might be referred to as a 20-base oligomer or a 20-mer, for short.

10.28 The genetic material of bacteriophage ΦX174 is single-stranded DNA. What base equalities or inequalities might we expect for single-stranded DNA?

Answer: We can make no predictions regarding the base content of single-stranded DNA. Any base-pair equality would depend on the overall sequence of the chromosome: A might be equal to T, but that need not be the case, given that is only one of many, many possibilities.

10.29 If a virus particle contains double-stranded DNA with 200,000 base pairs, how many complete 360° turns occur in this molecule?

Answer: Since there are 10 base pairs per complete 360° turn of a B-form double-stranded DNA molecule, one expects there to be 200,000/10 = 20,000 complete turns in the viral DNA.

10.30 A double-stranded DNA molecule is 100,000 base pairs (100 kilobases) long.

a. How many nucleotides does it contain?
b. How many complete turns are there in the molecule?
c. How long is the DNA molecule?

Answer:

a. Each base pair has 2 nucleotides, so the molecule has 200,000 nucleotides.
b. There are 10 base pairs per complete 360° turn, so that there will be 100,000/10 = 10,000 complete turns in the molecule.

c. There are 0.34 nm between the centers of adjacent base pairs. There will be 100,000 x 0.34 nm = 3.4×10^4 nm = 34 μm.

10.31 If nucleotides were arranged at random in a single-stranded RNA 10^6 nucleotides long, and if the base composition of this RNA was 20 percent A, 25 percent C, 25 percent U and 30 percent G, how many times would you expect the specific sequence 5'-GUUA-3' to occur?

Answer: The chance of finding the sequence 5' GUUA 3' is (0.30 x 0.25 x 0.25 x 0.20) = 0.00375. In a molecule 10^6 nucleotides long, there are nearly 10^6 groups of four bases: the first group of four is bases 1, 2, 3, and 4, the second group is bases 2, 3, 4, and 5, etc. Thus, the number of times this sequence is expected to appear is 0.00375 x 10^6 = 3,750.

CHAPTER 11

THE ORGANIZATION OF DNA IN CHROMOSOMES

I. CHAPTER OUTLINE

STRUCTURAL CHARACTERISTICS OF BACTERIAL AND VIRAL CHROMOSOMES
- Bacterial Chromosomes
- T-Even Phage Chromosomes
- Bacteriophage ΦX174 Chromosome
- Bacteriophage λ Chromosome

STRUCTURAL CHARACTERISTICS OF EUKARYOTIC CHROMOSOMES
- The Karyotype
- Chromosomal Banding Patterns
- Cellular DNA Content and the Structural or Organizational Complexity of the Organism
- The Molecular Structure of the Eukaryotic Chromosome
- Centromeres and Telomeres

UNIQUE-SEQUENCE AND REPETITIVE-SEQUENCE DNA IN EUKARYOTIC CHROMOSOMES
- Unique-sequence DNA
- Repetitive-sequence DNA

II. REVIEW OF KEY TERMS, SYMBOLS AND CONCEPTS

Without consulting the text and in your own words, write a brief definition of each term in the groups below. Then, either using a short phrase or a simple diagram, identify the relationship(s) between specific pairs of terms within a set. Finally, consult the text (and perhaps a friend who has also done the exercise) to check your answers.

1	2	3	4
relaxed circular DNA	*E. coli* chromosome	karyotype	linear chromosome
supercoiled circular DNA	nucleoid region	G-banding	circularly permute
topoisomerase I, II	HU, H proteins	Q-banding	denature
negative, positive supercoiling	DNA domains	AT-rich region	renature
DNA gyrase	histone proteins	FISH	reanneal
	nonhistone proteins	*in situ* hybridization	T-even phage
			T-odd phage
			headful packaging

5	6	7	8
double-stranded DNA chromosome	C-value	chromatin	histones
single-stranded DNA chromosome	C-value paradox	euchromatin	nonhistones
ΦX174	structural complexity	heterochromatin	positively charged
T-even phage	phylum	constitutive heterochromatin	negatively charged
λ chromosome		facultative heterochromatin	basic
cos sites		centromere	acidic
ter enzyme		telomere	phosphodiester backbone
DNA ligase			H1, H2A, H2B, H3, H4
concatamers			nucleosome
			nucleosome core particle
			histone octomer

9	10	11
nucleosome	centromere	unique-sequence DNA
10 nm nucleofilament	yeast CEN sequences	moderately repetitive DNA
30 nm chromatin fiber	consensus sequence	highly repetitive DNA
H1-depleted chromatin	telomere	denaturation
looped domain	simple telomeric sequences	renaturation
nonhistone protein scaffold	telomere associated sequences	$C_o t$ curves
nuclear envelope	chromosome stabilization	$C_o t_{1/2}$
nuclear matrix	chromosome segregation	protein coding genes
MARs		satellite DNA
metaphase chromosome		% GC content
		SINES & LINES
		Alu repeats

III. THINKING ANALYTICALLY

As you begin to understand the details of how DNA is organized in chromosomes of different organisms, look for patterns to identify general principles underlying DNA packaging. Try to understand how these patterns and principles relate to the "biology of the organism."

DNA packaging occurs in three dimensions, and so it will be helpful to try to visualize this process by making sketches and drawings of how DNA is packaged. You do not have to be a great artist — the point is to convey how a molecule, that if stretched end-to-end, would be larger than most cells, can fit and retain functionality, in a subcellular organelle.

One of the most challenging topics (especially for those of us who do not always visualize well in three dimensions) is to visualize DNA supercoiling. Try to visualize some of the principles involved by taking a spiral telephone cord, attaching it at the ends to circularize it, introducing additional twists and examining how the molecule becomes a coiled coil.

IV. QUESTIONS FOR PRACTICE

A. Multiple Choice Questions

Choose the correct answer or answers for the following questions.

1. The chromosome of *E. coli* is packaged in the nucleoid region in a
   a. nuclear membrane.
   b. semicircular form.
   c. relaxed form.
   d. supercoiled form.

2. Topoisomerases are enzymes that do all of the following except
   a. cleave a phosphodiester backbone.
   b. twist one DNA strand about another.
   c. introduce positive supercoils into relaxed DNA.
   d. introduce negative supercoils into relaxed DNA.
   e. change the topological form of the DNA, but not the DNA sequence.

3. Circularly permuted chromosomes are found in
   a. T2 and T4 phage.
   b. certain bacteria only.
   c. any phage that is packaged from a concatamer.
   d. all eukaryotes.
   e. lambda phage.

4. The DNA of viruses ΦX174 and Qβ are
   a. single stranded.
   b. double stranded.
   c. circularly permuted.
   d. extensively supercoiled.

5. The total amount of DNA in the haploid genome of any organism is
   a. its karyotype.
   b. its C-value.
   c. its $C_ot_{1/2}$ value.
   d. an indication of its structural complexity.

6. Facultative heterochromatin
   a. is always inactive.
   b. is inactive only in certain cells.
   c. contains only middle-repetitive DNA.
   d. can contain unique sequence DNA.

7. The fundamental unit of chromatin packaging in eukaryotes is
   a. a histone protein.
   b. a nucleosome.
   c. a 30 nm chromatin fiber.
   d. a looped domain.

8. Which of the following are not true about both centromeres and telomeres?
   a. They are associated with constitutive heterochromatin.
   b. They are associated with consensus sequence elements.
   c. They are essential for eukaryotic chromosomal function.
   d. Their DNA has unusual G–G base pairing.

9. Consider a 100 bp stretch of double-stranded DNA having 50 A–T base pairs followed by 50 G–C base pairs. What will happen if it is heated slowly in solution?
   a. At a particular temperature, the molecule will denature from one double-stranded into two single-stranded DNA molecules.
   b. As the temperature increases, the A–T region will denature first. At a higher temperature, the G–C region will denature, leading to a two-stage strand dissociation.
   c. As the temperature increases, the G–C region will denature first. At a higher temperature, the A–T region will denature, leading to a two-stage strand dissociation.
   d. As the temperature increases, phosphodiester bonds will break, leading to the production of G–C and A–T base pairs.

10. In a particular renaturation experiment, SV40 DNA is seen to have a $C_ot_{1/2}$ value of 9 x 10^{-3}. SV40 has 5,000 base pairs in its genome. Under identical conditions, λ DNA has a $C_ot_{1/2}$ value of 9 x 10^{-2}. About how many bases does λ DNA have?
   a. 500
   b. 5,000
   c. 50,000
   d. 500,000

Answers: 1d; 2c; 3a; 4a; 5b; 6b & d; 7b; 8d; 9b; 10c

B. Thought Questions

1. The histones, H1, H2A, H2B, H3, and H4, are said to be the most highly conserved of all proteins. From this fact, it is proposed that they serve a function that is basic to life for all eukaryotes. Comment on the implications of this assertion with regard to evolutionary theory.

2. The histones are highly basic proteins that interact with DNA (specifically, with the acidic phosphodiester backbone). What kinds of proteins are the nonhistones? How might they differ from the histones in their interactions with DNA?

3. How do you account for the presence of both unique and repetitive sequences in the genomes of eukaryotes? Why are the latter lacking in prokaryotes?

4. Explain one of the fundamental puzzles of life: An amphibian that lives in a swamp has 100-fold more DNA than a human that lives in on the 48th floor of a luxury apartment complex. In what is the amphibian richer?

5. What different techniques exist for staining eukaryotic chromosomes, and how do they work? How might some of these be used to correlate cytological maps of eukaryotic chromosomes with meiotic map positions?

6. Summarize the structural composition and organization of the chromosomes of eukaryotic cells. How do they compare with the chromosomes of viruses and of prokaryotes?

7. Consider three phages: λ, T7 and T4. What different consequences would there be on chromosome packaging if each phage underwent a deletion comprising 20 percent of its genome? What if the mutation were a 20 percent insertion?

V. SOLUTIONS TO TEXT PROBLEMS

11.1 Two double-stranded DNA molecules from a population of T2 phages were heat-denatured to produce the following four single-stranded DNAs:

1 TAGCTCC → 3 GCTCCTA →

and

2 ATCGAGG ← 4 CGAGGAT ←

These separated strands were allowed to renature. Diagram the structures of the renatured molecules most likely to appear when (a) strand 2 renatures with strand 3, and (b) strand 3 renatures with strand 4. Mark the strands and indicate the sequences and the polarity.

Answer: a. The sequence C G A G G in molecule 2 is complementary to the sequence G C T C C in molecule 3. When these pair up, one has:

3   G C T C C T A →
←
2 A T C G A G G

Each strand has two unpaired bases sticking out. These bases are complementary to each other, so that if the molecule bends, one has:

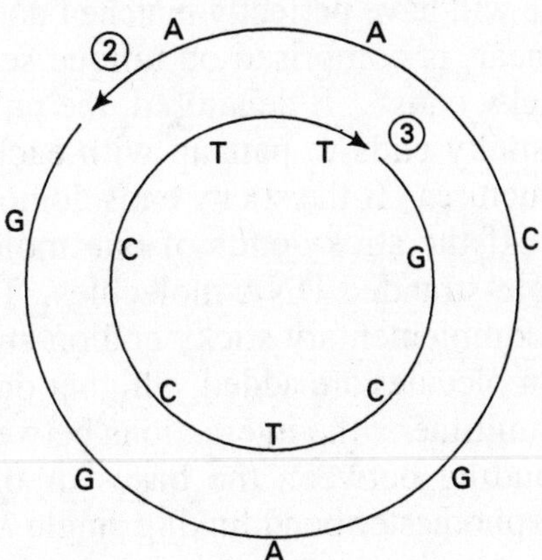

b. The sequence in molecule 3 is complementary to the sequence in molecule 4. It also has opposite polarity, so that the two strands can pair up. One has:

3 G C T C C T A →
←
4 C G A G G A T

11.2 Capital letters represent regions in the chromosome of phage T4. A particular *E. coli* cell was infected by a single T4 chromosome with the sequence ABCDEFAB, but before this chromosome could replicate it suffered a deletion of the E region. This did not interfere with phage replication. What would you expect to be the chromosome sequence(s) of the progeny phage produced upon lysis of this cell? Explain your reasoning.

Answer: T4 chromosomes are circularly permuted and terminally redundant. Since progeny T4 chromosomes are cut by the "headful" from a long concatamer, the deletion of E will produce a shorter unique sequence, and result in longer terminal redundancy. Thus, we should see ABCDFABC, BCDFABCD, CDFABCDF, DFABCDFA and FABCDFAB.

11.3

a. If you were to denature and renature a population of normal T4 chromosomes, what kinds of structures would form?
b. How would the results differ if the chromosomes were from T7?
c. From λ?

Answer:

a. The chromosomes are circularly permuted, terminally redundant molecules. Upon denaturation and renaturation, there would be few perfectly matched double-stranded molecules. Most molecules would have single-stranded ends. Molecules with very long single-stranded ends might form complex partially circularized structures, such as those shown for T2 phage in text Figure 11.5b, p. 323.
b. T7 chromosomes are linear, and are unique sequence. If dissociated and allowed to renature, unique sequences on one strand will pair with complementary unique sequences on the other. When renaturation is complete, one will have perfectly matched double-stranded molecules.
c. λ DNA is linear, is comprised on unique sequences, and has short, single-stranded "sticky ends". If denatured, the unique sequences would renature, leaving the sticky ends to pair up with each other. This could result in a variety of structures. If the sticky ends do not interact, one will have a linear λ-monomer. If the sticky-ends of one molecule pair up, one would have circular double-stranded DNA molecules. If a sticky end on one molecule pairs with a complementary sticky end on another, one will have a λ-dimer. If additional molecules are added, a higher order λ-multimer will form. Note that in such multimers, the interactions between λ chromosomes are those of hydrogen-bonding between the bases of the cos-sequence. There is no covalent phosphodiester bond linking single λ chromosomes.

11.4 What are topoisomerases?

Answer: Topoisomerases are a class of enzymes that convert one topological form of DNA to another. In *E. coli*, topoisomerase II (a.k.a. DNA gyrase) untwists relaxed DNA to produce negative supercoils. Topoisomerase I twists negatively supercoiled DNA to convert it to a more relaxed state. The two enzymes together allow DNA to be converted between negatively supercoiled DNA and a relaxed state. (Notice that positively supercoiled DNA does not exist because Topoisomerase I cannot induce relaxed DNA into a positively supercoiled state.) See text pp. 319-320.

11.5 In typical human fibroblasts in culture, the G_1 period of the cell cycle lasts about 10 h, S lasts about 9 h, G_2 takes 4h, and M takes 1h. Imagine you were to do an experiment in which you added radioactive (^{3}H) thymidine to the medium and left it there for 5 min (pulse), and then washed it out and put in ordinary medium (chase).

a. What percentage of cells would you expect to become labeled by incorporating the ^{3}H-thymidine into their DNA?
b. How long would you have to wait after removing the ^{3}H medium before you would see labeled metaphase chromosomes?
c. Would one or both chromatids be labeled?
d. How long would you have to wait if you wanted to see metaphase chromosomes containing ^{3}H in the regions of the chromosomes that replicated at the beginning of the S period?

Answer: a. Assuming that the fibroblast culture is completely asynchronous, and that cells are distributed in all stages of the cell cycle, one would expect that 10/24 of the cells would be in G_1, 9/24 of the cells would be in S, 4/24 of the cells would be in G_2, and 1/24 of the cells would be in M. Only cells in S are replicating their DNA, so only 9/24 of the cells will be capable of incorporating ^{3}H-thymidine into their DNA. (9/24)(100%) = 37.5%.

b. If a cell that was in the very last stages of S phase incorporated ^{3}H-thymidine into its DNA, it would have to proceed through G_2 (four hours) before entering M. Not until M can one see metaphase chromosomes. Therefore, one would need to wait a bit over four hours before labeled metaphase chromosomes could be seen.

c. DNA replication is semi-conservative, meaning that each of the two double-strands is used as a template for the synthesis of a new, complementary strand. When ^{3}H-thymidine is incorporated into the newly synthesized DNA, it will be incorporated into each new complementary strand, and so be in each chromatid that is synthesized.

d. Cells that took up ^{3}H-thymidine into their DNA at the beginning of S phase would need to complete the S phase, and then go through G_2 before label could be seen in metaphase chromosomes. Thus, one would have to wait a bit over 13 hours.

The experiment that is outlined in this problem gives one an elegant means to determine which chromosomal region replicates early in S phase and which chromosomal region replicates late in S phase. Proceed by systematically collecting cells at various time points after the ^{3}H-thymidine pulse. Spread their chromosomes on slides, cover the chromosomes with a photographic emulsion that is able to detect decay of ^{3}H particles. After a period of time, develop the emulsion to detect sites on chromosomes that were replicating at a given stage of S-phase. This kind of experimentation supported a view that there are early and late-replicating regions of eukaryotic chromosomes. The late replicating regions include centromeric regions. For another look at this information, see problems 11.6 and 12.25.

11.6 Assume you did the experiment in Question 11.5, but left the radioactive medium on the cells for 16 h instead of 5 min. How would your answers to the above questions change?

Answer: a. 16 hours is longer than any single stage of the cell cycle. In particular, a cell that just entered G_2 at the beginning of the labeling period would, at the end of a 16 hour labeling period, be in early S phase. Therefore, it would be labeled. Indeed, every cell would be labeled.

b. Labeled metaphase chromosomes would already be present by the time the ^{3}H-thymidine was removed, as some of the cells in G_1 at the beginning of the labeling period would go through S and G_2, and be in M after 16 hours.

c. Both chromatids would be labeled, as in a 5 min pulse of ^{3}H-thymidine.

d. This would be difficult to do. If you waited a bit over 13 hours, you might catch a few cells that had just been labeled at the beginning of S. You would also pick up cells that had been labeled at the end of one S as well as the beginning of the next.

11.7 Karyotype analysis performed on cells cultured from an amniotic fluid sample reveals that the cells contain 47 chromosomes. The stained chromosomes are classified into groups and the arrangement shows 6 chromosomes in A, 4 in B, 16 in C, 6 in D, 6 in E, 4 in F, and 5 in the G group. Based on the above information:

a. What could be the genotype of the fetus? If more than one possibility exists, give all.

b. How would you proceed to distinguish between the possibilities?

Answer: Chromosomes are placed in groups according to their size (see text Figure 11.12, p. 330). The table below summarizes members of each group in a normal and the abnormal individual.

Group	Normal Chromosome Number	Normal Chromosome Type	Abnormal Chromosome Number
A	6	1, 2, 3	6
B	4	4, 5	4
C	15 or 16	6, 7, 8, 9, 10, 11, 12, X (XX in females)	**16**
D	6	13, 14, 15	6
E	6	16, 17, 18	6
F	4	18, 20	4
G	4 or 5	21, 22, (Y, in males)	**5**

Examination of which group has a potentially unusual number of chromosomes suggests either group C or G. The most likely explanations are that either the individual is XXY, or female and trisomy 21 (XX, 21, 21, 21). Other explanations are possible (e.g., XY, with trisomy for any of 6, 7, 8, 9, 10, 11, 12, or XX with trisomy for 22) but less likely as such aneuploids typically are spontaneously aborted before the stage (12 weeks) at which amniocentesis would be possible. To distinguish between these possibilities, use a staining technique that affords more resolution of the chromosomal banding patterns, such as Q- or G-banding (Figure 11.13, p. 331), and/or use chromosome-specific probes and perform FISH (Figure 11.14, p. 331) to determine which chromosomes are present.

11.8 What is the relationship between cellular DNA content and the structural or organizational complexity of the organism?

Answer: There is, paradoxically, no simple relationship between the haploid DNA content of a cell and its structural or organizational complexity. This is the "C-value paradox." See text Figure 11.15, p. 333 and Table 11.3, p. 332, and the discussion on those pages.

11.9 Match the DNA type with the chromatin type in the following table. (More than one DNA type may match a given chromatin type.)

DNA TYPE	CHROMATIN TYPE
Barr body (inactivated DNA)	Euchromatin
Centromere	Facultative heterochromatin
Telomere	Constitutive heterochromatin
Most expressed genes	

Answer: DNA type matched with chromatin type:

DNA TYPE	CHROMATIN TYPE
Barr body (inactivated DNA)	Facultative heterochromatin
Centromere	Constitutive heterochromatin
Telomere	Constitutive heterochromatin
Most expressed genes	Euchromatin

11.10 Eukaryotic chromosomes contain (choose the best answer):

a. protein.
b. DNA and protein.
c. DNA, RNA, histone, and nonhistone protein.
d. DNA, RNA, and histone.
e. DNA and histone.

Answer: c.

11.11 In a particular eukaryotic chromosome (choose the best answer):

a. Heterochromatin and euchromatin are regions where genes make functional gene products (i.e., where they are active).
b. Heterochromatin is active but euchromatin is inactive.
c. Heterochromatin is inactive but euchromatin is active.
d. Both regions are inactive.

Answer: c.

11.12 List four major features that distinguish eukaryotic chromosomes from prokaryotic chromosomes.

Answer: There are many differences. Here are five:

1. DNA of eukaryotic chromosomes is in linear form. Prokaryotic chromosomes, such as that of *E. coli,* can be closed circles.
2. The proteins that associate with the two kinds of chromosomes are different. Histones and non-histones are associated with eukaryotic chromosomes.

Prokaryotic chromosomes can be associated with a variety of proteins, but do not use histones.

3. Centromeres and telomeres are integral and essential parts of eukaryotic chromosomes. Centromeres are needed for appropriate segregation and telomeres are necessary to protect the ends from degradation. Prokaryotic chromosomes have neither.
4. The kinds and amounts of repetitive DNA sequences differ. Eukaryotic chromosomes can have highly repetitive and middle repetitive DNA, while in general, prokaryotic chromosomes have neither. Although there are often some genes that are present in more than one copy in prokaryotes (e.g., in *E. coli*, the genes for ribosomal RNA are present seven times), there is not significant repetitive DNA in prokaryotes.
5. Prokaryotic cells typically have a single chromosome, while eukaryotic cells typically have multiple chromosomes. They are also typically larger.

11.13 From the list below, identify

a. three features that both eukaryotic and bacterial chromosomes have in common.
b. four features that eukaryotic chromosomes have but which are not found in bacterial chromosomes.
c. one feature that bacterial chromosomes have but which are not found in eukaryotic chromosomes.

<u>Words for matching</u>

A.	centromeres	F.	nonhistone protein scaffolds
B.	hexose sugars	G.	DNA
C.	amino acids	H.	nucleosomes
D.	supercoiling	I.	circular chromosome
E.	telomeres	J.	looping

Answer: a. D (supercoiling), G (DNA), J (looping).

b. A (centromeres), E (telomeres), F (nonhistone protein scaffolds), H (nucleosomes).

c. I (circular chromosome).

11.14 A nucleosome core particle is composed of

a. Two molecules each of H2A, H2B, H3 and H4.
b. One molecule each of H1, H2A, H2B, H3, and H4.
c. Two molecules each of H1, H2, H3 and H4.

Answer: a.

11.15 Discuss the structure and role of nucleosomes.

Answer: Nucleosomes are the fundamental unit of DNA packaging in eukaryotic chromosomes. In a nucleosome, about 146 base pairs of DNA is wrapped one and three-quarters times around a protein octomer consisting of two copies of each of four histone proteins (H2A, H2B, H3, H4). This nucleosome core particle packs the DNA into a flattened disk about 5.7 x 11 nm. The double-stranded DNA is packed into nucleosomes at regular intervals, with about 80 base pairs between particles. This packing effectively condenses the DNA into a smaller size, and forms the basis for subsequent packaging by non-histone chromosomal proteins. The interactions between the histones and the DNA are not sequence specific, but rather are based on the basic, positively charged histone proteins interacting with the acidic, negatively charged sugar-phosphate backbone of the DNA. See text pp. 335-338.

11.16 Arrange the following figures in increasing order of eukaryotic chromosome condensation. Designate the simplest chromosome organization as number 1 and the most complex chromosome organization as number 6.

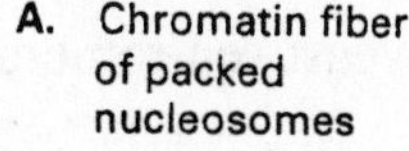

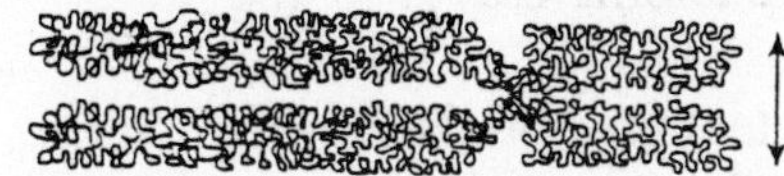

C. DNA double helix

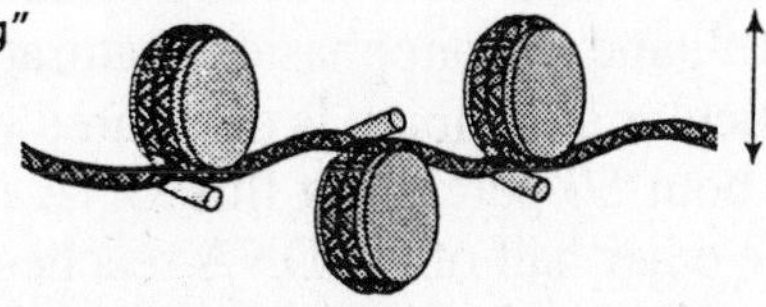

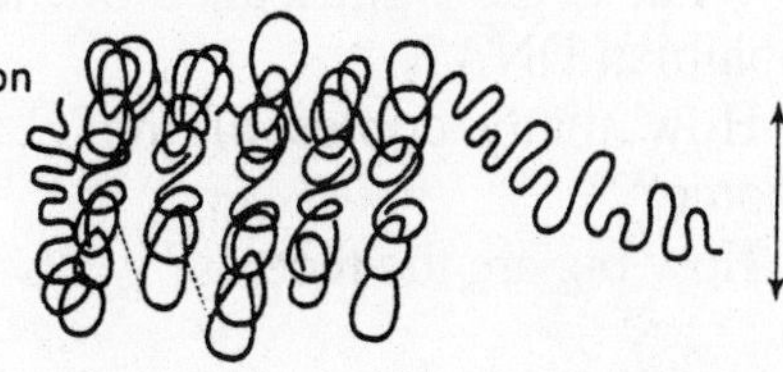

Answer: $1 = $ **c**, DNA double helix
↓
$2 = $ **e**, "Beads-on-a-string" form of chromatin
↓
$3 = $ **a**, chromatin fiber of packed nucleosomes
↓
$4 = $ **d**, extended section of chromosome
↓
$5 = $ **f**, condensed section of chromosome
↓
$6 = $ **b**, metaphase chromosome

11.17 What are the main molecular features of yeast centromeres?

Answer: See text pp. 339-340.

11.18 What are telomeres?

Answer: See text pp. 340-341.

11.19 Would you expect to find most protein coding genes in unique-sequence DNA, in moderately repetitive DNA, or in highly repetitive DNA?

Answer: In unique-sequence DNA. See text p. 344.

11.20 Would you expect to find ribosomal RNA genes in unique-sequence DNA, in moderately repetitive DNA, or in highly repetitive DNA?

Answer: Moderately repetitive DNA. See text pp. 344-345.

11.21 A particular virus has a genome consisting of 10^5 bp of double stranded DNA. When this DNA is denatured and renatured, it reaches 50 percent renaturation at a C_ot of 3×10^{-1}, and a monophasic renaturation curve is seen. In contrast, when the DNA of a particular amphibian is denatured and renatured, a biphasic renaturation curve is seen. About 50 percent of this DNA reaches 50 percent renaturation at a C_ot of 3, while the other half of the DNA reaches 50 percent renaturation at a C_ot of 3×10^4.

a. What is the size of the amphibian genome?
b. What is the significance of the biphasic renaturation curve shown by the amphibian DNA?
c. How many copies of the C_ot 3 species are present in the amphibian genome?
d. How big are the repeats?

Answer: Renaturation curves are very informative, but one must be quite careful in analyzing them. The answers that are requested here are not plainly obvious from a casual inspection of the data. Therefore, before trying to answer the individual questions, thoroughly analyze the data that is given.

One is told that the amphibian DNA has a biphasic renaturation curve, with 50 percent of the amphibian DNA reaching 50 percent renaturation at a C_ot of 3, while the other 50 percent of the DNA reaches 50 percent renaturation at a C_ot of

3×10^4. The biphasic nature of the renaturation curve means that the amphibian DNA has two components: a repetitive component and a unique component. The amounts of DNA in each component give an indication of how much of the amphibian genome is repetitive, and how much is unique: half of the DNA is repetitive, and half is unique.

The half of the DNA that is repetitive consists of DNA that is a set of sequences repeated many times. If one just considered the repeated sequences themselves, they would have a certain complexity. That is, they would consist of a number of individual DNA sequences of a specified number of base pairs having a certain sequence. Their complexity is far less than the complexity of the fraction of unique DNA, where all of the sequences are unique.

The relationship between the complexity of the repetitive and unique fractions is given by the $C_ot_{1/2}$ values, which are the 50 percent renaturation values for each component. Hence, the repetitive component of the amphibian genome has a $C_ot_{1/2}$ value of 3, while the unique component has a $C_ot_{1/2}$ value of 3×10^4. The $C_ot_{1/2}$ value is related directly to the complexity of the DNA being examined. As the complexity increases, the $C_ot_{1/2}$ value increases: a 10-fold larger $C_ot_{1/2}$ value corresponds to a 10-fold increase in complexity.

The $C_ot_{1/2}$ value for the viral genome provides a standard to assess the complexity of each component of the amphibian genome. The viral genome shows a monophasic renaturation curve, indicating that it consists of entirely unique sequences. Since its size is known to be 10^5 bp, one can infer that a $C_ot_{1/2}$ value of 0.3 corresponds to a complexity of 10^5 bp. Therefore, the $C_ot_{1/2}$ value of 3 for the repetitive fraction of the amphibian genome corresponds to a complexity of 10^6 bp, and the $C_ot_{1/2}$ value of 3×10^4 for the unique portion of the amphibian genome corresponds to a complexity of 10^{10} bp.

Since the amphibian genome consists of half repetitive and half unique DNA, and the unique portion has a complexity of 10^{10} bp, the repetitive portion must also have 10^{10} bp. Since the repetitive portion has a complexity of 10^6 bp, there must be $10^{10}/10^6 = 10^4$ repeats.

a. 2×10^{10} bp.
b. The biphasic curve indicates two components with different complexities. One component consists entirely of unique sequence DNA, while the other consists of **sequences** that are repeated many times.
c. 10^4.
d. 10^6 bp.

The relationship between the complexity of the repetitive and unique fractions [illegible] 50 percent [illegible] repetitive component [illegible] $C_0t_{1/2}$ value of [illegible]. The $C_0t_{1/2}$ value is related directly to the complexity of the DNA [illegible] the $C_0t_{1/2}$ value increases [illegible] 10-fold increase in complexity.

The $C_0t_{1/2}$ value for the [illegible] genome [illegible] complexity of each component of the amphibian genome. [illegible] consists of entirely unique sequences. [illegible] $C_0t_{1/2}$ value of [illegible]. Therefore, the $C_0t_{1/2}$ value of 3 for the repetitive fraction [illegible] the $C_0t_{1/2}$ value [illegible] of the amphibian genome [illegible].

CHAPTER 12

DNA REPLICATION AND RECOMBINATION

I. CHAPTER OUTLINE

DNA REPLICATION IN PROKARYOTES
- Early Models for DNA Replication
- The Meselson-Stahl Experiment
- Enzymes Involved in DNA Synthesis
- Molecular Details of DNA Replication in Prokaryotes

DNA REPLICATION IN EUKARYOTES
- Molecular Details of DNA Synthesis in Eukaryotes
- Replicating the Ends of Chromosomes
- Assembly of New DNA into Nucleosomes
- Genetics of the Eukaryotic Cell Cycle

DNA RECOMBINATION
- Crossing-over: Breakage and Rejoining of DNA
- The Holliday Model for Recombination
- Gene Conversion and Mismatch Repair

II. REVIEW OF KEY TERMS, SYMBOLS AND CONCEPTS

Without consulting the text and in your own words, write a brief definition of each term in the groups below. Then, either using a short phrase or a simple diagram, identify the relationship(s) between specific pairs of terms within a set. Finally, consult the text (and perhaps a friend who has also done the exercise) to check your answers.

1	2	3	4
semiconservative DNA replication	DNA polymerase I, II, III	origin of replication	semidiscontinuous DNA replication
conservative DNA replication	dNTP, dNMP	DNA helicase	replication fork
dispersive DNA replication	Kornberg enzyme	DNA primase	Okazaki fragments
Meselson-Stahl Experiment	*polA1* mutant	primosome	bidirectional replication
CsCl equilibrium gradient	conditional mutant	RNA primer	leading strand
^{15}N, ^{14}N	3'-to-5' exonuclease activity	SSB protein	lagging strand
buoyant density	5'-to-3' exonuclease activity	replication bubble	5'-to-3' synthesis
satellite DNA	proofreading activity	replication fork	
	5'-to-3' synthetic activity		

5	6	7	8
Okazaki fragments	rolling circle model	harlequin chromosomes	recombination
DNA polymerase I, III	twisting problem	BUdR	Holliday Model
RNA primer	supercoiled DNA	base analog	recognition & alignment
DNA ligase	semidiscontinuous-DNA replication	semi-conservative DNA replication	Holliday intermediate
5'-to-3' exonuclease activity		replicon	heteroduplex
lagging strand elongation		ARS element	branch migration
		telomerase	cleavage & ligation
			patched duplex
			spliced duplex

9
3:1 tetrad
gene conversion
mismatch repair
co-conversion

III. THINKING ANALYTICALLY

The material that is presented in this chapter requires very close attention to detail. As you read the chapter, make frequent reference to the text figures. Gain a clear comprehension of each section before going on to the next. After each section, try closing the text, and sketching out and summarizing the aspect of replication or recombination that has just been presented. By reproducing the text figures and diagrams, you will gain insight into the details of the processes, and force yourself to confront unclear concepts.

Sketching may not be enough to grasp the Holliday model for recombination. Try constructing a Holliday intermediate using different colored yarns. Rotate the strands to discover the consequences of branch migration. Visualizing this three-dimensional process is easy for some, but for many, quite challenging. If it is challenging for you, take the time to construct a model. It will pay off.

IV. QUESTIONS FOR PRACTICE

A. Multiple Choice Questions

Choose the correct answer or answers for the following questions.

1. Consider the Meselson and Stahl experiment where *E. coli* were grown in ^{15}N-media. Which of the following models of DNA replication is eliminated by the results of analyzing DNA after exactly one round of replication in ^{14}N-media?
   a. semi-conservative
   b. dispersive
   c. conservative

2. The enzymes that are most directly concerned with catalyzing the synthesis of DNA are the DNA
   a. ligases.
   b. exonucleases.
   c. polymerases.
   d. primases.

3. Which of the following is not essential for the *in vitro* synthesis of DNA?
   a. magnesium ions
   b. DNA polymerase I
   c. DNA primase
   d. a DNA fragment

4. As the parent strands of DNA separate during replication, they must untwist. The untwisting process is catalyzed by which enzyme?
   a. DNA primase
   b. DNA helicase
   c. DNA gyrase
   d. DNA ligase

5. Which of the following *do not provide evidence for* semi-conservative DNA replication?
   a. The Meselson-Stahl experiment
   b. Harlequin chromosomes
   c. The existence of ARS elements.
   d. Okazaki fragments

6. Which of the following *E. coli* polymerases have proofreading activity?
   a. DNA polymerase I
   b. DNA polymerase II
   c. DNA primase
   d. DNA polymerase III
   e. RNA polymerase

7. DNA replication in certain viruses, such as the circular ϕX174, and in the *E. coli F*-factor during conjugation is achieved

a. using a rolling circle model and semidiscontinuous DNA replication.
b. using a rolling circle model and continuous DNA replication.
c. without using an origin of replication.
d. using DNA fragmentation.

8. The first stage of genetic recombination according to the Holliday model is
   a. the first gap phase (G_1).
   b. chromosome attachment.
   c. branch migration.
   d. recognition and alignment.
   e. strand invasion.

9. Mismatch repair of heteroduplex DNA can result in
   a. restoration of allelic segregation.
   b. nullification of allelic segregation.
   c. a non-Mendelian segregation of alleles.
   d. genetic diversity.
   e. gene conversion.

10. During chromosome replication, new histones
   a. are synthesized during G_1 and S, and assembled into nucleosomes that are distributed conservatively.
   b. are synthesized during G_1 and S, and assembled into nucleosomes that are distributed semi-conservatively.
   c. are synthesized during G_1, S and G_2, and assembled into nucleosomes that are distributed conservatively.
   d. are synthesized during G_1, S and G_2, and assembled into nucleosomes that are distributed semi-conservatively.

Answers: 1c; 2c; 3c; 4b; 5c; 6a, b & d; 7a; 8d; 9c & e; 10b

B. Thought Questions

1. Describe the mechanism of semidiscontinuous DNA replication.

2. Does the fact that DNA replication is semidiscontinuous determine that it must replicate semi-conservatively? Does the fact that it replicates semi-conservatively exclude the possibility of replication being continuous on both strands?

3. In his investigation of the requirements for *in vitro* synthesis of DNA, Kornberg found that an absolute minimum of four components was necessary. What were these components, and why in particular were each and all necessary? Also, what limitations did his *in vitro* method have as compared to *in vivo* synthesis, which involves other components, such as helicase, primase and ligase?

4. Describe how eukaryotic DNA is assembled into nucleosomes after DNA replication.

5. How is gene conversion related to mismatch repair of heteroduplex DNA?

6. In *Neurospora*, gene conversion events can be seen in a 6:2 (= 3:1) spore ratio in the 8-spored ascus. How would you explain a 5:3 or a 7:1 spore ratio? (Hint: What would you expect to see if only one member of a pair of mitotic products immediately following meiosis were converted?)

7. What phenotype would you expect to see in a mutant in the *E. coli polA* gene that showed heat-sensitive 3'-to-5' exonuclease activity, but normal 5'-to-3' exonuclease activity at all temperatures?

V. SOLUTIONS TO TEXT PROBLEMS

12.1 Compare and contrast the conservative and semiconservative models for DNA replication.

Answer: In semiconservative DNA replication, the two strands of the double helix separate and each serves as a template for new DNA synthesis. Replication is "semiconservative" because only one strand is retained from the parental molecule when daughter molecules are produced. Each daughter double helix consists of one parental and one newly synthesized strand. In a conservative model of DNA replication, an entirely new double helix is synthesized while leaving the parental molecule intact, or "conserved", into the next generation. The result of DNA replication is two helices, one of which is the original parental double helix, and the other of which consists of two new strands.

12.2 Describe the Meselson and Stahl experiment, and explain how it showed that DNA replication is semiconservative.

Answer: By growing *E. coli* for many generations on media containing "heavy" nitrogen, or ^{15}N, bacterial DNA was uniformly labeled with heavy nitrogen. Such ^{15}N-DNA has a greater buoyant density than ^{14}N-DNA, and can be differentiated from ^{14}N-DNA by its banding position in a CsCl density gradient. Semi-conservative DNA replication was demonstrated by placing *E. coli* with ^{15}N-DNA into ^{14}N-media, and following the density of the DNA that was present in cells after growth proceeded through each of several successive generations. If DNA replication were conservative, after one cell division and one round of DNA replication, one would expect to see two distinct, equally dense bands, one corresponding to ^{14}N-DNA, and one corresponding to ^{15}N-DNA. This was not seen. If DNA replication were either semi-conservative or dispersive, after one cell division and one round of DNA replication, one would expect to see one band, consisting of DNA of density half-way between ^{14}N-DNA and ^{15}N-DNA. This was seen. To further distinguish between semi-conservative and dispersive replication, the consequences of another round of cell division and DNA replication were examined. In semi-conservative replication, one would expect two bands, one of density half-way between ^{14}N-DNA and ^{15}N-DNA, and one having ^{14}N-DNA density. In dispersive replication, only DNA with both ^{14}N-DNA and ^{15}N-DNA would be seen (although in two different proportions). The density of the two bands seen (one at the level of ^{14}N-DNA and one at the level of half ^{14}N-DNA and half ^{15}N-DNA) supported the semiconservative model. See text pp. 353-256, particularly Figure 12.2.

12.3 In the Meselson and Stahl experiment ^{15}N-labeled cells were shifted to ^{14}N medium, at what we can designate as generation 0.

a. For the semiconservative model of replication, what proportion of ^{15}N-^{15}N, ^{15}N-^{14}N, and ^{14}N-^{14}N would you expect to find at generations 1, 2, 3, 4, 6, and 8?

b. Answer the above question in terms of the conservative model of DNA replication.

Answer:

GENERATION	SEMI-CONSERVATIVE MODEL ^{15}N-^{15}N	^{15}N-^{14}N	^{14}N-^{14}N	CONSERVATIVE MODEL ^{15}N-^{15}N	^{15}N-^{14}N	^{14}N-^{14}N
1	0	1	0	1/2	0	1/2
2	0	1/2	1/2	1/4	0	3/4
3	0	1/4	3/4	1/8	0	7/8
4	0	1/8	7/8	1/16	0	15/16
6	0	1/32	31/32	1/64	0	63/64
8	0	1/128	127/128	1/256	0	255/256

12.4 Suppose *E. coli* cells are grown on an ^{15}N medium for many generations. Then they are quickly shifted to an ^{14}N medium, and DNA is extracted from the samples taken after one, two, and three generations. The extracted DNA is subjected to equilibrium density gradient centrifugation in CsCl. In the figure below, using the reference positions of pure ^{15}N and pure ^{14}N DNA as guides, indicate where the bands of DNA would equilibrate if replication were semiconservative or conservative.

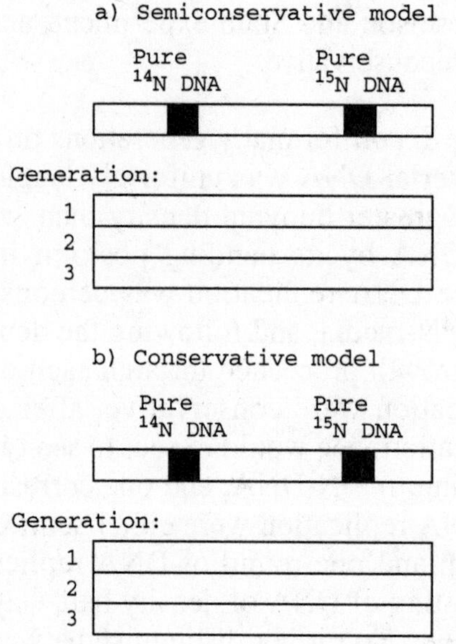

Answer:

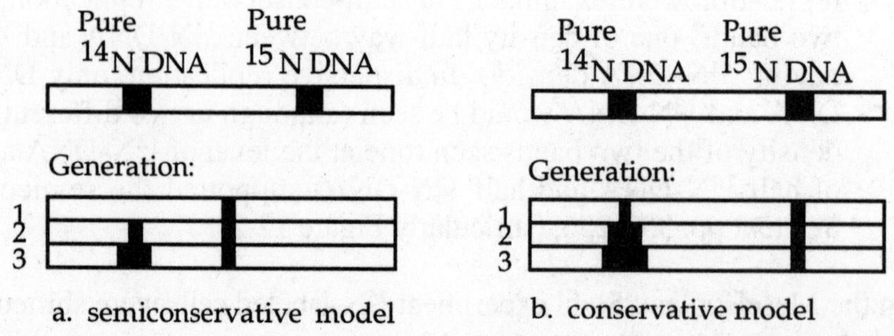

a. semiconservative model

b. conservative model

12.5 A spaceship lands on earth and with it a sample of extraterrestrial bacteria. You are assigned the task of determining the mechanism of DNA replication in this organism.

You grow the bacteria in unlabeled medium for several generations, then grow it in presence of ^{15}N exactly for one generation. You extract the DNA and subject it to CsCl centrifugation. The banding pattern you find is shown in the figure below.

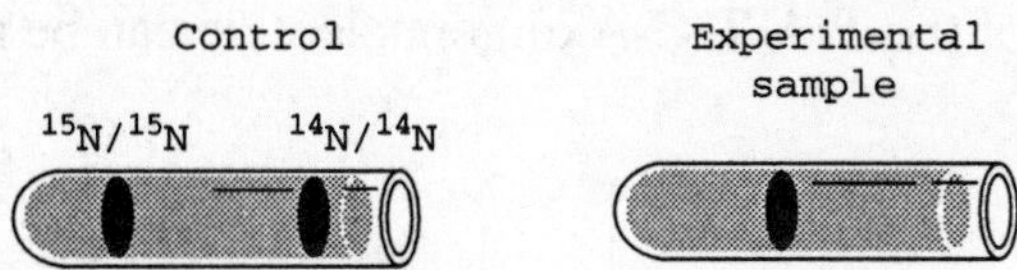

It appears to you that this is evidence that DNA replicates in the semiconservative manner, but this result does not prove that this is so. Why? What other experiment could you perform (using the same sample and technique of CsCl centrifugation) that would further distinguish between semiconservative and dispersive modes of replication?

Answer: The CsCl centrifugation result eliminates the possibility of the conservative model of replication, but is still consistent with either semiconservative or dispersive models of DNA replication. To distinguish between these two possibilities using the same sample and the technique of CsCl centrifugation, one could denature the DNA, and then subject the single-stranded sample to CsCl centrifugation. This could be done in practice by using an alkaline CsCl gradient, as the two DNA strands will denature at high pH. The expected results are shown below.

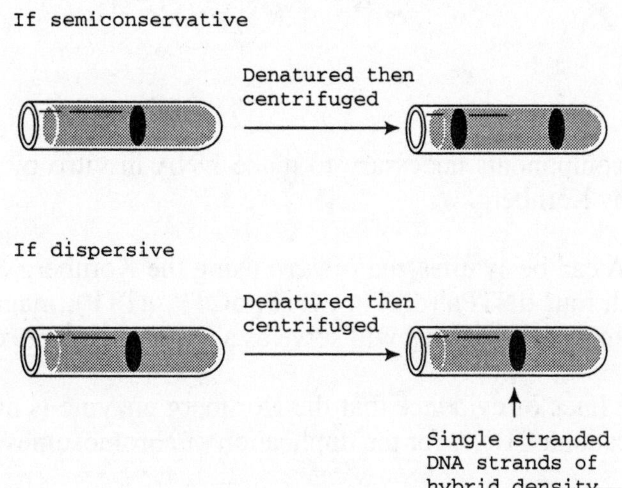

12.6 Assume you have a DNA molecule with the base sequence T-A-T-C-A going from the 5' to the 3' end of one of the polynucleotide chains. The building blocks of the DNA are drawn as in the following figure. Use this shorthand system to diagram the completed double-stranded DNA molecule, as proposed by Watson and Crick.

A G C T

PPP OH PPP OH PPP OH PPP OH

Answer: The deoxyribonucleotides have a triphosphate group at the 5' end of the molecule, and a hydroxyl group at the 3' end of the molecule. This information establishes the polarity of each of the deoxyribonucleotides (dATP, dGTP, dTTP and dCTP). Since the phosphodiester bond will form between the 5' and 3' ends, the 5'-to-3' bonding of the T-A-T-C-A oligonucleotide can be represented as:

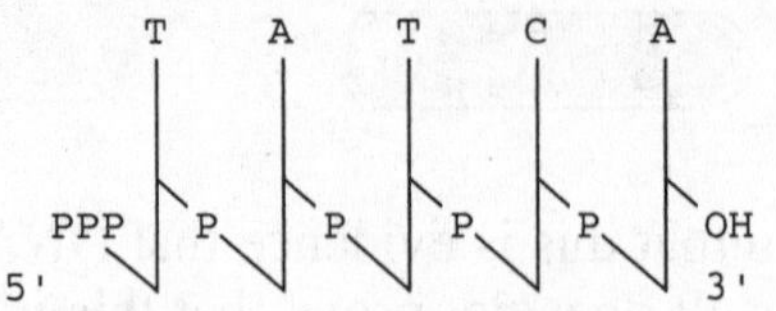

```
        T     A     T     C     A

PPP    P     P     P     P     OH
5'                               3'
```

The complementary strand will base-pair with antiparallel orientation:

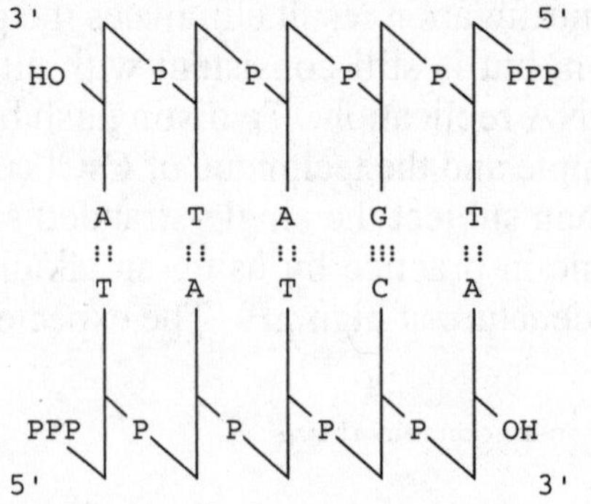

```
3'                               5'
HO     P     P     P     P     PPP

        A     T     A     G     T
        ::    ::    ::    :::   ::
        T     A     T     C     A

PPP    P     P     P     P     OH
5'                               3'
```

12.7 List the components necessary to make DNA in vitro by using the enzyme system isolated by Kornberg.

Answer: DNA can be synthesized *in vitro* using the Kornberg enzyme (DNA polymerase I), all four dNTPs (dATP, dGTP, dCTP, dTTP), magnesium ions (Mg^{2+}) and a fragment of DNA that will serve as a template. See text pp. 356-357.

12.8 Give two lines of evidence that the Kornberg enzyme is not the enzyme involved in the replication of DNA for the duplication of chromosomes in growth of *E. coli.*

Answer: Evidence that the Kornberg enzyme is not the enzyme involved in the replication of DNA came from an analysis of the growth and biochemical phenotypes of mutations in the *E. coli* gene encoding DNA polymerase I, *polA1*. Mutants lacking *polA1* function are viable, suggesting that DNA polymerase I is not essential to cell function. Furthermore, conditional mutations in *polA1*, in which DNA polymerase I activity is temperature-sensitive, multiply at the restrictive temperature of 42°C at same rate as strains with wild-type *E. coli.* Additional biochemical analyses using cell extracts of these mutants showed that they had almost no detectable DNA polymerase activity, even though DNA synthesis occurred at a normal rate. Thus, the ability of *polA1* mutants to grow, and the biochemical analysis of them indicated that there must be an additional enzyme or enzymes that are used for the duplication of chromosomes. See text pp. 357-358.

12.9 Kornberg isolated DNA polymerase I from *E. coli.* DNA polymerase I has an essential function in DNA replication. Which of the following is that function?

a. Filling gaps left by the removal of RNA primer
b. Filling in gaps where introns are removed
c. The formation of stem loops in tRNA
d. Recognition of rho factor for the initiation of transcription
e. Production of poly(A) tails on eukaryotic mRNAs

Answer: a. When DNA polymerase III reaches an RNA primer and dissociates from the DNA, DNA polymerase I continues to synthesize the DNA in a 5'-to-3' direction. It simultaneously removes the RNA primer using its 5'-to-3' exonuclease activity, and replaces the RNA with DNA nucleotides. See text pp. 363-364.

12.10 Base analogs are compounds that resemble the natural bases found in DNA and RNA, but are not normally found in those macromolecules. Base analogs can replace their normal counterparts in DNA during in vitro DNA synthesis. Four base analogs were studied for their effects on in vitro DNA synthesis using the *E. coli* DNA polymerase. The results were as follows, with the amounts of DNA synthesized expressed as percentages of that synthesized from normal bases only.

	NORMAL BASES SUBSTITUTED BY THE ANALOG			
ANALOG	A	T	C	G
A	0	0	0	25
B	0	54	0	0
C	0	0	100	0
D	0	97	0	0

Which bases are analogs of adenine? of thymine? of cytosine? of guanine?

Answer: None are analogs for adenine, B and C are analogs of thymine, C is an analog of cytosine, and A is an analog of guanine.

12.11 Describe the semidiscontinuous model for DNA replication. What is the evidence showing that DNA synthesis is discontinuous on at least one template strand?

Answer: The semidiscontinuous model of DNA replication is best described using both diagrams and words, especially as is shown in text Figures 12.6 (p. 363) and 12.10 (p. 366). The main features of the model are that

i. DNA replication proceeds only in the 5'-to-3' direction.
ii. Because DNA replication proceeds only in one direction, it is continuous on the 3'-to-5' template strand (the leading strand), while it is discontinuous on the 5'-to-3' template strand (the lagging strand). This is why it is *semi*discontinuous.
iii. Replication initiates at a replication origin, where a complex of proteins and enzymes forms at a replication fork. Initially, helicase untwist the DNA and SSB (single-stranded binding proteins) bind the single-stranded DNA to protect it from nuclease attack and keep it unwound.
iv. Since DNA polymerase can only polymerize DNA if a 3'-OH is provided, an RNA primer must be synthesized by a primase-enzyme complex (the primosome). The primosome binds to the single-stranded DNA and synthesizes a short RNA primer in the 5'-to-3' direction.

v. Once an RNA primer is synthesized, DNA polymerase III (a complex of proteins) adds onto the 3'-OH, using the unwound DNA as a template and dNTPs to synthesize DNA in the 5'-to-3' direction. The newly synthesized DNA is thus synthesized onto an RNA primer. This is referred to as an Okazaki fragment. On the lagging strand, synthesis moves "away" from the fork, and synthesis must be reinitiated frequently. On the leading strand, synthesis proceeds "into" the fork. The proteins involved in replication are associated into a replisome, so that the DNA polymerase III of the lagging strand is complexed with DNA polymerase III on the leading strand. The replisome that forms at a replication fork moves as a unit along the DNA, enabling new DNA to be synthesized efficiently on both the leading and lagging strands. See Figure 12.10, p. 366.

vi. On the lagging strand, DNA polymerase III will (after about 1,000 bp) encounter the 5' end of an RNA primer of an earlier synthesized Okazaki fragment. It then dissociates, and is replaced by DNA polymerase I, which uses its 5'-to-3' exonuclease activity to excise the RNA primer and replace it with DNA.

vii. Once the RNA primer is replaced with DNA, the gap between adjacent DNA fragments is sealed by DNA ligase.

The primary evidence that DNA synthesis is discontinuous on at least one strand is the existence of Okazaki fragments, short segments of DNA that have short RNA primers on their 5' ends.

12.12 Distinguish between a primer strand and a template strand.

Answer: A primer strand is a nucleic acid sequence that is extended at the 3'-end by 5'-to-3' DNA synthesis. A template strand directs the base sequence of the DNA strand being made (an A on the template results in the insertion of a T in the new chain). The template strand has antiparallel orientation (3'-to-5') to the newly synthesized strand.

12.13 The length of the *E. coli* chromosome is about 1,100 μm.

a. How many base pairs does the *E. coli* chromosome have?

b. How many complete turns of the helix does this chromosome have?

c. If this chromosome replicated unidirectionally and if it completed one round of replication in 60 minutes, how many revolutions per minute would the chromosome be turning during the replication process?

d. The *E. coli* chromosome, like many others, replicates bidirectionally. Draw a simple diagram of a replicating *E. coli* chromosome that is halfway through the round of replication. Be sure to distinguish new and old DNA strands.

Answer: a. There are 0.34 nm between base pairs, so there will be

$$1{,}100\ \mu\text{m} \times \frac{1000\ \text{nm}}{1\ \mu\text{m}} \times \frac{1\ \text{bp}}{0.34\ \text{nm}} = 3.24 \times 10^6\ \text{bp}$$

b. There are 10 base pairs per turn in a B-form DNA helix. There will be $(3.24 \times 10^6$ bp$)/(10$ bp/turn$) = 3.24 \times 10^5$ turns.

c. If replication is unidirectional in 60 minutes, one has: $(3.24 \times 10^5$ turns$)/(60$ minutes$) = 5{,}400$ turns/minute.

d. If a circular chromosome replicates bidirectionally starting from a single origin, a theta (θ) structure is formed. In the figure below, the dotted lines represent newly synthesized DNA, the intact lines parental DNA, and, O, the origin of replication.

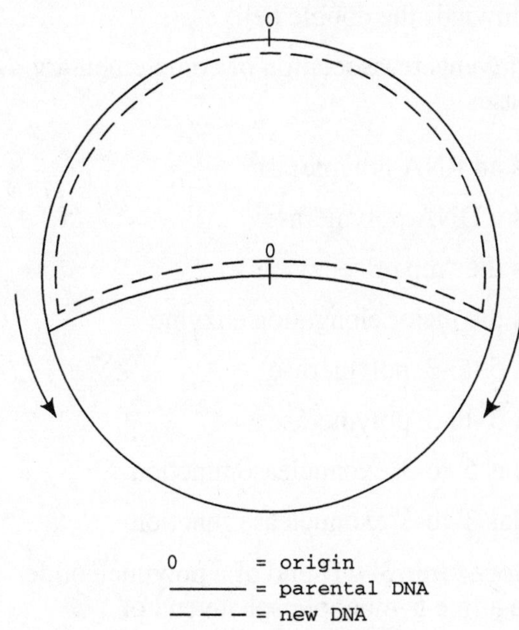

12.14 In *E. coli* the replication fork moves forward at 500 nucleotide pairs per second. How fast is the DNA ahead of the replication fork rotating?

Answer: Since there are 10 bp per turn, one has

$$\frac{500 \text{ bp}}{\text{second}} \times \frac{1 \text{ turn}}{10 \text{ bp}} \times \frac{60 \text{ seconds}}{1 \text{ minute}} = \frac{300 \text{ revolutions}}{\text{minute}}$$

12.15 A diploid organism has 4.5×10^8 base pairs in its DNA. This DNA is replicated in 3 minutes. Assuming all replication forks move at a rate of 10^4 base pairs per minute, how many replicons (replication units) are present in this organism's genome?

Answer: If there were only a single replicon, the time it would take to replicate its DNA

$$= 4.5 \times 10^8 \text{ bp} \times \frac{1 \text{ minute}}{10^4 \text{ bp}} = 4.5 \times 10^4 \text{ minutes.}$$

Since it only takes 3 minutes, there must be $[(4.5 \times 10^4)/3] = 1.5 \times 10^4$ replicons. If one reads the statement of the problem to mean that there are 4.5×10^8 base pairs in the entire diploid genome, then this is the number of replicons in the diploid genome. The number of replicons in a haploid complement would be $(1.5 \times 10^4)/2 = 7.5 \times 10^3$.

12.16 The following events, steps or reactions occur during *E. coli* DNA replication. For each entry in Column A, select the appropriate entry in Column B. Each entry in A may have more than one answer, and each entry in B can be used more than once.

		COLUMN A		COLUMN B
____	a.	Unwinds the double helix	A	Polymerase I
____	b.	Prevents reassociation of complementary bases	B	Polymerase III
			C	Helicase
____	c.	Is an RNA polymerase	D	Primase
____	d.	Is a DNA polymerase	E	Ligase
____	e.	Is the "repair" enzyme	F	SSB protein
____	f.	Is the major elongation enzyme	G	Gyrase
____	g.	A 5'-to-3' polymerase	H	None of these
____	h.	A 3'-to-5' polymerase		
____	i.	Has 5'-to-3' exonuclease function		
____	j.	Has 3'-to-5' exonuclease function		
____	k.	Bonds free 3'-OH end of a polynucleotide to a free 5' monophosphate end of polynucleotide		
____	l.	Bonds 3'-OH end of a polynucleotide to a free 5' nucleotide triphosphate		
____	m.	Separates daughter molecules and causes supercoiling		

Answer: a. C
b. F
c. D
d. A, B
e. A
f. B
g. A, B, D
h. H
i. A, B
j. A, B
k. E
l. A, B, (D, after the first two bases of the RNA primer are positioned)
m. G

12.17 Compare and contrast the three *E. coli* DNA polymerases with respect to their enzymatic activities.

Answer: All three *E. coli* DNA polymerases can synthesize DNA using a template and a primed strand, which provide a free 3'-OH on which to add bases in the 5'-to-3' direction. All three also have proofreading activity, as they have 3'-to-5' exonuclease activity. However, only DNA polymerase I has 5'-to-3' exonuclease activity, the activity needed to excise the RNA primer at the 5' end of Okazaki fragments. See text pp. 356-359 and Table 12.1 on p. 359.

12.18 In *E. coli*, distinguish between the activities of primase; single-stranded binding protein; helicase; DNA ligase; DNA polymerase I; and DNA polymerase III in DNA replication.

Answer:

PROTEIN/ENZYME	ACTIVITY
primase	Primase synthesizes short RNA primer in the 5'-to-3' direction, providing a free 3'-OH to which DNA will be added.
single-stranded binding protein	SSBs attach to the single-stranded DNA that is formed after helicase unwinds the DNA at the replication fork. It keeps the DNA from reassociating with its complementary strand, allowing the primosome access to single-stranded DNA.
helicase	Helicase unwinds the DNA at the replication fork.
DNA ligase	Ligase seals the gaps left between the DNA portion of Okazaki fragments after the RNA primer is removed.
DNA polymerase I	Pol I excises the RNA primer at the 5'-end of the Okazaki fragment, replacing it with DNA.
DNA polymerase III	Pol III adds DNA onto the RNA primer in a 5'-to-3' direction. It is the major synthetic enzyme.

12.19 Describe the molecular action of the enzyme DNA ligase. What properties would you expect an *E. coli* cell to have if it had a temperature-sensitive mutation in the gene for DNA ligase?

Answer: DNA ligase catalyzes the formation of a phosphodiester bond between the 3'-OH and the 5'-phosphate groups on either side of a gap, sealing the gap. See Figure 12.9, p. 365. Temperature-sensitive ligase mutants would be unable to seal such gaps at the restrictive (high) temperature, leading to fragmented lagging strands, and presumably cell death. If a biochemical analysis were to be performed on the DNA synthesized after *E. coli* were shifted to a restrictive temperature, one should see an accumulation of DNA fragments the size of Okazaki fragments. This would provide additional evidence that DNA replication must be discontinuous on one strand.

12.20 Chromosome replication in *E. coli* commences from a constant point, called the origin of replication. It is known that DNA replication is bidirectional. Devise a biochemical experiment to prove that the *E. coli* chromosome replicates bidirectionally. (Hint: Assume that the amount of gene product is directly proportional to the number of genes.)

Answer: Assume the amount of a gene's product is directly proportional to the number of copies of the gene present in the *E. coli* cell. Assay the enzymatic activity of genes positioned at various positions in the *E. coli* chromosome during the replication period. One should observe that some genes (those immediately adjacent to the origin) will double their activity very shortly after replication begins. One can relate the map position of genes having doubled activity to the amount of time that has transpired since replication was initiated. If replication is bidirectional, there should be a doubling of the gene products both clockwise and counterclockwise from the origin.

12.21 What property of DNA replication was indicated by the presence of Okazaki fragments?

Answer: Okazaki fragments indicate that DNA replication occurs in a discontinuous fashion.

12.22 A space probe returns from Jupiter and brings with it a new microorganism for study. It has double-stranded DNA as its genetic material. However, studies of replication of the alien DNA reveal that, while the process is semiconservative, DNA synthesis is continuous on both the leading-strand and the lagging-strand templates. What conclusion(s) can you make from that result?

Answer: Clearly DNA replication in the Jovian bug does not occur as it does in *E. coli*. Two explanations may be offered. First, if the double-stranded DNA is antiparallel as it is in *E. coli*, the Jovian DNA polymerases would appear to be able to synthesize DNA in both the 5'-to-3' direction (on the leading strand) as well as in the 3'-to-5' direction (on the lagging strand). This would be unlike any DNA polymerase on earth. Second, if the double-stranded DNA is not organized in an antiparallel, but in a parallel, fashion, then the DNA could still be synthesized in only a 5'-to-3' direction on both template strands as the replication fork moves.

12.23 Compare and contrast eukaryotic and prokaryotic DNA polymerases.

Answer: See text Table 12.1, p. 359, and Table 12.4, p. 374.

12.24 Draw a eukaryotic chromosome as it would appear at each of the following cell cycle stages. Show both DNA strands, and use different line styles for old and newly synthesized DNA.

a. G_1
b. anaphase of mitosis
c. G_2
d. anaphase of meiosis I
e. anaphase of meiosis II

Answer:

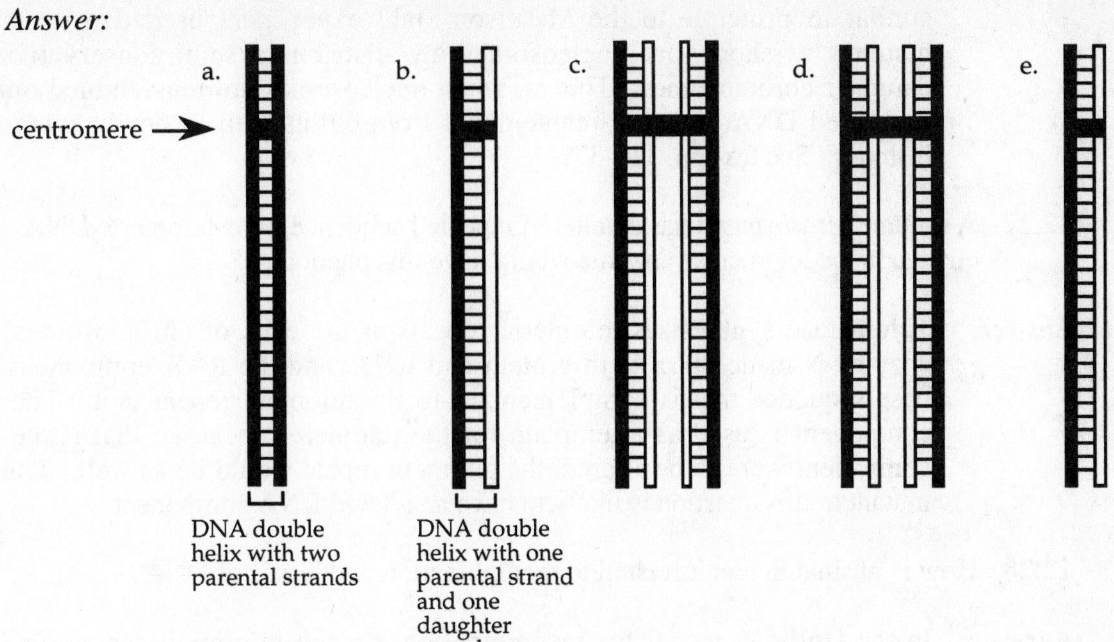

12.25 Autoradiography is a technique which allows radioactive areas of chromosomes to be observed under the microscope. The slide is covered with a photographic emulsion, which is exposed by radioactive decay. In regions of exposure the emulsion forms silver grains upon being developed. The tiny silver grains can be seen on top of the (much larger) chromosomes. Devise a method for finding out which regions in the human karyotype replicate during the last 30 min of the S period. (Assume a cell cycle in which the cell spends 10 h in G_1, 9 h in S, 4 h in G_2 and 1 h in M.)

Answer: If cells spend 4 hours in G_2, then there will be 4.5 to 5 hours between the last 30 minutes of S and metaphase in M. To see which chromosomal regions replicate during the last 30 minutes of S, add ^{3}H-thymidine for a 5 minute pulse, chase it with (cold, unlabeled) thymidine, wait 4.5 - 5 hours and prepare a spread of metaphase chromosomes on a glass slide. (Cells can be blocked at metaphase by adding colchicine, a drug that blocks the separation of the chromatids by interfering with microtubule function.) Immerse the slide in photographic emulsion, and after a suitable exposure period, develop the exposed silver grains and counterstain the chromosomes. Examine which regions of the chromosomes have silver grains lying over them to infer where ^{3}H-thymidine was incorporated.

12.26 When the eukaryotic chromosome duplicates, the nucleosome structures must duplicate. Discuss the synthesis of histones in the cell cycle, and discuss the model for the assembly of new nucleosomes at the replication forks.

Answer: New histones are synthesized to complex with newly synthesized DNA. Histone genes are transcribed starting near the end of G_1, just prior to S. The mRNAs that are produced are translated throughout S, and when DNA synthesis terminates at the end of S, histone transcription ceases, and histone mRNAs are degraded.

Nucleosomes assemble and disassemble during replication. There is a 200-300 bp zone around the replication fork free of nucleosomes. Experiments similar in principle to the Meselson-Stahl experiment used density-labeled histones to show that nucleosomes are distributed semi-conservatively to daughter chromosomes. That is, when nucleosomes are reassembled onto the replicated DNA, they are reassembled from old as well as newly synthesized histones. See text pp. 376-377.

12.27 A mutant *Tetrahymena* has an altered repeated sequence in its telomeric DNA. What change in the telomerase enzyme would have this phenotype?

Answer: Telomerase synthesizes telomeric repeats at the ends of chromosomes. The enzyme is made up of both protein and RNA, and the RNA component has a base sequence that is complementary to the telomere repeat unit. The RNA component is used as a template for the telomere repeat, so that if the RNA component were to be altered, the telomere repeat would be as well. Thus, the mutant in this question is likely to have an altered RNA component.

12.28 How is mismatch repair related to recombination and repair of DNA?

Answer: In the Holliday model for recombination, branch migration can result in the generation of heteroduplexes that contain mismatched base pairs subject to repair. The mismatched sequences are recognized and removed using an exonuclease that excises a segment of one DNA strand, a DNA polymerase to replace the excised strand with newly synthesized DNA, and DNA ligase to seal the gap between the new DNA and the preexisting DNA. See text pp. 379-382, especially Figures 12.21 and 12.23.

12.29 Crosses were made between strains, each of which carried one of three different alleles of the same gene, *a*, in yeast. For each cross, some unusual tetrads resulted at low frequencies. Explain the origin of each of these tetrads:

Cross:	$a1\ a2^+$	$a1\ a3^+$	$a2\ a3^+$
	x	x	x
	$a1^+\ a2$	$a1^+\ a3$	$a2^+\ a3$
Tetrads:	$a1^+\ a2$	$a1^+\ a3$	$a2^+\ a3$
	$a1^+\ a2^+$	$a1^+\ a3$	$a2^+\ a3^+$
	$a1\ a2^+$	$a1^+\ a3^+$	$a2\ a3^+$
	$a1\ a2^+$	$a1\ a3^+$	$a2\ a3^+$

Answer: For the $a1\ a2^+ \times a1^+\ a2$ cross, the tetrad shows 2:2 segregation of both alleles. However, since $a1$ and $a2$ are alleles at the same gene, one does not expect to see the $a1^+ a2^+$ and $a1\ a2$ spores very often. They could have arisen via a rare intragenic crossover between the $a1$ and $a2$ alleles. Mismatch repair might also have occurred. Recombination may have occurred much as is diagrammed in Figure 12.23 (through step 2). A gene conversion event in which an $a2/a2^+$ mismatch was repaired to $a2^+$ on one strand and to $a2$ on the other would result in this kind of tetrad.

In the remaining tetrads, there is evidence that the segregation of one of the alleles in the tetrad has resulted from gene conversion caused by mismatch repair of the heteroduplex DNA. The tetrad arising from the $a1^+\ a3 \times a1\ a3^+$ cross shows 2:2 segregation of $a3{:}a3^+$, but 3:1 segregation of $a1^+{:}a1$. This could have resulted from gene conversion of an *a1* allele to *a1*+. The tetrad arising from the *a2 a3*+ x *a2*+ *a3* cross shows 2:2 segregation of *a2:a2*+, but 3:1 segregation of *a3*+*:a3*. This could have resulted from gene conversion of an *a3* allele to *a3*+.

12.30 From a cross of *y1 y2*+ x *y1*+ *y2*, where *y1* and *y2* are both alleles of the same gene in yeast, the following tetrad type occurs at very low frequencies:

y1+ *y2*
y1 y2
y1 y2
y1 y2+

Explain the origin of this tetrad at the molecular level.

Answer: The tetrad show evidence of gene conversion of both alleles: the wild-type *y1*+ allele has undergone conversion to *y1*, and the wild-type *y2*+ allele has undergone conversion to *y2*. Remember that when a heteroduplex is formed that contains a mismatch needing repair, the mismatch can be repaired to either the wild-type, or the mutant allele. If strand exchange and branch migration within the *y* gene generated mismatches at both *y1* and *y2*, the strands could have been repaired to give the indicated gene conversion events.

12.31 In *Neurospora* the *a*, *b*, and *c* loci are situated in the same arm of a particular chromosome. The location of *a* is near the centromere, *b* is near the middle, and *c* is near the telomere of the arm. Among the asci resulting from a cross of *ABC* x *abc*, the following ascus was found (the 8 spores are indicated in the order in which they were arranged in the ascus): *ABC, ABC, ABc, ABc, aBC, aBC, abc, abc*. How might this ascus have arisen?

Answer: A recombination event has occurred between *a* and *c*. *A*/*a* shows 4:4 *A:a* segregation, *B*/*b* shows 3:1 *B:b* segregation and *C*/*c* shows 2:2:2:2 segregation. Thus, *A*/*a* shows first division segregation and *C*/*c* shows second division segregation. An exchange in the vicinity of *B*/*b* could result in a recombination intermediate that could be resolved so that the two recombinant chromatids would contain base mismatches. If repair of the mismatches occurred in the same direction in both chromatids, the 3*B*:1*b* segregation pattern would be produced. This is diagrammed below.

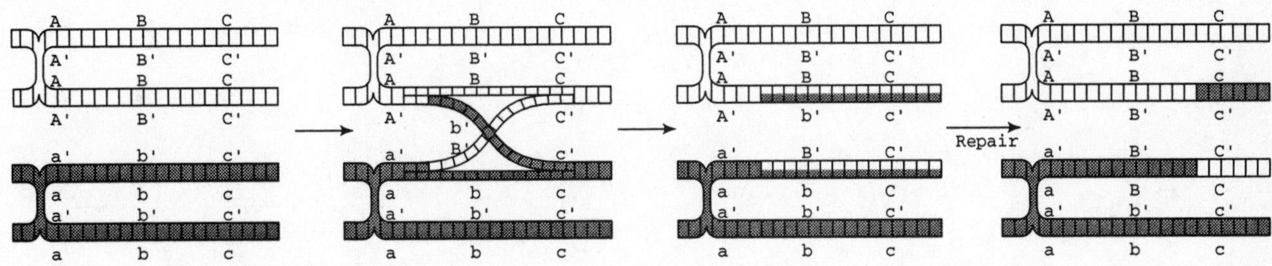

12.32 In the population of asci produced in question 12.31 an ascus was found containing, in this order, the spores *ABC, ABC, ABc, Abc, aBC, aBC, abc, abc*. How could this ascus have arisen?

Answer: The segregation patterns seen in this ascus are: for *A/a*, 4:4; for *B/b*, 3:1:2:2; for *C/c*, 2:2:2:2. Consider that the same series of events diagrammed in question 12.30 may have occurred. If mismatch repair did not take place in one of the recombinant chromatids, the mismatched strands would be segregated into spores 3 and 4. This would give rise to the observed pattern.

CHAPTER 13

TRANSCRIPTION, RNA MOLECULES, AND RNA PROCESSING

I. CHAPTER OUTLINE

THE TRANSCRIPTION PROCESS
- RNA Synthesis
- Classes of RNA and the Genes That Code for Them

TRANSCRIPTION OF PROTEIN-CODING GENES
- Prokaryotes
- Eukaryotes
- mRNA Molecules

TRANSCRIPTION OF OTHER GENES
- Ribosomal RNA and Ribosomes
- Transfer RNA

II. REVIEW OF KEY TERMS, SYMBOLS AND CONCEPTS

Without consulting the text and in your own words, write a brief definition of each term in the groups below. Then, either using a short phrase or a simple diagram, identify the relationship(s) between specific pairs of terms within a set. Finally, consult the text (and perhaps a friend who has also done the exercise) to check your answers.

1	2	3	4
Central Dogma	gene regulatory elements	mRNA	sucrose density gradient
replication	RNA polymerase	tRNA	Svedberg unit (S unit)
transcription	5'-to-3', 3'-to-5'	rRNA	23S, 16S, 5S rRNAs
primary transcript	coding strand	hnRNA	
precursor RNA	template strand	snRNA	
translation	antitemplate strand	RNA polymerase I, II, and III	
gene expression		α-amanitin	
		nucleolus	
		nucleus	

5	6	7	8
promoter closed, open promoter complex RNA polymerase holoenzyme core enzyme vs. sigma factor Pribnow box –35 region	terminator *rho*-dependent termination *rho*-independent termination Type I, II terminators hairpin loop	regulatory elements transcription factors regulatory factors promoter elements TATA, CAAT, GC boxes Goldberg-Hogness box enhancer elements silencer elements upstream activator sequences (UAS) RNA polymerase II TFIIA, B, C, E, F, S transcription initiation complex	transcription unit leader sequence coding sequence trailer sequence post-transcriptional modification coupled transcription and translation

9	10	11	12
pre-mRNA hnRNA introns exons R-loops mature mRNA 5' 7-methyl guanosine cap 5'-to-5' bond 3' polyadenylation RNA endonuclease poly(A) polymerase	mRNA splicing branch point sequence 3'-, 5'-splice sites snRNA snRNP (*snurps*) spliceosome U1, U2, U4, U5, U6 lariat structure 2'-to-5' bond	30S, 50S ribosome subunits 70S ribosome 23S, 16S, 5S rRNA 40S, 60S ribosome subunits 80S ribosome 28S, 18S, 5.8S, 5S rRNA precursor rRNA rDNA repeat unit spacer sequence nontranscribed spacer (NTS) external transcribed spacer (ETS)	self-splicing group I introns ribozyme

11	12
RNA polymerase III 5S rRNA genes tRNA genes internal control region (ICR) pre-tRNA molecule gene redundancy	tRNA anticodon post-transcriptional modification pseudouridine RNase P RNase Q RNA ligase

III. THINKING ANALYTICALLY

The wealth of important information in this chapter can be tackled successfully by organizing it. Organization will arrange the information contextually, so it can be remembered.

As a beginning, go through the list of terms with a view toward identifying their chemical nature. Is the term a region of DNA, RNA or a part of a protein? After this, relate the term to its function in the overall process of gene expression. If it is DNA, how does it relate to the structure of a gene or a transcription unit? If it is RNA, how does it relate to the RNA being synthesized or being processed? If it is a part of a protein, how is it involved in transcription or RNA processing?

As you relate a term to its function, try to develop an image of a transcription unit and how the transcript of a gene is processed. Diagram this process and relate the DNA structure of a gene to the structure of its primary transcript. Then, relate the structure of the primary transcript to the structure of the processed, mature transcript.

Next, distinguish between processes and characteristics in prokaryotes and eukaryotes. While there are some similarities between prokaryotes and eukaryotes, there are many, many differences. The fact that three RNA polymerases are active in eukaryotes results in three sets of data. For each RNA polymerase, there are distinct requirements for initiation and termination. Different DNA sequences are used, and different protein factors are involved in each case. It may help to prepare a list of analogous sites and factors, pairing prokaryotic elements with those doing similar jobs in eukaryotes.

IV. QUESTIONS FOR PRACTICE

A. Multiple Choice Questions

Choose the correct answer or answers for the following questions.

1. Match each item in Column A with one of the items in Column B and one of the items from Column C.

	COLUMN A	COLUMN B	COLUMN C
1.	_____ sigma factor	a. DNA	i. prokaryotic
2.	_____ promoter	b. RNA	ii. eukaryotic
3.	_____ terminator	c. protein	iii. i & ii.
4.	_____ transcription factor	d. a &b	
5.	_____ regulatory factor	e. b & c	
6.	_____ Pribnow box	f. a & c	
7.	_____ transcription initiation complex	g. a, b, & c	
8.	_____ spacer sequences		
9.	_____ NTS		
10.	_____ hairpin loop		
11.	_____ snRNP		
12.	_____ spliceosome		
13.	_____ rho		
14.	_____ promoter element		
15.	_____ RNA polymerase core enzyme		
16.	_____ ribozyme		
17.	_____ consensus sequence		
18.	_____ enhancer		
19.	_____ U1		

Answer: 1 c, i; 2 a, iii; 3 a, iii; 4 c, iii; 5 c, iii; 6 a, i; 7 f, iii; 8 d, iii; 9 a, ii; 10 b, iii; 11 e, ii; 12 e, ii; 13 c, i; 14 a, iii; 15 c, i; 16 b, i; 17 d, iii; 18 a, ii; 19 b, ii.

2. Which two statements below are correct?
   a. During transcription, the template strand is read in a 3'-to-5' direction.
   b. During transcription, the template strand is read in a 5'-to-3' direction.
   c. During transcription, an RNA is transcribed in the 3'-to-5' direction.
   d. During transcription, an RNA is transcribed in the 5'-to-3' direction.

3. How are RNA polymerases different than DNA polymerases?
    a. RNA polymerase is made of RNA, while DNA polymerase is made of DNA.
    b. Only DNA polymerases can proofread.
    c. Some RNA polymerases, but not DNA polymerase, are sensitive to α-amanitin.
    d. Only RNA polymerase can initiate the formation of a polynucleotide.
    e. RNA polymerases synthesize in a 3'-to-5' direction, while DNA polymerases synthesize in a 5'-to-3' direction.
    f. RNA polymerases use an antiparallel template, while DNA polymerases use a parallel template.

4. Which of the following is used to recognize an *E. coli* promoter?
    a. sigma factor
    b. core RNA polymerase
    c. TFIID
    d. Pribnow box

5. Which of the following is used to recognize a eukaryotic promoter?
    a. sigma factor
    b. core RNA polymerase
    c. TFIID
    d. Pribnow box

6. Which of the following statements are true?
    a. Both prokaryotic and eukaryotic mRNAs have 5' untranslated leader sequences.
    b. Only prokaryotic mRNAs are polyadenylated at the 3' end.
    c. In prokaryotes, transcription is coupled to translation.
    d. In eukaryotes, RNA splicing occurs after the mRNA is transported into the cytoplasm.
    e. All RNA splicing requires the formation of a spliceosome.
    f. Consensus sequences can be found in DNA as well as in RNA.

Answers: 2a & d; 3b, c & d; 4a; 5c; 6a, c, & f

B. Thought Questions

1. Compare and contrast each of the following pairs of terms with respect to location, function, and host organism (prokaryote or eukaryote).
    a. Pribnow box & TATA element
    b. promoter elements, enhancer elements, silencer elements, UAS
    c. transcription in T7 and T4 bacteriophage
    d. The promoters bound by RNA polymerase I, II, III
    e. Promoters located internally within a gene
    f. rRNA genes

2. Do prokaryotes or eukaryotes have more RNA polymerase (per unit total protein)? Why should the two groups of organisms differ?

3. How does *rho*-dependent termination differ from *rho*-independent termination? What is the role of the hairpin loop in each of these events? Is the hairpin loop in DNA or RNA?

4. What takes the place of sigma factor in eukaryotes?

5. Re-consider the C-value paradox. Would you expect a relationship to exist between the number of rDNA repeat units and the structural or organizational complexity of the organism? The amount of hnRNA? The complexity of hnRNA? The amount of cytoplasmic mRNA? The complexity of cytoplasmic mRNA? Why or why not?

6. The promoters for RNA polymerase II have been highly conserved, and thus differ little among eukaryotes. However, the promoters for RNA polymerase I show significant differences among eukaryotes. Why might this be the case? (Hint: Consider the number of genes read by each polymerase.)

7. Consider what will happen to an mRNA once it is processed and then address the following questions. How precise does RNA splicing need to be? How precise does transcription initiation need to be? How precise does polyadenylation need to be?

8. What post-transcriptional modifications are routinely made to a eukaryotic mRNA? Where in the cell do they occur?

V. SOLUTIONS TO TEXT PROBLEMS

13.1 Describe the differences between DNA and RNA.

Answer: DNA contains deoxyribose and thymine, while RNA contains ribose and uracil. DNA is frequently double-stranded, while RNA is usually single-stranded.

13.2 Compare and contrast DNA polymerases and RNA polymerases.

Answer: Both DNA polymerases and RNA polymerases catalyze the synthesis of nucleic acids in the 5'-to-3' direction. Both use a DNA template and synthesize a nucleic acid polynucleotide that is complementary to the template. However, DNA polymerases require a 3'-OH to add on to, while RNA polymerases do not. That is, RNA polymerases can initiate chains without primers, while DNA polymerases cannot. Furthermore, RNA polymerases usually require specific base-pair sequences as signals to initiate transcription.

13.3 All base pairs in the genome are replicated during the DNA synthesis phase of the cell cycle, but only *some* of the base pairs are transcribed into RNA. How is it determined *which* base pairs of the genome are transcribed into RNA?

Answer: The DNA sequences that are transcribed into RNA are determined using two general principles. First, there are signals in the DNA base sequences that identify a specific region as one that can be transcribed. Regions not bounded by transcription initiation and transcription termination signals are not transcribed. In regions that are bounded by these signals, only one strand is ordinarily transcribed, so that transcripts are formed in a 5'-to-3' direction using a single DNA template strand. Second, transcription within a defined region occurs only if additional transcription-inducing molecules are present. Some of these molecules recognize the signals within the DNA that define the transcription unit and are used in the transcription process.

13.4 Discuss the structure and function of the *E. coli* RNA polymerase. In your answer, be sure to distinguish between RNA core polymerase and RNA core polymerase-sigma factor complex.

Answer: RNA polymerase in *E. coli* consists of a core enzyme bound to an additional polypeptide, the sigma-factor. The core enzyme itself consists of four polypeptides. The complex of all five polypeptides is referred to as the holoenzyme, or the complete enzyme. All five polypeptides are required to bind to a promoter and initiate transcription. Without the sigma-factor, the core enzyme binds to DNA in various places, but fails to initiate transcription. The holoenzyme binds to a promoter in two distinct steps: After loosely binding a sequence about 35 bp before transcription initiation (the -35 region), it changes configuration, tightly binding a region about 10 bp before transcription initiation (the -10 region) and melting about 17 bp of DNA around this region. The two-step binding to the promoter orients the polymerase on the DNA and facilitates transcription initiation in the 5'-to-3' direction. After about 8-9 bases are formed in a new transcript, sigma factor dissociates from the holoenzyme, and the core enzyme completes the transcription process. See text pp. 393-397, especially Figure 13.4.

13.5 Discuss the similarities and differences between the *E. coli* RNA polymerase and eukaryotic RNA polymerases.

Answer: Both eukaryotic and *E. coli* RNA polymerases transcribe RNA in a 5'-to-3' direction using a 3'-to-5' DNA template strand. There are many differences between the enzymes, however. In *E. coli*, a single RNA polymerase core enzyme is used to transcribe genes. In eukaryotes, there are three types of RNA polymerase molecules, RNA polymerase I, II, and III. RNA Pol I transcribes the 28S, 18S, and 5.8S rRNA genes and is found in the nucleolus. RNA Pol II transcribes genes that encode hnRNA, mRNA and some snRNAs and is nuclear. RNA pol III transcribes tRNA genes, the 5S rRNA genes and some snRNAs and is nuclear. In addition, the mechanisms that are used to regulate the transcription activity of the *E. coli* RNA polymerase differ substantially from those employed by the eukaryotic RNA polymerases. See text pp. 391 onward.

13.6 Discuss the molecular events involved in the termination of RNA transcription in prokaryotes.

Answer: Termination of transcription in *E. coli* is signaled by controlling elements (sequences within the DNA) called terminators. Two classes of terminators exist, rho-independent (type I) and rho-dependent (type II). Both Type I and Type II termination events lead to the cessation of RNA synthesis, the release of the RNA chain and the RNA polymerase from the DNA.

Type I terminators utilize sequences with two-fold symmetry lying about 16-20 bp upstream of the termination point to signal the termination site. See text figure 13.5, p. 396. The hairpin loop that forms when a two-fold symmetric sequence is transcribed, plus a string of Us lead to termination, perhaps by destabilizing the RNA-DNA hybrid in the terminator region. Type II terminators lack the structure of Type I terminators, and instead use an ATP-activated rho-protein that binds to recognition sequences in the transcribed termination region. The binding of rho leads to the hydrolysis of the ATP, and the release of the transcript and the RNA polymerase from the DNA template.

13.7 Which classes of RNA do each of the three eukaryotic RNA polymerases synthesize? What are the functions of the different RNA types in the cell?

Answer:

RNA POLYMERASE	PRODUCT	FUNCTION
I	28S, 18S, 5.8S rRNA	structural and functional ribosomal components; the ribosome functions during translation
II	mRNA	provide protein coding information to ribosome
	some snRNAs	small nuclear RNAs function in a variety of nuclear processes, including RNA splicing and processing
III	tRNA	transfer RNA carries an amino acid to the ribosome during translation
	some snRNAs	small nuclear RNAs function in a variety of nuclear processes
	5S rRNA	functional and structural component of the ribosome 60S large subunit

13.8 What is the Pribnow box? The Goldberg-Hogness box (TATA element)?

Answer: The Pribnow box is the -10 element of the prokaryotic promoter (see solution to problem 13.4). It is located at -10 bp relative to the starting point of transcription and has the consensus sequence 5'-TATAAT-3'.

The Goldberg-Hogness box is a eukaryotic counterpart of the Pribnow box for many, but not all protein-coding genes. In higher eukaryotes, it is located at -25 to -30 bp before the starting point of transcription and has the consensus sequence 5'-TATAAA-3'. It is also called the TATA box or TATA element. In yeast, TATA elements range from 40 to 120 bp upstream of the transcription starting point. See text pp. 394 and 397-398.

13.9 What is an enhancer element?

Answer: An enhancer element is a DNA sequence that increases the amount of transcription of a nearby gene. Specific enhancer elements have consensus sequences. Enhancers can be bound by protein factors that, by interactions with other proteins and transcription factors, can stimulate transcription. Enhancer elements do not necessarily have to be a fixed position relative to the start of transcription, and also need not be in a particular orientation relative to the direction of transcription. That is, they can be effective in either orientation and in varying position relative to the start of transcription. See text pp. 398-400, particularly Figure 13.7.

13.10 A piece of mouse DNA was sequenced as follows (a space is inserted after every 10th base for ease in counting; "..." means a lot of unspecified bases):

```
AGAGGGCGGT   CCGTATCGGC   CAATCTGCTC   ACAGGGCGGA
TTCACACGTT   GTTATATAAA   TGACTGGGCG   TACCCCAGGG
TTCGAGTATT   CTATCGTATG   GTGCACCTGA   CT[(...)
GCTCACAAGT   ACCACTAAGC   (...)
```

What can you see in this sequence to indicate it might be all or part of a transcription unit?

Answer: The DNA sequence that is given reads 5'-to-3', and represents the antitemplate strand, the strand that is complementary to the 3'-to-5' template strand used for transcription. A region that is transcribed will produce RNA having an identical sequence, except that U will replace T.

Approach this problem by examining the sequence for the GC, CAAT and TATA box consensus sequences described on pp. 397-398 of the text. Some interesting findings are illustrated below. A bonus finding is that a conceptual mRNA that could be produced from this sequence has two signals that might be read by the ribosome to determine where translation would initiate and terminate. One of these is a triplet sequence that would encode a methionine (5'-AUG-3'), the amino acid that initiates a polypeptide chain. Some distance from this potential translation start site, a triplet sequence is found that would encode a chain termination sequence (5'-UAA-3'). These features of translation are described in Chapter 14.

```
              -80                     -65                        -50
          GC element               CAAT box                  GC element
5'  AGAGGGCGGT   CCGTATCGGC   CAATCTGCTC   ACAGGGCGGA

                              -30
                           TATA box
    TTCACACGTT   GTTATATAAA   TGACTGGGCG   TACCCCAGGG

    +1 (approx., conceptual)
    transcription initiation potential translation start
    TTCGAGTATT   CTATCGTATG   GTGCACCTGA   CT(...)
mRNA 5'uauu      cuaucguaug   gugcaccuga   cu(...)

    potential translation stop
    GCTCACAAGT   ACCACTAAGC   (...)
    gcucacaagu   accacuaagc   (...)
```

13.11 Compare and contrast the structures of prokaryotic and eukaryotic mRNAs.

Answer: Prokaryotic and eukaryotic mRNAs differ in several regards. First, they are processed quite differently. Prokaryotic mRNAs are not substantively changed between the time they are synthesized and the time they are decoded in translation. Eukaryotic mRNAs, on the other hand, are processed extensively prior to the time they are transported from the nucleus to the cytoplasm, where they will be translated. Eukaryotic mRNAs are capped at their 5'-ends with a 7-^{methyl}G attached via a unique 5'-5' linkage (see Figure 13.11, p. 402), have introns removed by RNA splicing (see pp. 403-408) and are polyadenylated at their 3'-ends (see Figure 13.13, p. 405).

Second, prokaryotic mRNAs and eukaryotic mRNAs differ in their structural organization. Prokaryotic mRNAs are typically polycistronic, and encode several different polypeptides by using consecutive translation reading frames. Eukaryotic mRNAs are typically monocistronic, and encode a single polypeptide. This reflects structural differences between the genes of prokaryotes and eukaryotes, (e.g., prokaryotes have operons), as well as the fact that transcription is coupled directly to transcription in prokaryotes, while transcription is temporally and spatially separated from translation in eukaryotes.

13.12 Compare the structures of the three classes of RNA found in the cell.

Answer: Cells have mRNA, rRNA and tRNA. Although all of these RNAs are single-stranded, they can differ substantially in the secondary and tertiary structures that can be achieved by intramolecular base-pairing interactions. These interactions, in the case of rRNA and tRNA, allow them to achieve their different functions.

mRNA typically has a 5' untranslated leader sequence, a protein-coding sequence and a 3' untranslated trailer sequence (see Figure 13.8, text p. 400). In eukaryotes, mRNAs are processed to have 5' caps and 3' poly-A tails, and to remove intronic sequences (see Figure 13.13, p. 405).

rRNAs are of three sizes in prokaryotes (5S, 16S, 23S) and four sizes in eukaryotes (5S, 5.8S, 18S, and 28S). They are single stranded, and three rRNAs are produced by processing a single rRNA precursor transcript (exception = eukaryotic 5S rRNA) (see Figure 13.19, p 411 and Figure 13.20, p. 412). They fold in a complex manner and have both a structural and functional role in the ribosome.

tRNA molecules are small, about 4S, and consist of a single chain of 75-90 nucleotides. They contain modified bases such as inosine, ribothymidine, pseudouridine, dihydrouridine methylguanosine, dimethylguanosine and methylinosine. There is extensive base-pairing between different parts of the tRNA molecule, allowing it to form a cloverleaf structure having three or four loops. The tRNA molecule functions as a carrier for amino acids to the ribosome. An amino acid is attached to its 3'-end in the cytoplasm, and it brings this amino acid to the ribosome. An anticodon in loop II binds to the mRNA, and the ribosome transfers the amino acid to a growing polypeptide chain. See text pp. 417-421, especially Figures 13.26 and 13.27.

13.13 Many eukaryotic mRNAs, but not prokaryotic mRNAs, contain introns. What is the evidence for the presence of introns in genes? Describe how these sequences are removed during the production of mature mRNA.

Answer: Evidence for introns stems from R-looping experiments in which a mature mRNA was hybridized with a double-stranded DNA segment that encoded the mRNA. One of the strands of the DNA is the template strand that is used to transcribe a precursor-mRNA. Since the mature mRNA is processed, this DNA strand has complementary sequences to only part of the mature mRNA used in the R-looping experiment. Because an RNA-DNA hybrid is more stable than a DNA-DNA hybrid, the RNA will bind to complementary sequences in the template strand and displace any sequences not bound as a single-stranded region. These sequences, although transcribed, were introns and removed by RNA splicing. Their location can be visualized by electron microscopy, as is shown in text Figure 13.12, p. 404.

Introns are removed from a pre-mRNA molecule by the action of a spliceosome. Consensus sequences at the intron-exon boundaries and within the intron itself are recognized by spliceosome components as the spliceosome assembles around the intronic sequence. First, U1 snRNP binds to the 5' splice site by base-pairing interactions between the U1 snRNA and the 5' splice site sequence. Then, U2 snRNP binds to a branch point region inside of the intron. Following the association of a pre-assembled U4/U6/U5 particle with the bound U1 and U2 snRNPs, U4 snRNP dissociates from the complex, allowing the formation of an active spliceosome. Splicing proceeds via cleavage of the 5' exon-intron junction, attaching the 5' end of the intron sequence to a branch-point sequence via an unusual 2'-to-5' bond, and then cleavage at the 3' exon-intron junction. When the two exons are ligated together, a lariat structure composed of intron sequences is released. See text Figure 13.14, p. 406 and Figure 13.16, p. 408.

13.14 The figure below shows the interpretative diagram of an electron micrograph (text p. 424) resulting from an R-looping experiment in which the mature chicken ovalbumin mRNA was hybridized with the ovalbumin gene. What would be the minimum number of phosphodiester bonds that would have to be cut to produce the mRNA from the pre-mRNA?

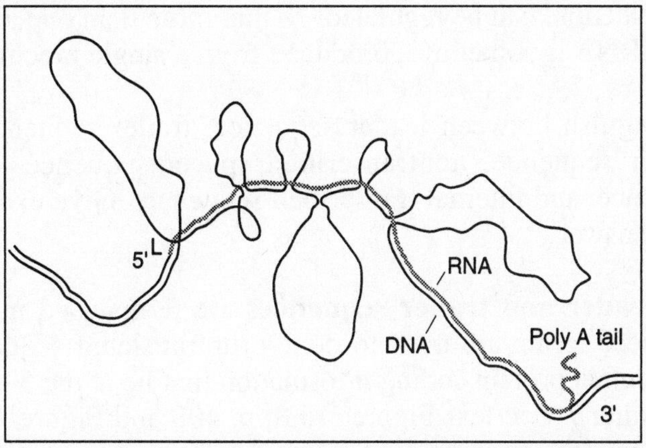

Answer: There are seven single-stranded regions in the figure, indicating that ovalbumin has seven introns. The removal of a single intron requires two phosphodiester bonds to be cut (see Figure 13.14, p. 406), so that the removal of seven introns would require 14 bonds to be cut. One additional cut must be made during polyadenylation (see Figure 13.13, p. 405), bringing the total to 15.

13.15 Discuss the posttranscriptional modifications that take place on the primary transcripts of tRNA, rRNA, and protein-coding genes.

Answer: tRNA genes can be found in single copies as well as in clusters. In prokaryotes, a cluster of tRNA genes may be transcribed to produce a single RNA transcript containing a number of tRNA sequences. The leader, trailer and spacer sequences in such transcripts will be removed by specific enzymes, including RNase P and RNase Q. In eukaryotes, some tRNA genes contain introns that are removed by specific endonucleases. The resulting RNA pieces are spliced together by RNA ligase. See text pp. 417-421.

In prokaryotes, a 30S pre-rRNA (p30S) is transcribed that contains a 5'-leader sequence, the 16S, 23S and 5S rRNA sequences separated by spacer sequences, and a 3'-trailer sequence. The p30S molecule is processed by RNase III so that the three rRNA precursors are produced while transcription is still occurring. See text Figure 13.19, p. 411. In eukaryotes, 18S, 28S and 5.8S rRNA genes are transcribed from the rDNA into a single pre-rRNA molecule. The pre-mRNA is processed by removing the internal and external spacer sequences, leading to the production of the mature rRNAs. See text Figure 13.20, p. 412. The eukaryotic 5S rRNA is transcribed separately to produce mature rRNA molecules that need no further processing. Eukaryotic 5S rRNAs are imported into the nucleolus where they are assembled with the other mature rRNAs and the ribosomal protein subunits to produce functional ribosomal subunits. Some *Tetrahymena* pre-rRNAs have self-splicing introns (group I) in the 28S rRNA. It is important to remember that the removal of the intron in the 28S rRNA via self-splicing is a separate process from the cleavage of spacer sequences from the mature rRNAs. See text Figure 13.23, p. 415.

Protein-coding eukaryotic transcripts synthesized by RNA polymerase II are extensively processed. A 5'-5' bonded m^7Gppp cap is added to the 5'-end of the nascent transcript when the chain is 20-30 nucleotides long. A poly(A) tail of variable length can be added to the 3'-end of the transcript, and spliceosomes can remove intronic sequences. The site of poly(A) addition, as well as splice-site selection, can be regulated, so that more than one alternatively processed mature mRNA is sometimes produced from a single precursor mRNA transcript.

13.16 Distinguish between leader sequence, trailer sequence, coding sequence, intron, spacer sequence, nontranscribed spacer sequence, external transcribed spacer sequence, and internal transcribed sequence. Give examples of actual molecules in your answer.

Answer: **Leader and trailer sequences** are terms used in several, related ways. First, these terms are used to denote untranslated sequences in mRNA that do not contain protein coding information that lie at the 5'-end (for leader) or 3'-end (for trailer). See text Figure 13.8, p. 400 and Figure 13.13, p. 405. Second, these terms are sometimes used to denote the sequences that lie at the 5'-end (for leader) or 3'-end (for trailer) of pre-rRNA transcripts in prokaryotes. See text Figure 13.19, p. 411.

Coding sequences are sequences in mRNA that are translated by the ribosome. These sequences encode information (in the form of a triplet code – see Chapter 14) that can be read by the ribosome to direct the synthesis of a polypeptide. **Introns** are sequences present in eukaryotic precursor RNA molecules that are spliced out in the mature, functional molecule. Introns are present in pre-mRNAs as well as in pre-rRNAs and pre-tRNAs.

Nontranscribed spacer sequences are sequences, such as those in the repeated rDNA genes, that lie between adjacent, repeating transcription units. **External transcribed spacer sequences** are found in pre-rRNA transcripts in eukaryotes (from transcription of 18S, 28S and 5.8S pre-rRNAs). They are located immediately upstream of the 5'-end of the 18S sequence and downstream of the 3'-end of the 28S sequence. The **internal transcribed spacer sequences** are located on either side of the 5.8S sequence. See text Figure 13.20, p. 412.

13.17 Describe the organization of the ribosomal DNA repeating unit of a higher-eukaryotic cell.

Answer: rDNA in a eukaryotic cell is organized into tandem arrays of repeating units, each of which have an 18S, 5.8S and 28S rRNA gene. In between each repeating unit is a nontranscribed spacer. The three genes within each unit are transcribed by RNA polymerase I as a 45S precursor rRNA molecule in the nucleolus. 5' and 3' flanking spacer sequences (external transcribed spacer sequences) as well as internal spacer sequences are removed, leading to the production of 18S, 5.8S and 28S rRNAs.

13.18 Which of the following kinds of mutations would be likely to be recessive lethals in humans? Explain your reasoning.

a. Deletion of the U1 genes.
b. Deletion within intron 2 of β-globin
c. Deletion of 4 bases at the end of intron 2 and 3 bases at the beginning of exon 3 in β-globin.

Answer:

a. One would expect deletion of U1 genes to be recessive lethal, as U1 snRNA is essential for the identification of the 5' splice site in RNA splicing. Incorrect splicing would lead to non-functional gene products for many genes, a non-viable situation.
b. If a deletion within intron 2 did not affect a region important for the removal of intron 2 (e.g., the branch point, the regions near the 5' or 3' splice sites), it would have no affect on the mature mRNA produced. Consequently, such a mutation would lack a phenotype. However, if splicing of intron 2 were affected and the mRNA altered, such a mutation could result in non-functional hemoglobin and death.
c. The deletion described would affect the 3' splice site of intron 2, leading to, at best, aberrant splicing of intron 2. This could lead to a non-functional protein being made and death.

13.19 The diagram in the figure below shows the transcribed region of a typical eukaryotic protein-coding gene:

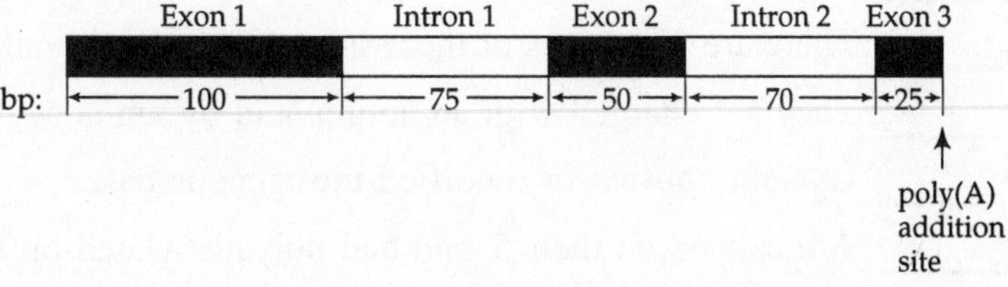

What is the size (in bases) of the fully processed, mature mRNA? Assume in your calculations a poly(A) tail of 200 As.

Answer: 1 (5' m^7G cap) + 100 (exon 1) + 50 (exon 2) + 25 (exon 3) + 200 (poly-A tail) = 376 bases.

13.20 Which of the following could occur in a single mutational event in a human? Explain.

a. Deletion of 10 copies of the 5S ribosomal RNA genes only.
b. Deletion of 10 copies of the 18S rRNA genes only.
c. Simultaneous deletion of 10 copies of the 18S, 5.8S, and 28S rRNA genes only.
d. Simultaneous deletion of 10 copies each of the 18S, 5.8S, 28S, and 5S rRNA genes.

Answer: a. Could occur. Since the 5S genes are clustered, 10 copies could be removed in one deletion.
b. Could not occur. 18S genes are part of a cluster of 18S, 5.8S and 28S genes. One cannot excise 18S genes without also excising others.
c. Could occur. The genes are clustered.
d. Could not occur. The 18S, 5.8S and 28S genes are clustered at a locus separate from the 5S genes.

13.21 During DNA replication in a mammalian cell a mistake occurs: 10 wrong nucleotides are inserted into a 28S rRNA gene, and this mistake is not corrected. What will likely be the effect on the cell?

Answer: The cell will most likely be just fine. The other 1,000 copies of the 28S rRNA genes should be normal, allowing this error to have little phenotypic consequence.

13.22 Give the correct answers, noting that each blank may have more than one correct answer, and that each answer (1 through 4) could be used more than once.

1. Eukaryotic mRNAs
2. Prokaryotic mRNAs
3. Transfer RNAs
4. Ribosomal RNAs

a. ______ Have a cloverleaf structure
b. ______ Are synthesized by RNA polymerases
c. ______ Display an anticodon each
d. ______ Are the template of genetic information during protein synthesis
e. ______ Contain exons and introns
f. ______ There are four types of these in eukaryotes and only three types in *E. coli*.
g. ______ They get charged with an amino acid by aminoacyl-tRNA synthetase.
h. ______ Contain unusual or modified nitrogenous bases
i. ______ Are capped on their 5' end and polyadenylated on their 3' end

Answer: a. 3
b. 1, 2, 3, 4
c. 3
d. 1 and 2
e. 1
f. 4
g. 3
h. 3
i. 1

CHAPTER 14

THE GENETIC CODE AND THE TRANSLATION OF THE GENETIC MESSAGE

I. CHAPTER OUTLINE

PROTEIN STRUCTURE
- Chemical Structure of Proteins
- Molecular Structure of Proteins

THE NATURE OF THE GENETIC CODE
- The Genetic Code Is a Triplet Code
- Deciphering the Genetic Code
- Nature and Characteristics of the Genetic Code

TRANSLATION OF THE GENETIC MESSAGE
- Aminoacyl-tRNA Molecules
- Initiation of Translation
- Elongation of the Polypeptide Chain
- Termination of Translation

PROTEIN SORTING IN THE CELL
- Proteins Distributed by the Endoplasmic Reticulum
- Proteins Transported into Mitochondria and Chloroplasts
- Proteins Transported into the Nucleus

II. REVIEW OF KEY TERMS, SYMBOLS AND CONCEPTS

Without consulting the text and in your own words, write a brief definition of each term in the groups below. Then, either using a short phrase or a simple diagram, identify the relationship(s) between specific pairs of terms within a set. Finally, consult the text (and perhaps a friend who has also done the exercise) to check your answers.

1	2	3	4
protein	protein	triplet code	reading frame
polypeptide	primary structure	codon	comma free code
amino acid	secondary structure	frameshift mutation	nonoverlapping code
α-carbon	tertiary structure	reversion	universal, degenerate code
peptide bond	quaternary structure	random copolymers	start, stop codon
acidic, basic amino acid	conformation	ribosome binding assay	sense, nonsense codon
neutral polar, nonpolar amino acid	α-helix	cell-free protein-synthesizing system	wobble hypothesis
C-terminus	β-pleated sheet		
N-terminus			

5	6	7	8
translation	fMet	elongation	termination factor
aminoacyl-tRNA	transformylase	peptidyl transferase	RF1, RF2, RF3
aminoacyl-tRNA synthetase	tRNA.fMet vs. tRNA.Met	P, A sites	selenocysteine
codon recognition	30S initiation complex	EF-Tu, EF-TS, EF-G	*selA, B, C, D*
initiation	ribosome binding site	translocation	UGA codon
elongation	Shine-Dalgarno sequence	polyribosome	UAG, UAA codons
termination	IF1, IF2, IF3	polysome	context effect
	cap binding factor		
	eIFs		

9
endoplasmic reticulum
membrane-bound ribosome
signal hypothesis
signal sequence
signal recognition particle (SRP)
docking protein
cotranslational transport
signal peptidase
Golgi complex
posttranslational transport
transit peptidase
nuclear localization sequences
glycoprotein

III. THINKING ANALYTICALLY

Much of the material presented in this chapter is detailed and descriptive. Initiation, elongation, and termination are all complex events. Each involves numerous enzymes and other factors that differ in prokaryotes and eukaryotes. As in several of the recent chapters, placing the information in context will help you learn it better.

There are several levels in which to organize the information in this chapter. First, consider the spatial organization. Associate the different types of RNA (rRNA, mRNA, tRNA) and factors involved in translation with the different components of the translation machinery in the cell. Then construct a temporal framework to follow these components through the stages of translation. What RNAs and factors are components of the initiation complex? What RNA and factors are present during elongation? What new factors are required for termination? Then come back to the spatial organization issue. Once proteins are synthesized, what factors are used to shuttle them to the appropriate location?

Keep in mind the polarity of DNA and RNA as you examine DNA sequences to infer which transcripts would be produced, and find reading frames that will be translated. Remember that transcripts are synthesized in the 5'-to-3' direction using a 3'-to-5' template strand, and that amino acids are synthesized using codons that start at the 5'-end of the mRNA. Pay close attention to keep correct mRNA, tRNA anticodon and DNA strand polarity.

One of the highlights of this chapter is the presentation of a set of elegant experiments used to decipher the genetic code. These technically demanding experiments addressed one of the fundamental questions in genetics: How is information in DNA used to code for proteins? Consider them carefully, and consider how interpreting them often uses probabilistic thinking. The product rule, which states that the frequency of two independent events occurring together is the product of their separate probabilities, allows one to deduce the appropriate codons that specify particular amino acids. Consider the following example:

(a) A synthetic mRNA is generated by randomly polymerizing adenine and guanine ribonucleotides. What codons would be formed, and what would be their relative frequency if you began with a limitless mixture of 3 adenine to 1 guanine in the synthesis?
(b) When the mRNAs generated in part (a) are added to a cell-free translation system, polypeptides are formed that, on average, have 56 percent lysine, 20 percent arginine, 20 percent glutamic acid, and 8 percent glycine. What can you infer about codon usage from these data (without consulting a table of the genetic code)?

Answer: First, determine what possible codons (three-nucleotide combinations) could be generated. Then, determine the frequency of that codon by multiplying together the probability of obtaining each base in one codon. Since there is a 3A : 1 G ratio, the chance of finding an A is 3/4, and the chance of finding a G is 1/4.

$p(\text{AAA}) = (3/4)^3 = 0.42$
$p(\text{AAG}) = (3/4)^2(1/4) = 0.14$
$p(\text{AGA}) = (3/4)^2(1/4) = 0.14$
$p(\text{AGG}) = (3/4)(1/4)^2 = 0.06$
$p(\text{GAA}) = (1/4)(3/4)^2 = 0.14$
$p(\text{GAG}) = (1/4)^2(3/4) = 0.06$
$p(\text{GGA}) = (1/4)^2(3/4) = 0.06$
$p(\text{GGG}) = (1/4)^3 = 0.02$

Since AAA accounts for 42 percent of the total codons that are translated, and only lysine constitutes at least 42 percent of the amino acid content of the polypeptides, AAA must encode lysine. By a similar sort of reasoning, GGG must encode glycine. Since there is only 8 percent glycine, and no codon alone accounts for 8 percent of the total codon usage, glycine must be encoded by at least two codons. GGG must be one of these, since if any other two codons encoded glycine, the amount of glycine would exceed 8 percent. The remaining glycine must be encoded by one of the GAG, GGA, or AGG codons, since they are represented 6 percent of the time. The remaining lysine must be encoded by either AAG, AGA, or GAA, as no combination of other codons would give 56 percent lysine. This leaves arginine to be encoded by one of the AAG, AGA, GAA codons and one of the AGG, GAG, GGA codons. Similarly, glutamic acid must be encoded by one of the AAG, AGA, GAA codons and one of the AGG, GAG, GGA codons. In order to make further inferences than those here, more experimentation (e.g., with different co-polymers) must be done.

IV. QUESTIONS FOR PRACTICE

A. Multiple Choice Questions

Choose the correct answer or answers for the following questions.

1. The term "secondary structure" refers to a polypeptide's
   a. sequence of amino acids.
   b. structure that results from local interactions between the residues of different amino acids.
   c. folding that results from local and distant interactions between the residues of different amino acids.
   d. interactions with a second polypeptide chain.

2. The eukaryotic equivalent of the Shine-Dalgarno sequence is the
   a. 60S large ribosomal subunit.
   b. 40S small ribosomal subunit.
   c. 5' cap of the mRNA.
   d. poly(A) tail.

3. In which stages of translation is GTP used?
   a. Initiation
   b. Elongation
   c. Translocation
   d. Termination

4. Which of the following are the same in both prokaryotes and eukaryotes?
   a. Elongation factors
   b. Stop codons
   c. The use of fMet.tRNA
   d. AUG initiation codons

5. Specificity in translation is maintained by
   a. using specific tRNA synthetases for each amino acid.
   b. using an anticodon.
   c. using a degenerate triplet code having codons for each amino acid.
   d. using RNA polymerase.
   e. allowing wobble in the anticodon.

6. Where can coding sequences be found?
   a. introns
   b. exons
   c. mRNA
   d. hnRNA
   e. rRNA

7. In an unprocessed, newly synthesized eukaryotic protein,
   a. the N-terminus is encoded by sequences in the 5'-region of the mRNA.
   b. the C-terminus is encoded by sequences in the 5'-region of the mRNA.
   c. the first amino acid is always methionine or f-methionine.
   d. the last amino acid is always encoded by the codon AAA.

8. During translation in *E. coli*,
   a. an intact 70S ribosome binds the mRNA during initiation.
   b. Shine-Dalgarno sequences present in the mRNA align it with the 23S rRNA.
   c. both initiation and elongation require energy in the form of GTP.
   d. termination occurs when the ribosome recognizes a chain termination codon with the aid of Rfs.

9. The RNAs that are used in eukaryotic translation include
   a. 5S rRNA.
   b. 28S rRNA.
   c. mRNA.
   d. tRNA.
   e. snRNA.

10. In eukaryotes, proteins destined to remain in the cell are distributed
   a. by packaging into secretory vesicles.
   b. after complexing with an SRP using an N-terminal signal sequence.
   c. by using posttranslational transport.
   d. by using transit sequences.

Answers: 1b; 2c; 3a, b, & c; 4b & d; 5a, b, & c; 6b, c, & d; 7a & c; 8c & d; 9a, b, c, & d; 10b, c & d.

B. Thought Questions

1. The following sequence of bases in *E. coli* DNA is part of a gene. Mark the region that, if transcribed, will form the first codon to be translated. Mark the region, that if transcribed, will form a Shine-Dalgarno sequence? Is the strand shown the template or nontemplate strand? (Hint: Find the initiation sequence. Be careful about direction!)

5'-CCGCATTCCTCCGGCGGGACCTACT-3'

2. Why is the code for methionine not degenerate? (Hint: what special role does methionine play in translation?)

3. What are the components of a cell-free protein synthesizing system? What is the function of each?

4. Describe three complementary methods by which the code was deciphered. Is any single method sufficient to completely decipher the code? What advantages and disadvantages does each method have?

5. The genetic code is nearly universal. To what extent does this indicate that the code was "settled on" very early in the evolution of cells?

6. Consider which bases can wobble to base pair with each other. What structural similarities exist between bases at the 3'-end of codon that can wobble base-pair with the base at the 5'-end of the anticodon?

7. What is the "context effect" in codon translation? Cite two examples.

8. What percentage of 70S ribosomes do you expect to be associated with mRNA in prokaryotes? What percentage of 80S ribosomes do you expect to be associated with mRNA in eukaryotes?

9. Discuss the various ways in which proteins within a cell are (1) targeted for secretion, (2) targeted to the mitochondria, (3) targeted to the nucleus. How are these targeting mechanisms similar? How are these targeting mechanisms different?

V. SOLUTIONS TO TEXT PROBLEMS

14.1 The form of genetic information used directly in protein synthesis is (choose the correct answer):

a. DNA
b. mRNA
c. rRNA
d. Ribosomes

Answer: b.

14.2 Proteins are (choose the correct answer):

a. Branched chains of nucleotides
b. Linear, folded chains of nucleotides
c. Linear, folded chains of amino acids
d. Invariable enzymes

Answer: c.

14.3 Various proteins are treated as specified. For each case indicate what level(s) of protein structure would change as the result of the treatment.

a. Hemoglobin is stored in a hot incubator at 80° C.
b. Egg white (albumin) is boiled.
c. RNAse (a single polypeptide enzyme) is heated to 100° C.
d. Meat in your stomach is digested (gastric juices contain proteolytic enzymes).
e. In the β-polypeptide chain of hemoglobin, the amino acid valine replaces glutamic acid at the number six position.

Answer: a. The hemoglobin would dissociate into its four component subunits, because the heat will destabilize the ionic bonds that stabilize the quaternary structure of the protein. The individual subunit's tertiary structure may also be altered, as the thermal energy of the heat may also destabilize the folding of each polypeptide subunit.

b. The protein will denature. Its tertiary structure is destabilized by heating, so that it does not retain a pattern of folding that allows it to be soluble.
c. The protein will denature when its tertiary structure is destabilized by heating. RNAse, unlike albumin, will renature if cooled slowly, and re-establish its normal, functional tertiary structure.
d. It is likely that the meat proteins' tertiary structure will first be denatured by the acid conditions of the stomach. Then, the primary structure of the proteins in the meat will be destroyed as the proteins are degraded into amino acid components by the proteolytic enzymes.
e. Valine is a neutral, non-polar amino acid, unlike the acidic glutamic acid (see Figure 14.2, p. 429). A change in the chemical properties of this amino acid may alter the function of the hemoglobin molecule by affecting one or more levels of protein structure. It could affect the local interactions between amino acids near the number 6 position and potentially alter its secondary structure. It could also affect the folding patterns of the protein and alter the tertiary structure of the β-subunit. Finally, if the number 6 amino acid residue were important for interactions between the subunits of the hemoglobin molecules, it could alter the quaternary structure of the protein. Mutations at this position are known to alter the function of the hemoglobin molecule (see Figures 9.12 and 9.13, p. 282) and result in sickle-cell anemia.

14.4 The process in which ribosomes engage is (choose the correct answer):

a. Replication
b. Transcription
c. Translation
d. Disjunction
e. Cell division

Answer: c.

14.5 What are the characteristics of the genetic code?

Answer:

1. The genetic code is triplet, with three nucleotides specifying either the insertion of an amino acid into a polypeptide chain, or a chain termination event.
2. An mRNA coding region is read in a continuous fashion without skipping nucleotides.
3. A coding region is read in a non-overlapping manner. Triplets are read sequentially.
4. The code is nearly universal. Nearly all organisms use the same code, with exceptions being found in mammalian mitochondria and the nuclear genomes of some protozoa.
5. The code is degenerate, meaning that more than one codon typically codes for a single amino acid.
6. The code has translation stop (chain termination) and translation start signals. AUG, which encodes methionine within a coding region, is usually used as a start codon. One of three nonsense (*not sensing* an amino acid) codons are used as chain termination codons.

7. Wobble occurs in the 5' base of the anticodon, so that some tRNAs can recognize multiple codons (coding for the same amino acid).
See text pp. 434-436.

14.6 Base-pairing wobble occurs in the interaction between the anticodon of the tRNAs and the codon of the mRNA. On the theoretical level, determine the minimum number of tRNAs needed to read the 61 sense codons.

Answer: Consider Figure 14.9, p. 435, in answering this question.

AMINO ACID	tRNAS NEEDED	RATIONALE
Ile	1	3 codons can use 1 tRNA (wobble)
Phe	1	2 codons can use 1 tRNA (wobble)
Tyr	1	" " " " "
His	1	" " " " "
Gln	1	" " " " "
Asn	1	" " " " "
Lys	1	" " " " "
Asp	1	" " " " "
Glu	1	" " " " "
Cys	1	" " " " "
Trp	1	1 codon
Met	2	Single codon, but need one tRNA for initiation and one tRNA for elongation
Val	2	4 codons: 2 can use 1 tRNA (wobble)
Pro	2	" " " " "
Thr	2	" " " " "
Ala	2	" " " " "
Gly	2	" " " " "
Leu	3	6 codons: 2 can use 1 tRNA (wobble)
Arg	3	" " " " "
Ser	3	" " " " "
Total	**32**	**61 codons**

14.7 Antibiotics have been very useful in elucidating the steps of protein synthesis. If you have an artificial messenger of the sequence of AUGUUUUUUUUUU..., it will produce the following polypeptide in a cell-free, protein-synthesizing system: fMet-Phe-Phe-Phe... In your search for new antibiotics you find one called putyermycin, which blocks protein synthesis. When you try it with your artificial mRNA in a cell-free system, the product is fMet-Phe. What step in protein synthesis does putyermycin affect? Why?

Answer: Since a dipeptide is formed, translation initiation if not affected, nor is the first step of elongation—the binding of a charged tRNA in the A site and the formation of a peptide bond. However, since *only* a dipeptide is formed, it appears that translocation is inhibited.

14.8 Describe the reactions involved in the aminoacylation (charging) of a tRNA molecule.

Answer: Aminoacylation of a tRNA occurs enzymatically using specific aminoacyl-tRNA synthetases. There is an aminoacyl-tRNA synthetase for each amino acid, so that specificity in charging a tRNA can be maintained. Without such specificity, the specificity of codons determining the insertion of specific amino acids would be circumvented. To attach a specific amino acid to a specific tRNA, the amino acid first reacts with ATP to form a aminoacyl-AMP complex. This reaction is catalyzed by an aminoacyl-tRNA synthetase. A second reaction, also catalyzed by the aminoacyl-tRNA synthetase, results in the amino acid becoming attached by its carboxyl group to the last ribose of the 3' end of the tRNA chain. See text Figure 14.12, p. 439.

14.9 Compare and contrast the following in prokaryotes and eukaryotes:

a. protein synthesis initiation
b protein synthesis termination

Answer:

a. See text pp. 440-441.
b. See text pp. 445-446.

14.10 Discuss the two species of methionine tRNA, and describe how they differ in structure and function. In your answer, include a discussion of how each of these tRNAs binds to the ribosome.

Answer: Different methionine tRNAs are used in translation initiation and elongation. In both eukaryotes and prokaryotes, a special methionine tRNA is used for translation initiation. In prokaryotic initiation, methionine is attached to a special tRNA, tRNA.fMet. After methionine-tRNA synthetase catalyzes the transfer of methionine to this tRNA, transformylase catalyzes the addition of a formyl group to the amino group of the methionine. The charged fMet-tRNA.fMet and the mRNA are assembled as a complex of initiation factors (IF1, IF2, IF3), GTP, and the 30S ribosomal subunit during the initiation of translation. In eukaryotes, a special initiator charged Met-tRNA.Met is similarly bound by a complex of the 40S subunit with eukaryotic initiation factors and GTP. Once translation has initiated, a separate tRNA.Met is used during translation elongation. Once charged, this Met-tRNA.Met will bind to the A-site in a complex with elongation factor Tu and GTP. In both initiation and elongation, the anticodon on the Met (or fMet)-bound tRNA recognizes the codon 5'-AUG-3'. See pp. 440-441.

14.11 Random copolymers were used in some of the experiments that revealed the characteristics of the genetic code. For each of the following ribonucleotide mixtures, give the expected codons and their frequencies, and give the expected proportions of the amino acids that would be found in a polypeptide directed by the copolymer in a cell-free, protein-synthesizing system:

a. 4 A:6 C
b. 4 G:1 C
c. 1 A:3 U:1 C
d. 1 A:1 U:1 G:1 C

Answer: One can determine the expected amino acids in each case by calculating the expected frequency of each kind of triplet codon that might be formed, and inferring from these what types and frequencies of amino acids would be used during translation.

a. 4 A:6 C gives $2^3 = 8$ codons, specifically AAA, AAC, ACC, ACA, CCC, ACA, CAC, CAA. Since there is 40% A and 60% C, one has
p(AAA) = 0.4 x 0.4 x 0.4 = 0.064, or 6.4% Lys
p(AAC) = 0.4 x 0.4 x 0.6 = 0.096, or 9.6% Asn
p(ACC) = 0.4 x 0.6 x 0.6 = 0.144, or 14.4% Thr
p(ACA) = 0.4 x 0.6 x 0.4 = 0.096, or 9.6% Thr (24% Thr total)
p(CCC) = 0.6 x 0.6 x 0.6 = 0.216, or 21.6% Pro
p(CCA) = 0.6 x 0.6 x 0.4 = 0.144, or 14.4% Pro (36% Pro total)
p(CAC) = 0.6 x 0.4 x 0.6 = 0.144, or 14.4% His
p(CAA) = 0.6 x 0.4 x 0.4 = 0.096, or 9.6% Gln

b. 4 G:1 C gives $2^3 = 8$ codons, specifically GGG, GGC, GCG, GCC, CGG, CGC, CCC, CCG. Since there is 80% G and 20% C, one has
p(GGG) = 0.8 x 0.8 x 0.8 = 0.512, or 51.2% Gly
p(GGC) = 0.8 x 0.8 x 0.2 = 0.128, or 12.8% Gly (64% Gly total)
p(GCG) = 0.8 x 0.2 x 0.8 = 0.128, or 12.8% Ala
p(GCC) = 0.8 x 0.2 x 0.2 = 0.032, or 3.2% Ala (16% Ala total)
p(CGG) = 0.2 x 0.8 x 0.8 = 0.128, or 12.8% Arg
p(CGC) = 0.2 x 0.8 x 0.2 = 0.032, or 3.2% Arg (16% Arg total)
p(CCC) = 0.2 x 0.2 x 0.2 = 0.008, or 0.8% Pro
p(CCG) = 0.2 x 0.2 x 0.8 = 0.032, or 3.2% Pro (4% Pro total)

c. 1 A:3 U:1 C gives $3^3 = 27$ different possible codons. Of these, one will be UAA, a chain terminating codon. Since there is 20% A, 60% U and 20% C, the probability of finding this codon is 0.6 x 0.2 x 0.2 = 0.024, or 2.4%. All of the remaining 26 (97.6%) codons will be sense codons. One can proceed in the same manner as in a and b to determine their frequency, and thus determine the kinds of amino acids expected. To take the frequency of nonsense codons into account, one must divide the frequency of obtaining a particular amino acid considering all 27 possible codons by the frequency of obtaining a sense codon. One has:
(0.8/0.976)% = 0.82% Lys
(3.2/0.976)% = 3.28% Asn
(12.0/0.976)% =12.3% Ile
(9.6/0.976)% = 9.84% Tyr
(19.2/0.976)% =19.67% Leu
(28.8/0.976)% =29.5% Phe
(4.0/0.976)% = 4.1% Thr
(0.8/0.976)% = 0.82% Gln
(3.2/0.976)% = 3.28% His
(4.0/0.976)% = 4.1% Pro
(12.0/0.976)% =12.3% Ser.

It is likely that the chains produced would be relatively short, due to the chain terminating codon.

d. 1 A:1 U:1 G:1 C will produce $4^3 = 64$ different codons, all of those possible in the genetic code. The probability of each codon is 1/64, and so there will be a 3/64 chance of a codon being chain terminating. With those exceptions, the relative proportion of amino acid incorporation is directly dependent on the codon degeneracy for each amino acid. Inspecting the table of the genetic code in Figure 14.9, and taking the frequency of nonsense codons into account, one expects there to be:

AMINO ACID	# CODONS	FREQUENCY
Trp	1	1/61 =1.64%
Met	1	1.64%
Phe	2	2/61 = 3.28%
Try	2	3.28%
His	2	3.28%
Gln	2	3.28%
Asn	2	3.28%
Lys	2	3.28%
Asp	2	3.28%
Glu	2	3.28%
Cys	2	3.28%
Ile	3	3/61 = 4.92%
Val	4	4/61 = 6.56%
Pro	4	6.56%
Thr	4	6.56%
Ala	4	6.56%
Gly	4	6.56%
Leu	6	6/61 = 9.84%
Arg	6	9.84%
Ser	6	9.84%

14.12 Other features of the reading of mRNA into proteins being the same as they are now (i.e., codons must exist for 20 different amino acids), what would the minimum WORD (CODON) SIZE be if the number of different bases in the mRNA were, instead of four:

a. two
b. three
c. five

Answer: a. Approach this by trial and error. If only two bases existed in mRNA and codons were only four bases long, there would only be $2^4 = 16$ possible codons, not enough to code for 20 amino acids. In order to encode 20 different amino acids, one would need to read codons five bases long. For such a length of codon, there would be $2^5 = 32$ possible codons.

b. If only three bases existed in mRNA, one would need to read codons 3 bases long ($3^3 = 27$ possible combinations).

c. If five bases existed in mRNA, one would need to read codons 2 bases long ($5^2 = 25$ possible combinations).

14.13 Suppose that at stage A in the evolution of the genetic code only the first two nucleotides in the coding triplets led to unique differences and that any nucleotide could occupy the third position. Then, suppose there was a stage B in which differences in meaning arose depending upon whether a purine (A or G) or pyrimidine (C or T) was present at the third position. Without reference to the number of amino acids or multiplicity of tRNA molecules, how many triplets of different meaning can be constructed out of the code at stage A? At stage B?

Answer: Stage A: $4^2 = 16$ different meaningful triplets
Stage B: $4^2 \times 2 = 32$ different meaningful triplets

14.14 A gene makes a polypeptide 30 amino acids long containing an alternating sequence of phenylalanine and tyrosine. What are the sequences of nucleotides corresponding to this sequence in the following:

a. The DNA strand which is read to produce the mRNA, assuming Phe = UUU and Tyr = UAU in mRNA
b. The DNA strand which is not read
c. tRNA

Answer:
a. 3' – AAA ATA AAA ATA AAA ATA ...–5'
b. 3' – TTT TAT TTT TAT TTT TAT ...–5'
c. 3' –AAA–5' in the anticodon for Phe, and 3' –AUA– 5' in the anticodon for Tyr

14.15 A segment of a polypeptide chain is Arg-Gly-Ser-Phe-Val-Asp-Arg. It is encoded by the following segment of DNA:

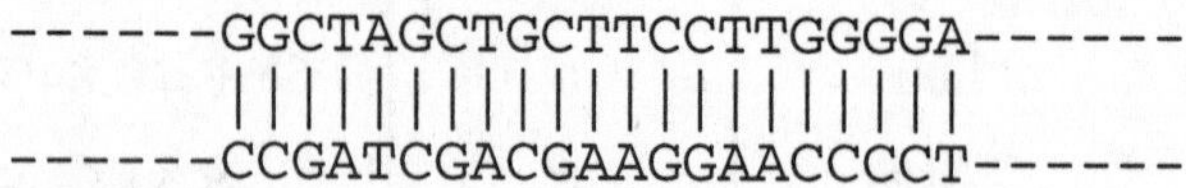

Which strand is the template strand? Label each strand with its correct polarity (5' and 3').

Answer: The template strand is that which is read to produce the mRNA. It is complementary to the mRNA, and of the opposite polarity. The non-template strand has the same 5'-to-3' polarity as the mRNA, and if U is replaced by T, the same sequence. Given a specified polypeptide segment, and a table of the genetic code, one can determine what possibilities exist for the first three codons of an mRNA (read 5'-to-3') and the sequence of the non-template DNA strand. Let (N= any nucleotide, Y = a pyrimidine (C or U) and R = a purine (G or A). In this case, one has:

amino acids:	Arg	- Gly -	Ser
potential codons:	5' AGR	GGN	AGY 3'
	or		or
	CGN		UCN
nontemplate strand sequence:	5' AGR	GGN	AGY 3'
	or		or
	CGN		TCN

Comparing the nontemplate strand sequence to that given, one finds that the top strand to be the non-template strand, with the 5' end on the right side. The template strand is the bottom strand, and transcription occurs from right to left. Note that one can also determine the C-terminus and the N-terminus of the polypeptide segment from this information.

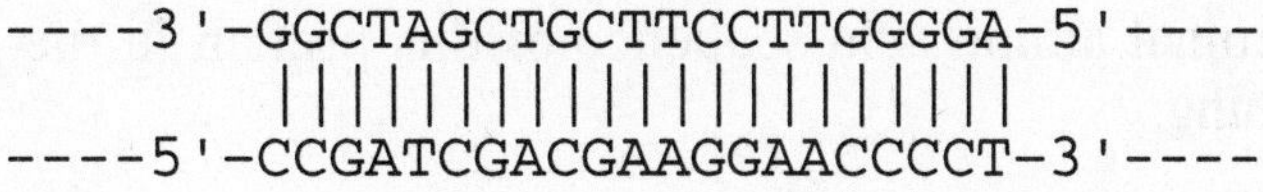

Transcribed into: 3' GGC UAG CUG CUU CCU UGG GGA 5'
Translated into: C -Arg-Asp-Val-Phe-Ser-Gly-Arg- N

14.16 Two populations of RNAs are made by the random combination of nucleotides. In population A the RNAs contain only A and G nucleotides (3A:1G), while in population B the RNAs contain only A and U nucleotides (3A:1U). In what ways *other than amino acid content* will the proteins produced by translating the population A RNAs differ from those produced by translating the population B RNAs?

Answer: In population A, the codons that can be produced encode Lys (AAA, AAG), Arg (AGG, AGA), Glu (GAG, GAA) and Gly (GGA, GGG). All of these are sense codons, and so long polypeptide chains will be synthesized containing these amino acids. In population B, the codons that can be produced encode Lys (AAA), Asn (AAU), Ile (AUA, AUU), Tyr (UAU), Leu (UUA), Phe (UUU) and stop (UAA). The frequency of the stop codon will be (1/4 x 3/4 x 3/4) = 9/64 = 0.14 or 14%. Thus, the polypeptides formed in population B will, on average, be shorter than those formed in population A. If a stop codon appears 14 percent of the time, one expects polypeptides that are, on average, about 7 amino acids long (1/0.14 = 7.14).

14.17 In *E. coli* a particular tRNA normally has the anticodon 5'-GGG-3', but because of a mutation in the tRNA gene, the mutant tRNA has the anticodon 5'-GGA-3'.

a. What amino acid would this tRNA carry?
b. What codon would the normal tRNA recognize?
c. What codon would the mutant tRNA recognize?
d. What would be the effect of the mutation on the proteins in the cell?

Answer:

a. The normal anticodon 5'-GGG-3' binds to the codon 5'-CCC-3', which encodes Pro (proline). Since the amino acid that is attached to the tRNA is unaffected by a mutant anticodon, the mutant tRNA will continue to carry proline to the ribosome.
b. The normal tRNA would recognize 5'-CCC-3'.

c. The mutant tRNA would recognize 5'-UCC-3'.

d. 5'-UCC-3' encodes Ser (serine). The mutant tRNA would therefore compete with tRNA.Ser for binding to UCC codons. In a percentage of UCC codons, proline would be inserted instead of serine into a polypeptide chain. If there is only one tRNA.Pro able to recognize the codon CCC (and if wobble is used to pair the 5' G in the anticodon with a 3' U and A in the codons CCU and CCA), the codon CCC (and perhaps CCU and CCA) would not be able to be read as a sense codon, and the ribosome would stall when the codon CCC (and perhaps CCU and CCA) were encountered. This could result in chain termination. Therefore, proteins with the wrong, or shortened amino acid sequences would be formed.

14.18 A particular protein found in *E. coli* normally has the N-terminal sequence Met-Val-Ser-Ser-Pro-Met-Gly-Ala-Ala-Met-Ser... . In a particular cell a mutation alters the anticodon of a particular tRNA from 5'-GAU-3' to 5'-CAU-3'. What would be the N-terminal amino acid sequence of this protein in the mutant cell? Explain your reasoning.

Answer: The anticodon 5'-GAU-3' recognizes the codon 5'-AUC-3', which encodes Ile. The mutant tRNA anticodon 5'-CAU-3' would recognize the codon 5'-AUG-3', which normally encodes Met. The mutant tRNA would therefore compete with tRNA.Met for the recognition of the 5'-AUG-3' codon, and if successful, insert Ile into a protein where Met should be. Since a special tRNA.Met is used for initiation, only codons other than the initiation AUG will be affected. Thus, this protein will have four different N-terminal sequences, depending on which tRNA occupies the A site in the ribosome when the codon AUG is present there:

Met-Val-Ser-Ser-Pro-**Ile**-Gly-Ala-Ala-**Ile**-Ser
Met-Val-Ser-Ser-Pro-**Met**-Gly-Ala-Ala-**Ile**-Ser
Met-Val-Ser-Ser-Pro-**Ile**-Gly-Ala-Ala-**Met**-Ser
Met-Val-Ser-Ser-Pro-**Met**-Gly-Ala-Ala-**Met**-Ser.

14.19 The gene encoding an *E. coli* tRNA containing the anticodon 5'-GUA-3' mutates so that the anticodon now is 5'-UUA-3'. What will be the effect of this mutation? Explain your reasoning.

Answer: The normal tRNA recognizes the codon 5'-UAC-3', and so must have carried the amino acid tyrosine. The altered anticodon will recognize the codon 5'-UAA-3', a chain termination codon. Consequently, a tyrosine will be inserted with the nonsense codon UAA on an mRNA is positioned in the A site of the ribosome. This will result in read-through of the mRNA some of the time (when the termination factor does not compete for binding to the chain termination codon), and the addition of amino acids onto the C-terminus of the protein. mRNAs having UAG and UGA chain termination codons will not be affected.

14.20 The normal sequence of the coding region of a particular mRNA is shown below, along with several mutant versions of the same mRNA. Indicate what protein would be formed in each case. (... = many [a multiple of 3] unspecified bases.)

normal: AUGUUCUCUAAUUAC(...)AUGGGGUGGGUGUAG
mutant *a*: AUGUUCUCUAAUUAG(...)AUGGGGUGGGUGUAG
mutant *b*: AGGUUCUCUAAUUAC(...)AUGGGGUGGGUGUAG

mutant *c*: AUGUUCUCGAAUUAC(...)AUGGGGUGGGUGUAG
mutant *d*: AUGUUCUCUAAAUAC(...)AUGGGGUGGGUGUAG
mutant *e*: AUGUUCUCUAAUUC(...)AUGGGGUGGGUGUAG
mutant *f*: AUGUUCUCUAAUUAC(...)AUGGGGUGGGUGUGG

Answer: First rewrite the sequences so that the codons can be readily seen, noting the mutations (underlined):

```
normal: AUG UUC UCU AAU UAC (...) AUG GGG UGG GUG UAG
     a: AUG UUC UCU AAU UAG (...) AUG GGG UGG GUG UAG
     b: AGG UUC UCU AAU UAC (...) AUG GGG UGG GUG UAG
     c: AUG UUC UCG AAU UAC (...) AUG GGG UGG GUG UAG
     d: AUG UUC UCU AAA UAC (...) AUG GGG UGG GUG UAG
     e: AUG UUC UCU AAU UC. ..)A UGG GGU GGG UGU AG.
     f: AUG UUC UCU AAU UAC (...) AUG GGG UGG GUG UGG
```

Mutants *a, b, c, d* and *f* are point mutations, in which one base has been substituted for another. Mutant *e* is a deletion of a single base that results in a shift in the reading frame of the mRNA (a frameshift mutation). Translating each sequence using the genetic code, one can determine the proteins that would be formed if these sequences were translated:

```
normal: AUG UUC UCU AAU UAC ... AUG GGG UGG GUG UAG
        met phe ser asn tyr ... met ala trp val stop

     a: AUG UUC UCU AAU UAG ... AUG GGG UGG GUG UAG
        met phe ser asn stop
```

Mutant *a* is a nonsense mutation, and results in premature chain termination.

```
     b: AGG UUC UCU AAU UAC ... AUG GGG UGG GUG UAG
                                met ala trp val stop
```

Mutant *b* mutates the initiation codon, so that a polypeptide would be formed (if formed at all) using a downstream initiation codon. It results in a polypeptide missing amino acids at its N-terminus.

```
     c: AUG UUC UCG AAU UAC ... AUG GGG UGG GUG UAG
        met phe ser asn tyr ... met ala trp val stop
```

Mutant *c* changes a base in the 3' end of the codon. This does not alter the amino acid that is inserted. It will be "silent" and have no phenotypic effect.

```
     d: AUG UUC UCU AAA UAC ... AUG GGG UGG GUG UAG
        met phe ser lys tyr ... met ala trp val stop
```

Mutant *d* changes a base in the 3' end of the codon. This does alter the amino acid that is inserted. It is a missense mutation, resulting in the insertion of a Lys instead of an Asn.

```
e: AUG UUC UCU AAU UC. ..A UGG GGU GGG UGU AG.
   met phe ser asn ser ... trp gly gly cys ?
```

Mutant *e* is a single base-pair deletion, and results in a frameshift mutation that alters the reading frame of the protein. All amino acids inserted following Asn are likely to be incorrect. It is conceivable that a stop codon could be read in the region that is indicated by the ..., leading to premature chain termination.

```
f: AUG UUC UCU AAU UAC ... AUG GGG UGG GUG UGG
   met phe ser asn tyr ... met ala trp val trp..
```

Mutant *e* changes a base in the chain terminating UAG codon so that the amino acid Trp will now be inserted. It will result in the addition of additional amino acids onto the C-terminus of the protein.

14.21 The normal sequence of a particular protein is given below, along with several mutant versions of it. For each mutant, explain what mutation occurred in the coding sequence of the gene.

Normal: Met-Gly-Glu-Thr-Lys-Val-Val-...-Pro
Mutant 1: Met-Gly
Mutant 2: Met-Gly-Glu-Asp
Mutant 3: Met-Gly-Arg-Leu-Lys
Mutant 4: Met-Arg-Glu-Thr-Lys-Val-Val-...-Pro

Answer: One approach to this problem is to infer the possible coding sequence(s) that could be used for the normal protein (using N = any nucleotide, R = purine, Y = pyrimidine), and then examine this sequence to deduce what possible mutations could have resulted in the mutant proteins.

Based on the normal coding sequence, one has:

amino acid sequence: `Met-Gly-Glu-Thr-Lys-Val-Val-...-Pro`
potential mRNA: `5'-AUG GGN GAR ACN AAR GUN GUN ... CCN-3'`

In mutant 1, a premature chain termination has occurred. This could have occurred if, in the DNA transcribed into the third (GAR (Glu)) codon, a GC base pair was changed to a TA base pair. This would lead to a UAR (stop) codon.

normal sequence: `Met-Gly-Glu-Thr-Lys-Val-Val-...-Pro`
normal mRNA: `5'-AUG GGN GAR ACN AAR GUN GUN ... CCN-3'`
mutant mRNA: `5'-AUG GGN UAR ACN AAR GUN GUN ... CCN-3'` (U underlined)
mutant sequence: `Met-Gly-stop`

It could also have occurred if a TA base pair insertion mutation occurred in the DNA, so that a U was transcribed in between the normal second and third codons, resulting in a UGA (stop) third codon.

amino acid sequence: `Met-Gly-Glu-Thr-Lys-Val-Val-...-Pro`
potential mRNA: `5'-AUG GGN GAR ACN AAR GUN GUN ... CCN-3'`
mutant mRNA: `5'-AUG GGN UGA RAC NAA RGU NGU N.. .CC-3'` (U underlined)
mutant sequence: `Met-Gly-stop`

In mutant 2, a premature chain termination has occurred after a wrong amino acid has been inserted. To explain both of these results as a consequence of a single mutational event, try either insertion or deletion mutations that would alter the reading frame. One possible explanation is that a GC base pair insertional mutation in the DNA resulted in a G being inserted after the third codon. If the N of the fourth codon were a U, then such a frameshifting insertion would change the Thr to Asp, and also introduce a chain termination codon into the fifth codon position.

amino acid sequence: Met-Gly-Glu-Thr-Lys-Val-Val-...-Pro
potential mRNA: 5'-AUG GGN GAR ACN AAR GUN GUN ... CCN-3'
mutant mRNA: 5'-AUG GGN GAR GAC UAA RGU NGU N.. .CC-3'
mutant sequence: Met-Gly-Glu-Asp-stop

In mutant 3, a similar situation to that in mutant 2 has occurred. This time however, several wrong amino acids are inserted before chain termination. To explain all of these consequences as the result of a single mutational event, check for the consequences of deletions or insertions in the region of the second and third codons. One possible explanation is that a deletion mutation in the DNA resulted in the N of the second codon being deleted. If, as in mutant 2, the N of the fourth codon is a U, and the R of the third codon is a G, the R of the fifth codon is an A, and the N of the sixth codon is an A, one would obtain the mutant sequence seen.

normal sequence: Met-Gly-Glu-Thr-Lys-Val-Val ... Pro
normal mRNA: 5'-AUG GGN GAR ACN AAR GUN GUN ... CCN-3'
normal mRNA: 5'-AUG GGN GAG ACU AAA GUA GUN ... CCN-3'
mutant mRNA: 5'-AUG GGG AGA CUA AAG UAG UN. ..C CN-3'
mutant sequence: Met-Gly-Arg-Leu-Lys-Stop

In mutant 4, the normal second amino acid (Gly) has been replaced with Arg. Arg is encoded by AGR or CGN, while Gly is encoded by GGN. If a GC base pair were substituted for a CG base pair in the DNA so that the first G of the second codon were replaced by a C, Arg would be inserted as the second amino acid.

normal sequence: Met-Gly-Glu-Thr-Lys-Val-Val ... Pro
normal mRNA: 5'-AUG GGN GAR ACN AAR GUN GUN ... CCN-3'
mutant mRNA: 5'-AUG CGN GAR ACN AAR GUN GUN ... CCN-3'
mutant sequence: Met-Arg-Glu-Thr-Lys-Val-Val ... Pro

14.22 In the recessive condition in humans known as sickle cell anemia, the β-globin polypeptide of hemoglobin is found to be abnormal. The only difference between it and the normal β-globin is that the 6th amino acid from the N-terminal is valine, whereas the normal β-globin has glutamic acid at this position. Explain how this occurred.

Answer: Both GAA and GAG code for glutamic acid, while GUU, GUC, GUA and GGG code for valine. The simplest explanation is that there was an AT to a TA change in the DNA, at the 17th base pair in the coding region of the gene. In this event, the 6th codon, instead of being GAA or GAG, would be GUA or GUG and encode valine.

14.23 Antibiotics have been useful in determining whether cellular events depend on transcription or translation. For example, actinomycin D is used to block transcription, and cycloheximide (in eukaryotes) is used to block translation. In some cases, though, surprising results are obtained after antibiotics are administered. The addition of actinomycin D, for example, may result in an increase, not a decrease, in the activity of a particular enzyme. Discuss how this result might come about.

Answer: Recall that some gene products can inhibit the activity of others. In this case, actinomycin D might block the transcription of a gene that codes for an inhibitor of an enzyme activity.

CHAPTER 15

RECOMBINANT DNA TECHNOLOGY AND THE MANIPULATION OF DNA

I. CHAPTER OUTLINE

GENE CLONING
- Restriction Enzymes
- Cloning Vectors and the Cloning of DNA

CONSTRUCTION OF GENOMIC LIBRARIES, CHROMOSOME LIBRARIES, AND cDNA LIBRARIES
- Genomic Libraries
- Chromosome Libraries
- cDNA Libraries

IDENTIFYING SPECIFIC CLONED SEQUENCES IN cDNA LIBRARIES AND GENOMIC LIBRARIES
- Identifying Specific Cloned Sequences in a cDNA Library
- Identifying Specific Cloned Sequences in a Genomic Library
- Identifying Specific DNA Sequences in Libraries Using Heterologous Probes
- Identifying Genes in Libraries by Complementation of Mutations
- Identifying Genes or cDNAs in Libraries Using Oligonucleotide Probes

ANALYSIS OF GENES AND GENE TRANSCRIPTS
- Restriction Enzyme Analysis of Cloned DNA Sequences
- Restriction Enzyme Analysis of Genes
- Analysis of Gene Transcripts

DNA SEQUENCE ANALYSIS
- Dideoxy (Sanger) DNA Sequencing
- Analysis of DNA Sequences

POLYMERASE CHAIN REACTION (PCR)

APPLICATIONS OF RECOMBINANT DNA TECHNOLOGY
- Analysis of Biological Processes
- Diagnosis of Human Genetic Diseases by DNA Analysis
- Isolation of Human Genes
- Human Genome Project
- DNA Typing
- Gene Therapy
- Commercial Products
- Genetic Engineering of Plants

II. REVIEW OF KEY TERMS, SYMBOLS AND CONCEPTS

Without consulting the text and in your own words, write a brief definition of each term in the groups below. Then, either using a short phrase or a simple diagram, identify the relationship(s) between specific pairs of terms within a set. Finally, consult the text (and perhaps a friend who has also done the exercise) to check your answers.

1	2	3	4
recombinant DNA	restriction endonuclease	cloning vector	plasmid vector
genetic engineering	isoschizomer	polylinker	λ vector
cloning vector	2-fold rotational symmetry	multiple cloning site	cosmid vector
molecular cloning	sticky, staggered end	*ori* vs. *cos* site	shuttle vector
restriction enzyme	5'-, 3'-overhang	selectable marker	*YAC*
restriction site	blunt end	unique restriction site	insert size
		blue-white selection	

5	6	7	8
genomic library	expression vector	restriction map	polymerase chain reaction (PCR)
partial digestion	library screening	preparative agarose gel electrophoresis	thermal cycler
cDNA	autoradiogram	Southern blot technique	amplification
cDNA library	autoradiography	northern blot analysis	primer
oligo(dT)	plaque lift	Maxam-Gilbert sequencing	ancient DNA
reverse transcriptase	nick translation labeling	Sanger (dideoxy) sequencing	thermostable DNA polymerase
RNase H	random primer labeling	dideoxy nucleotide	
linker	Klenow fragment	computer database	
chromosome library	nonradioactive labeling		
flow cytometry	digoxigenin		
	heterologous probe		
	oligonucleotide probe		

9	10	11
genetic counseling	Human Genome Project	*Agrobacterium tumefaciens*
restriction fragment length polymorphism (RFLP)	DNA typing (fingerprinting)	crown gall disease
in situ hybridization	highly polymorphic markers	monocots vs. dicots
CF (cystic fibrosis) gene	variable number of tandem repeats (VNTRs)	*Ti* plasmid
anonymous probe	gene therapy	*vir* (virulence) region
chromosome walking	somatic cell therapy	T-DNA
jumping library	germ-line therapy	plant cell transformation
CpG island	transgenic cell	transgenic cell
triplet repeat mutation	transgene	
Huntington's disease (HD) gene		

III. THINKING ANALYTICALLY

The material in this chapter is illustrative of the state of modern genetics. It presents many of the areas and issues that are on the cutting edge of progress in the discipline. As such, it requires one to consider core genetic concepts with state of the art technological advances. For example, identifying the chromosomal location and then molecularly cloning human genes utilizes the core concepts of genetic linkage with sophisticated RFLP and VNTR mapping, and chromosome walking. As you learn about the molecular methods, try to see how they rely on fundamental genetic principles.

The material in this chapter illustrates the key role played by technology in new discoveries and developments. It also shows how insights gained from exploration of basic scientific questions can be put to applied uses. In this chapter, illustrations can be found of the role of basic science in the diagnosis and treatment of human diseases, in the application of DNA typing to forensic issues, in the development of commercial products and in the application of genetic engineering in plants.

Thoroughly learning this material requires a clear understanding of not only the fundamental genetic principles underlying the application of the technology, but also of the logic and process of many different techniques. While it would of course be helpful to do or at least observe some of the techniques, this is not always an option. Indeed, the repertoire of techniques is now so great and so rapidly advancing that it is extremely unlikely that an individual scientist has first-hand experience in all methods. Still, individual scientists can think about how to use these methods constructively. As you approach a new method, think through the steps of the method very carefully, and address the point of each step. Try to address how different methods can be employed together, or how alternate methods could be used to accomplish a specific goal.

Much of the material in this chapter is presented at a quite sophisticated level of abstraction. It is a level that more often is seen in the physical sciences—chemistry and physics—than in biology. In its simplest form, the key question is how one manipulates things that one cannot sense directly (with the eyes, ears, nose, touch etc.). Each of the techniques explained in this chapter is based on a "model" of reality. They are based on our understanding of the molecular structure, function and activity of DNA, RNA, proteins, cells and viruses under certain conditions of temperature, growth media, and other environmental parameters. Each time one performs one of the techniques in this chapter and obtains reasonable results, those results validate (and sometimes expand) our view of the structure and action of various molecules and cellular processes. For example, we have never seen the actual base sequence in a fragment of DNA, but we have constructed a model of its structure based on prior experimental evidence. *If* that model is correct, then a technique such as the dideoxy-sequencing method should work. As you consider each of the methodologies in this chapter, ask yourself which basic models are being tested. Address what would occur if the model was incorrect in some fundamental aspect, that is, how the results of using the technique would be altered.

IV. QUESTIONS FOR PRACTICE

A. Multiple Choice Questions

For questions 1-11 match each of the techniques on the right with the statement that describes how or for what purpose it is used. Some of the techniques may be used more than once.

Purpose	Technique
1. To isolate a particular cDNA clone in a cDNA library.	a. Restriction mapping
2. To separate mRNA from a mixture of mRNA, rRNA and tRNA prior to construction of cDNA.	b. Partial digestion with restriction enzyme that recognizes 4 bp site.
3. To determine whether a particular mRNA is present in a cell at a specific stage of development.	c. Polymerase chain reaction
4. To determine gene homology.	d. Use expression vectors to produce a particular protein.
5. To determine the organization of introns in genes.	e. Perform a northern blot analysis.
6. To produce relatively large DNA fragments appropriate for cloning in a genomic library.	f. Blue/white selection
7. To identify plasmids that have inserted cDNA sequences.	g. Chromosomal walking
8. To identify phage having an insert homologous to a defined probe sequence.	h. Use oligo(dT) to isolate RNA with poly(A) tails.
9. To amplify a specific genomic DNA sequence.	i. RFLP mapping
10. To localize a disease gene to a specific chromosomal region.	j. Plaque lifts
11. To determine the size of a fragment of DNA inserted into a plasmid vector.	k. Agarose gel electrophoresis
	l. Southern blotting

12. Antibiotic resistance markers are important in the use of plasmid-cloning vectors
    a. because the plasmid must show resistance in order to accept inserted DNA.
    b. so that any previously sensitive bacterium that has taken up such a plasmid can be recognized.
    c. so that one can be sure of the presence of both the *ori* and the *cos* site.
    d. so that the resistance gene can be cut by a restriction enzyme.

13. When a particular PCR reaction is performed using genomic DNA as a template, a 1.5 kb product is amplified. When the same reaction is performed using cDNA as a template, a 0.8 kb product is amplified. The products are different because
    a. primers always bind to different sequences in different templates.
    b. there is a RFLP in the genomic DNA.
    c. there is an intron in the gene.
    d. the cDNA is degraded.

14. What are the reasons for choosing to screen a genomic library constructed in a cosmid instead of a phage?

a. One can screen fewer inserts.
b. One can screen more inserts.
c. Phage libraries only contain cDNA, not genomic DNA.
d. Cosmid libraries can be screened with an antibody.

15. The restriction enzyme *Bam*HI cleaves a phosphodiester bond between two GC base pairs at a six base pair site. The 5'-to-3' sequence of one of the sites is GGATCC. Which of the following are true?
a. *Bam*HI leaves a 5'-overhang.
b. *Bam*HI leaves a 3'-overhang.
c. *Bam*HI leaves a blunt end.
d. *Bam*HI recognizes a two-fold rotationally symmetric site.

Answers: 1d; 2h; 3e; 4a, l; 5a (k); 6b; 7f; 8j; 9c; 10i, g; 11k, a; 12b; 13c; 14a; 15a, d.

B. Thought Questions

1. Contrast the *cos* site with the *ori* site.
2. For what purposes would you choose to use each of a plasmid, lambda, cosmid or YAC library? When you could use either, why would you choose one over the other?
3. What kinds of organisms naturally make restriction enzymes? Of what use are they to the organism where they are naturally made?
4. Address the following questions to explore the kinds of results that could be obtained on a Southern blot made with genomic DNA.
   a. You digest the DNA with a restriction enzyme that cuts at a defined 6 bp sequence having 2 GC base pairs and 4 AT base pairs. You know that the genome is 40% GC. What is the *average* fragment size?
   b. You separate the cleaved DNA by size using agarose gel electrophoresis and stain the gel to observe the location of the DNA fragments. You observe a smear. Why do you observe a smear, and not just one band (or a few bands)?
   c. You transfer the DNA in the manner of Southern to a membrane, and allow sequences on the membrane to hybridize with a unique sequence probe made from a fragment that is 4 kb in length. You see three bands having sizes 3, 0.5 and 9 kb. How do you interpret this result?
   d. You allow the same blot to hybridize with a probe that has homology to middle repetitive DNA and has been made from a fragment that is 1.2 kb in size. You see about 40 bands on the blot, ranging in size from 3 to 17 kb. How do you interpret this result?
5. Construct a restriction map of a 10 kb DNA fragment using the following data:

ENZYMES USED	SIZES OF FRAGMENTS (in kb)
*Eco*RI	1, 4, 5
*Bam*HI	4 , 6
*Hin*dIII	0.8, 1.5, 7.7
*Eco*RI and *Bam*HI	1, 2, 3, 4
*Eco*RI and *Hin*dIII	0.5, 0.8, 1, 3.2, 4.5
*Bam*HI and *Hin*dIII	0.8, 1.5, 2.5, 5.2
*Bam*HI, *Eco*RI, and *Hin*dIII	0.5, 0.8, 1, 2, 2.5, 3.2

6. What are the aims of the human genome project? What kinds of information will be gathered, and how will (should) it be used? Who should have access to this information? Concurrent with the human genome project, the genomes of a number of other "model" organisms are being mapped, and possibly sequenced (e.g., the fly *Drosophila,* the plant *Arabidopsis,* the nematode *Caenorhabditis*, the yeast *Saccharomyces*). How will studying the organization and sequences of these genomes advance analysis of the human genome?

7. Many of the techniques used in molecular biology are exquisitely sensitive. Does this just complicate, or rule out entirely, their use in forensic investigations?

V. SOLUTIONS TO TEXT PROBLEMS

15.1 A new restriction endonuclease is isolated from a bacterium. This enzyme cuts DNA into fragments that average 4,096 base pairs long. Like all other known restriction enzymes, the new one recognizes a sequence in DNA that has twofold rotational symmetry. From the information given, how many base pairs of DNA constitute the recognition sequence for the new enzyme?

Answer: The average length of the fragments is an indication of how often, on average, one finds a particular sequence. If one assumes that DNA is equally composed of A, T, G, and C, then one can calculate the probability of finding a particular sequence at a particular point in a polynucleotide chain. The chance of finding a particular base pair (AT, TA, GC, or CG) in a specific position is 1/4. The chance of finding two base pairs in a specific position is $(1/4) \times (1/4) = (1/4)^2 = 1/16$. In general, the chance of finding n base pairs in a particular position is $(1/4)^n$. Here, one sees a particular sequence every 4,096 bases, so that the chance of seeing the sequence is 1/4,096. Noting that $4{,}096 = 4^6$, one can deduce that the enzyme recognizes a 6 bp sequence.

15.2 An endonuclease called *Avr*II ("a-v-r-two") cuts DNA whenever it finds the sequence
5'-CCTAGG-3'
3'-GGATCC-5'. About how many cuts would *Avr*II make in the human genome, which is about 3×10^9 base pairs long and about 40% GC?

Answer: The enzyme recognizes a sequence that has two GC base pairs, two CG base pairs, one AT base pair and one TA base pair in a particular order. Since the genome has 40% GC, the chance of finding a GC or CG base pair is 0.20, and the chance of finding an AT or a TA base pair is 0.30. The chance of finding six base pairs with this sequence is $(0.20)^4(0.3)^2 = 0.000144$. A genome with 3×10^9 base pairs will have about 3×10^9 different groups of six-base pair sequences. Thus, the number of sites in the human genome is $(0.000144) \times (3 \times 10^9) = 432{,}000$.

15.3 About 40 percent of the base pairs in human DNA are GC. On the average, how far apart (in terms of base pairs) will the following sequences be?

a. two *Bam*HI sites
b. two *Eco*RI sites
c. two *Not*I sites
d. two *Hae*III sites

Answer: Since the human genome has 40 percent GC, the probability of finding a GC or CG base pair is 0.20, and the probability of finding a TA or AT base pair is 0.30. [This assumes that in any region of the genome, one will find, on average, 40 percent GC or CG base pairs.] This information can be used to determine the probability of finding a particular restriction enzyme recognition sequence as follows:

ENZYME	RECOGNITION SEQUENCE	PROBABILITY OF FINDING SEQUENCE	AVERAGE DISTANCE BETWEEN SITES
*Bam*HI	5'-GGATCC-3' 3'-CCTAGG-5'	$(0.2)^4(0.3)^2 =$ 0.000144	1/0.000144 = 6,944 bp
*Eco*RI	5'-GAATTC-3' 3'-CTTAAG-5'	$(0.2)^2(0.3)^4 =$ 0.000324	1/0.000324 = 3,086 bp
*Not*I	5'-GCGGCCGC-3' 3'-CGCCGGCG-5'	$(0.2)^8 =$ 0.00000256	1/0.00000256 = 390,625 bp
*Hae*III	5'-GGCC-3' 3'-CCGG-5'	$(0.2)^4 = 0.0016$	1/0.0016 = 625 bp

15.4 What are the features of plasmid cloning vectors that make them useful for constructing and cloning recombinant DNA molecules?

Answer: Plasmids need three essential features to be utilized as cloning vectors:

1. A bacterial *ori*, or origin of replication sequence, to allow it to replicate in *E. coli.*
2. A dominant selectable marker, such as antibiotic resistance, to allow selection of cells harboring the plasmid.
3. At least one unique restriction enzyme cleavage site, so that DNA sequences cut with that enzyme can be spliced into the plasmid.

Modern plasmid cloning vectors have been engineered to possess additional features that facilitate easier use as cloning vectors.

1. They are present in a high copy number, which facilitates purification of plasmid DNA.
2. They contain many unique restriction sites in a *polylinker* or *multiple cloning site*, to facilitate cloning fragments of DNA obtained after cleavage with a variety of different restriction enzymes.
3. The polylinker is inserted in the 5'-end of the *lacZ* gene, which encodes β-galactosidase. Cells harboring a plasmid with an intact *lacZ* gene, when grown on media with the β-galactosidase substrate X-gal, form blue colonies. Cells harboring a plasmid whose *lacZ* gene has been interrupted by a cloned segment of DNA will not express functional β-galactosidase, and be white. Thus, one can tell if a bacteria colony harbors a plasmid with DNA insertion by the color of the colony.
4. They contain phage promoters flanking each side of the polylinker. These promoters are used to make *in vitro* RNA copies of the cloned DNA. These promoters can then be used for a variety of purposes, including making radioactively-labeled RNA probes.

15.5 Genomic libraries are important resources for isolating genes of interest and for studying the functional organization of chromosomes. List the steps you would use to make a genomic library of yeast in a lambda vector.

Answer: A genomic library made in a λ vector is a collection of λ phage that have had a portion of their genome replaced by different yeast genomic DNA sequences. Like two volumes of book series, two λ phage will have identical external protein coat and λ vector sequences, but different yeast DNA inserts. Such a library is made as follows:

1. Isolate high molecular weight yeast genomic DNA by isolating nuclei, lysing them, and gently purifying their DNA.
2. Cleave the DNA into fragments that are an appropriate size for the λ vector being used. This can be done by cleaving the DNA with *Sau*3A for a limited time (i.e., performing a *partial* digest), and then selecting fragments of an appropriate size by either sucrose density centrifugation or agarose gel electrophoresis.
3. Remove the central portion of the λ vector by digestion with *Bam*HI.
4. Anneal, and then ligate the left and right arms of the λ vector to the yeast DNA. The sticky ends that are left by the *Sau*3A and *Bam*HI are complementary.
5. Package the recombinant DNA molecules *in vitro* into λ particles.
6. Infect *E. coli* cells with the λ phage population, and collect progeny phage produced by cell lysis. These phage have different yeast DNA inserts, and represent the yeast genomic library.

15.6 The human genome contains about 3×10^9 bp of DNA. How many 40 kb pieces would you have to clone into a library if you wanted to be 90 percent certain of including a particular sequence.

Answer: There are two ways to solve this problem.

(1) Deduce the probability relationships from first principles.

p(identifying sequence in <u>one</u> 40 kb piece) $= 40{,}000 \text{ bp}/3 \times 10^9 \text{ bp}$
$= 13.333 \times 10^{-6}$

p(<u>not</u> finding sequence in <u>one</u> 40kb piece) $= [1 - (13.333 \times 10^{-6})]$

p(<u>not</u> finding sequence in <u>*n*</u> clones) $= [1 - (13.333 \times 10^{-6})]^n$

If, one wants to be 90 percent certain of finding the sequence, one is willing to allow a 10 percent chance that the sequence will not be found. Thus,

$0.10 = [1 - (13.333 \times 10^{-6})]^n$
$n = \ln(0.10)/\ln[1 - (13.333 \times 10^{-6})]$
$n = 172{,}693$ pieces.

(2) Recall from the text presentation that the probability of having any sequence represented in a genomic library is given by $N = \ln(1 - p)/\ln(1 - f)$, where N = necessary number of DNA molecules, p = probability of finding the sequence, f = fractional proportion of the genome in a single recombinant DNA molecule. Here, $p = 0.90$, $f = 40{,}000/(3 \times 10^9)$, so $N = 172{,}693$.

15.7 What is a cDNA library and from what cellular genetic material is it derived? How is a cDNA library used in cloning particular genes?

Answer: A cDNA library contains inserts made from DNA complementary to mRNA. The cDNA is synthesized by annealing an oligo-dT primer to the poly-A tail of mRNA, and using reverse-transcriptase to synthesize a DNA copy of the mRNA strand. The mRNA is partially degraded by RNase H, leaving a single-stranded complementary DNA with a short mRNA fragment attached. DNA polymerase I is then used to synthesize a second, complementary DNA strand, using the short mRNA as a primer. The resulting double-stranded DNA is cloned into a vector and propagated. See text pp. 470 - 471, especially Figures 15.12 and 15.13.

If a cDNA library is constructed using mRNA isolated from a particular tissue, the cDNA inserts represent partial copies of genes transcribed in that tissue. Thus, each clone can be used to identify a gene expressed in that tissue. If cDNA inserts are cloned into an expression vector, the cDNAs can be transcribed and translated. In this way, the protein products of the cloned cDNAs can be produced in a bacterial cell. If an antibody is available that binds to a protein expressed in a particular tissue, the antibody can be radioactively-labeled and used as a probe to identify clones expressing the protein. Such clones have cDNAs that encode the protein. This method of screening an expression vector library is described in detail on text pp. 471-472.

15.8 Suppose you wanted to produce human insulin (a peptide hormone) by cloning. Assume that this could be done by inserting the human insulin gene into a bacterial host, where, given the appropriate conditions, the human gene would be transcribed then translated into human insulin. Which do you think it would be best to use as your source of the gene, human genomic insulin DNA or a cDNA copy of this gene? Explain your choice.

Answer: It would be preferable to use cDNA. Human genomic DNA contains introns, while cDNA synthesized from cytoplasmic poly-A$^+$ mRNA does not. Prokaryotes do not process eukaryotic precursor mRNAs having intron sequences, so genomic clones will not give appropriate translation products. Since cDNA is a complementary copy of a functional mRNA molecule, the mRNA transcript will be functional, and when translated human (pro-)insulin will be synthesized.

15.9 You are given a genomic library of yeast prepared in a bacterial plasmid vector. You are also given a cloned cDNA for human actin, a protein which is conserved in protein sequences among eukaryotes. Outline how you would use these resources to attempt to identify the yeast actin gene.

Answer: One can radioactively- (or non-radioactively) label the human actin cDNA, and use it as a heterologous probe to screen the yeast genomic library. To radioactively label the human actin cDNA using a random primer method, denature the double-stranded cDNA by boiling, allow "random" short oligonucleotides (primers) to anneal to a cDNA strand, and synthesize DNA complementary to the cDNA using one or more labeled dNTPs and the Klenow fragment of DNA polymerase I. Once a probe has been made, clones harboring plasmids with inserts encoding the yeast actin gene can be identified as follows:

Plate the library onto bacterial media in petri plates at a density that allows separate colonies to be identified. Overlay the bacterial colonies with a positively charged membrane that can bind DNA, and then lift it off so that some of each bacterial colony is attached to the membrane. Lyse the bacterial cells bound to the membrane *in situ* using an alkaline solution, so that the plasmid DNA harbored by the cells binds to the membrane in single-stranded form. Allow the probe to hybridize with sequences on the membrane. Detect the location of the bound probe (e.g., use autoradiography if the probe is radioactively labeled), and then align the membrane with the original Petri plate from which colonies were lifted. Identify the clone having the hybridizing cDNA insert. This clone will have a yeast genomic DNA insert that has sequence homology to the human actin gene probe, and most likely, is a yeast actin gene.

15.10 Restriction endonucleases are used to construct restriction maps of linear or circular pieces of DNA. The DNA is usually produced in large amounts by recombinant DNA techniques. The generation of restriction maps is similar to the process of putting the pieces of a jigsaw puzzle together. Suppose we have a circular piece of double-stranded DNA that is 5,000 base pairs long. If this DNA is digested completely with restriction enzyme I, four DNA fragments are generated: fragment *a* is 2,000 base pairs long; fragment *b* is 1,400 base pairs long; *c* is 900 base pairs long; and *d* is 700 base pairs long. If, instead, the DNA is incubated with the enzyme for a short time, the result is incomplete digestion of the DNA, not every restriction enzyme site in every DNA molecule will be cut by the enzyme, and all possible combinations of adjacent fragments can be produced. From an incomplete digestion experiment of this type, fragments of DNA were produced from the circular piece of DNA, which contained the following combinations of the above fragments: *a-d-b, d-a-c, c-b-d, a-c, d-a, d-b* and *b-c*. Lastly, after digesting the original circular DNA to completion with restriction enzyme I, the DNA fragments were treated with restriction enzyme II under conditions conducive to complete digestion. The resulting fragments were: 1,400, 1,200, 900, 800, 400, and 300. Analyze all the data to locate the restriction enzyme sites as accurately as possible.

Answer: Sort out the results for the first enzyme by tabulating the results:

COMPLETE DIGESTION		PARTIAL DIGESTION	
FRAGMENT	SIZE (BP)	FRAGMENT	SIZE (BP)
a	2,000	*d-a-c*	3,600
b	1,400	*a-d-b*	3,100
c	900	*c-b-d*	3,000
d	700	*a-c*	2,900
total plasmid	5,000	*d-a*	2,700
		b-c	2,300
		d-b	2,100

Consider the following: if an enzyme cuts a circular molecule once, it will produce one fragment. If an enzyme cuts a circular molecule twice, it will produce two fragments. (Diagram these situations to convince yourself of this.) Since four fragments are produced when enzyme I completely cleaves the plasmid, enzyme I must cut the plasmid at four sites. A partial digestion occurs when all four sites are not cut. For the partial digestion fragments that contain three fragments, two cuts were made in neighboring sites. Thus, the *d-a-c* fragment was released when two cuts were made at sites flanking fragment *b*, the *a-d-b* fragment was released when two cuts were made at sites flanking fragment *c*, and the *c-b-d* fragment was released when two cuts were made a sites flanking fragment *a*. For the partial digestion fragments containing only two fragments, cuts were made that flank both fragments. Thus, *a* is next to *c*, *d* is next to *a*, *b* is next to *c* and *d* is next to *b*. This information can be used to order the fragments in the plasmid. Since *a* is next to *c* and *d*, the order must be *c-a-d*. Since *b* is next to *c* and *d*, the order must be *d-b-c*. Since *d* is next to *a* and *b*, the order must be *a-d-b*. Thus, the order of fragments in the plasmid is *c-a-d-b*.

When the plasmid is cleaved with both enzyme I and enzyme II, six fragments are produced. This indicates that the two enzymes together recognize six sites. Since enzyme I cleaves at four sites, enzyme II must cleave at two sites. Since the 1,400 and 900 bp fragments produced when enzyme I cleaves the plasmid remain intact in the double digestion, and the 2,000 and 700 bp fragments do not remain intact in the double digestion, enzyme II must cleave at sites within the 2,000 and 700 bp fragments (*a* and *d*). Given the 1,200 and 800 bp fragments produced with the double digestion, enzyme II must cleave *a* 800 bp from an enzyme I site. Given the 400 and 300 bp fragments produced with the double digestion, enzyme II must cleave *d* 300 bp from an enzyme I site. This gives the following map:

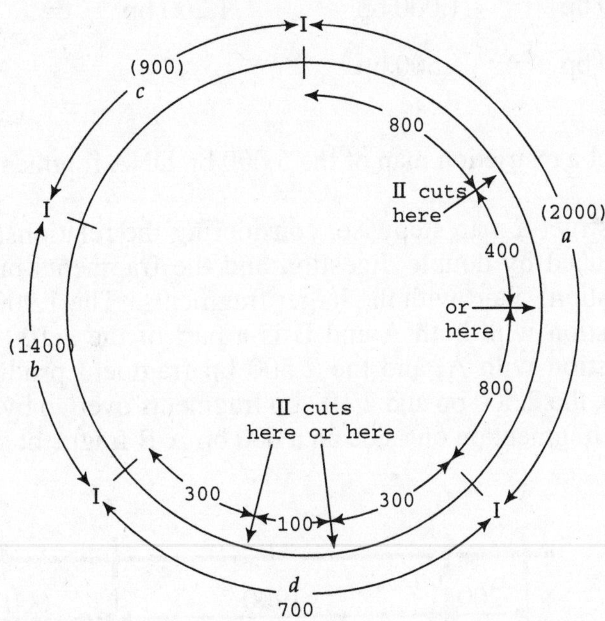

The ambiguities in enzyme II site positions could be resolved if one knew what size fragments were produced when the plasmid was digested to completion with enzyme II alone.

15.11 A piece of DNA 5,000 bp long is digested with restriction enzymes A and B, singly and together. The DNA fragments produced were separated by DNA electrophoresis and their sizes were calculated, with the following results:

Digestion With		
A	B	A + B
2,100 bp	2,500 bp	1,900 bp
1,400 bp	1,300 bp	1,000 bp
1,000 bp	1,200 bp	800 bp
500 bp		600 bp
		500 bp
		200 bp

Each A fragment was extracted from the gel and digested with enzyme B, and each B fragment was extracted from the gel and digested with enzyme A. The sizes of the resulting DNA fragments were determined by gel electrophoresis, with the following results.

A Fragment		Fragments Produced by Digestion with B	B Fragment		Fragments Produced by Digestion with A
2,100 bp	→	1,900, 200 bp	2,500 bp	→	1,900, 600 bp
1,400 bp	→	800, 600 bp	1,300 bp	→	800, 500 bp
1,000 bp	→	1,000 bp	1,200 bp	→	1,000, 200 bp
500 bp	→	500 bp			

Construct a restriction map of the 5,000 bp DNA fragment.

Answer: Construct a map stepwise, considering the relationship between the fragments produced by double digestion and the fragments produced by single enzyme digestion. Start with the larger fragments. The 1,900 bp fragment produced by digestion with both A and B is a part of the 2,100 bp fragment produced by digestion with A, and the 2,500 bp fragment produced by digestion with B. Thus, the 2,500 bp and 2,100 bp fragments overlap by 1,900 bp, leaving a 200 bp A-B fragment on one side an a 600 bp A-B fragment on the other. One has:

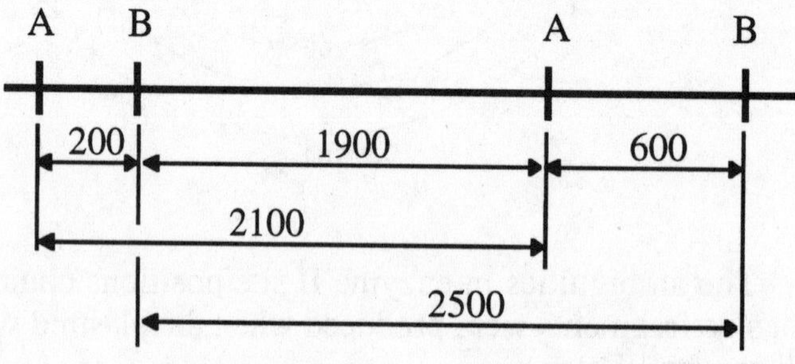

The map is extended in a stepwise fashion, until all fragments are incorporated into the map. The restriction map is:

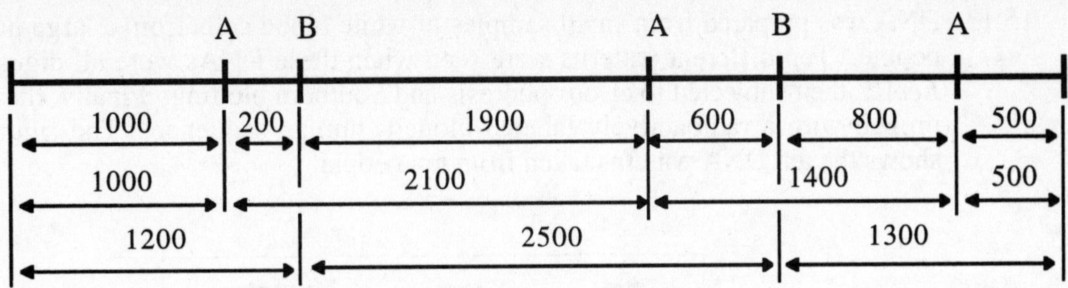

15.12 Draw the banding pattern you would expect to see on a DNA-sequencing gel if you annealed the primer 5'-C-T-A-G-G-3' to the following single-stranded DNA fragment and carried out a dideoxy sequencing experiment. Assume the dNTP precursors were all labeled.

3'-G-A-T-C-C-A-A-G-T-C-T-A-C-G-T-A-T-A-G-G-C-C-5'.

Answer: The primer will anneal to the fragment, and be extended at its 3'-end in four separate reactions, each with small amounts of a different dideoxynucleotide. In each reaction, some chains will be prematurely terminated when the dideoxynucleotide is incorporated. By using labeled dNTP precursors, all extension products will be labeled, and be able to be observed as distinct, labeled bands after separation on a denaturing polyacrylamide gel and signal detection.

fragment: `3'-GATCCAAGTCTACGTATAGGCC-5'`
primer: `5'-CTAGG`

GEL ANALYSIS
ddNTP Added

extension products:		ddA	ddG	ddC	ddT
5'-CTAGGTTCAGATGCATATCCG**G**	**G**		—		
5'-CTAGGTTCAGATGCATATCC**G**	**G**		—		
5'-CTAGGTTCAGATGCATATC**C**	**C**			—	
5'-CTAGGTTCAGATGCATAT**C**	**C**			—	
5'-CTAGGTTCAGATGCATA**T**	**T**				—
5'-CTAGGTTCAGATGCAT**A**	**A**	—			
5'-CTAGGTTCAGATGCA**T**	**T**				—
5'-CTAGGTTCAGATGC**A**	**A**	—			
5'-CTAGGTTCAGATG**C**	**C**			—	
5'-CTAGGTTCAGAT**G**	**G**		—		
5'-CTAGGTTCAGA**T**	**T**				—
5'-CTAGGTTCAG**A**	**A**	—			
5'-CTAGGTTCA**G**	**G**		—		
5'-CTAGGTTC**A**	**A**	—			
5'-CTAGGTT**C**	**C**			—	
5'-CTAGGT**T**	**T**				—
5'-CTAGG**T**	**T**				—

15.13 DNA was prepared from small samples of white blood cells from a large number of people. Ten different patterns were seen when these DNAs were all digested with *Eco*RI, then subjected to electrophoresis and Southern blotting. Finally, the blot was probed with a radioactively labeled cloned human sequence. The figure below shows the ten DNA patterns taken from ten people.

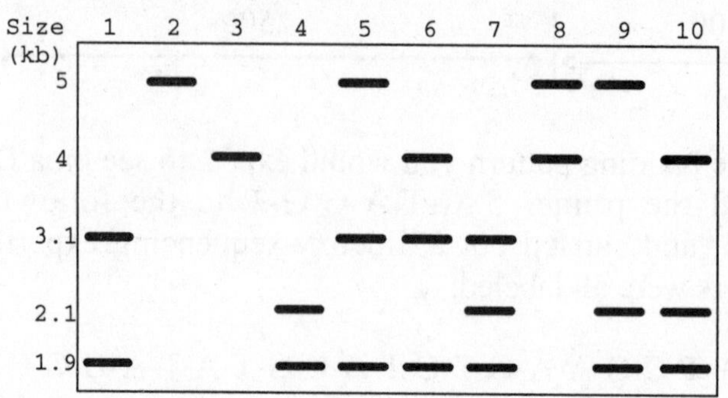

a. Explain the hybridization patterns seen in the ten people in terms of variation in *Eco*RI sites.
b. If the individuals whose DNA samples are in lanes 1 and 6 on the blot were to produce offspring together, what bands would you expect to see in DNA samples from these offspring?

Answer: a. The probe hybridizes to the same genomic region in each of the ten individuals. Different patterns of hybridizing fragments are seen because of polymorphism of the *Eco*RI sites in the region. If a site is present in one individual, but absent in another, different patterns of hybridizing fragments will be seen. This provides evidence of restriction fragment length polymorphism. To distinguish between sites that are invariant and those that are polymorphic, analyze the pattern of bands that appear. Notice that the sizes of the hybridizing bands in individual 1 add up to 5 kb, the size of the band in individual 2 and the largest hybridizing band. This indicates that there is a polymorphic site within a 5 kb region. This is indicated in the diagram below, where the asterisk over site *b* depicts a polymorphic *Eco*RI site:

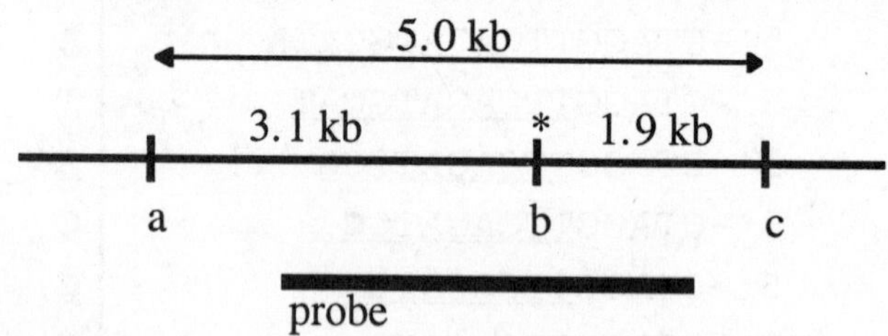

Notice also that the size of the band in individual 3 equals the sum of the sizes of the bands in individual 4. Thus, there is an additional polymorphic site in this 5 kb region. Since the 1.9 kb band is retained in individual 4, the additional site must lie within the 3.1 kb fragment. This site, denoted *x*, is incorporated into the diagram below. Notice that since the 1.0 kb fragment flanked by sites *a* and *x* is not seen on the Southern blot, the probe does not extend into this region.

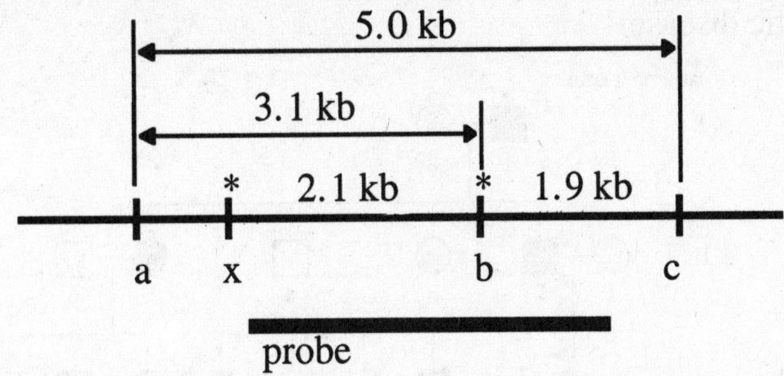

Depending on whether *x* and/or *b* are present, one will see either 5 kb, 3.1 and 1.9 kb, 2.1 and 1.9 kb, or 4 kb bands. In addition, if an individual has chromosomes with different polymorphisms, one can see combinations of these bands. Thus, individual 5 has one chromosome that lacks sites *x* and *b*, and one chromosome that has site *b*. The chromosomes in each individual can be tabulated as follows:

INDIVIDUAL	SITES ON EACH HOMOLOGUE	HOMOZYGOTE OR HETEROZYGOTE?
1	*a, b, c*	homozygote
2	*a, c*	homozygote
3	*x, c*	homozygote
4	*x, b, c*	homozygote
5	*a, c/a, b, c*	heterozygote
6	*x, c/a, b, c*	heterozygote
7	*a, b, c/x, b, c*	heterozygote
8	*a, c/x, c*	heterozygote
9	*a, c/x, b, c*	heterozygote
10	*x, c/x, b, c*	heterozygote

b. Since individual 1 is homozygous, chromosomes with sites at *a, b,* and *c* will be present in all of the offspring, giving bands at 3.1 and 1.9 kb. Individual 6 will contribute chromosomes of two kinds, one has sites at *x* and *c* and one has sites at *a, b,* and *c.* Thus, if this analysis is performed on their offspring, two equally frequent patterns will be observed: a pattern of bands at 3.1 and 1.9 kb, and a pattern of bands at 4, 3.1 and 1.9 kb. This is just like the patterns seen in the parents.

15.14 Filled symbols in the pedigree below indicate people with a rare autosomal dominant genetic disease.

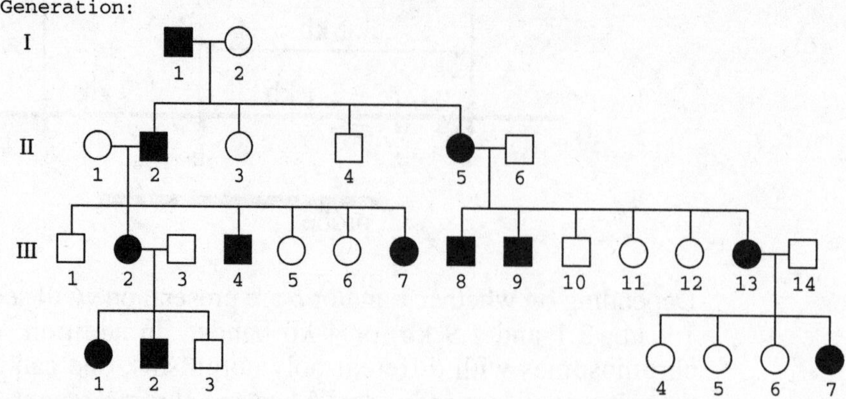

DNA samples were prepared from each of the individuals in the pedigree. The samples were restricted, electrophoresed, blotted, and probed with a cloned human sequence called DS12-88, with the results shown in the figure below.

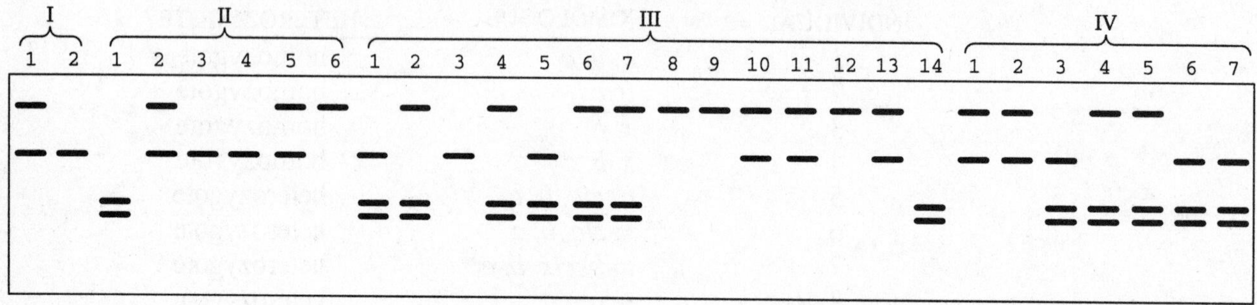

Do the data in these two figures support the hypothesis that the locus for the disease that is segregating in this family is linked to the region homologous to DS12-88? Make your answer quantitative, and explain your reasoning.

Answer: Assign the rare and autosomal dominant trait the symbol *D,* and the normal allele the symbol *d.* Since the trait is rare and autosomal dominant, and especially since the affected individuals in the pedigree (except for I-1) are known to have one normal parent, affected individuals are heterozygous (*D*/*d*) for the trait. Inspection of the Southern blot results shows that three different haplotypes are apparent. Homozygotes for each haplotype are found in individuals I-2 (one intermediate sized band), II-1 (two small bands) and II-6 (one large band). Call the haplotype seen in individual II-6 haplotype A, the haplotype seen in individual I-2 haplotype B, and the haplotype seen in individual II-1 haplotype C. Redraw the pedigree, including haplotype and genotype information. Wherever possible, include information on alleles contributed from the parents of each individual. This is done in the diagram below by using a slash to separate alleles contributed by one parent from those contributed by the other. Here, a slash does not necessarily mean that the alleles lie on the same chromosome.

Generation:

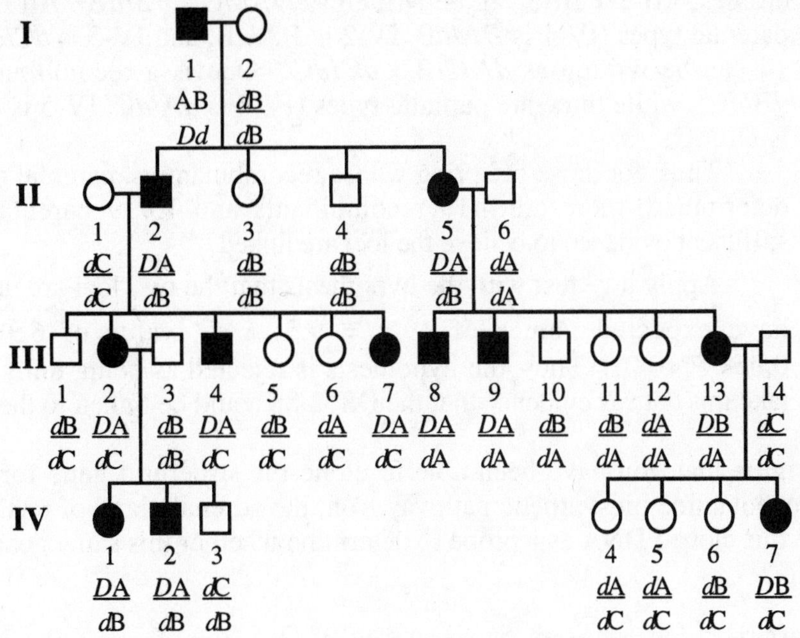

If the locus is linked to the disease trait, then a haplotype in an affected individual should be linked to the disease allele. In quantitative terms, it should show less than 50 percent recombination with the disease allele. Just as one would analyze data of a two-point cross, one can analyze how frequently progeny are recombinant or parental types. This is made somewhat complex by the fact that the recombinant or parental type status of some offspring cannot be determined. Thus, restrict the analysis to those offspring whose genotypes can be clearly determined.

Consider the first generation and their progeny. Individual I-1 is *Dd* and AB, individual I-2 is *dd* and BB. We cannot determine what the nature of a parental or recombinant type gamete is for individual I-1. Put another way, if the *D*/*d* and the DS12-88 loci are linked, we do not know whether I-1 had the *D* and A alleles contributed by one parent and the *d* and B alleles contributed by the other parent , or had the *D* and B alleles contributed by one parent and the *d* and B alleles contributed by the other parent. Thus, we cannot infer whether I-1 gives his progeny recombinant or parental type gametes. However, because individual I-2 is homozygous for both the *d* and B traits, we can be sure that individual I-2 contributes both *d* and B to her offspring. We can therefore infer that both II-2 and II-5 are *d*B/*D*A. Since we know their genotypes, and they mate with homozygotes, we can determine the alleles they contribute to their progeny and assess their progeny for receipt of recombinant and parental type chromosomes.

The progeny of the second generation are produced by two different crosses. II-1 x II-2 can be written as *d*C/*d*C x *d*B/*D*A. Of the six progeny, 1 is a recombinant type (III-6 is *d*C/***d*A**), while five are parental types (III-1 is *d*B/*d*C, III-2 is *D*A/*d*C, III-4 is *D*A/*d*C, III-5 is *d*B/*d*C, and III-7 is *D*A/*d*C). II-5 x II-6 can be written as *d*B/*D*A x *d*A/*d*A. Of the six progeny, two are recombinant types (III-12 is ***d*A**/*d*A, and III-13 is ***D*B**/*d*A), while four are parental types (III-8 is *D*A/*d*A, III-9 is *D*A/*d*A, III-10 is *d*B/*d*A, and III-11 is *d*B/*d*A).

The progeny of the third generation are also produced by two different crosses. III-2 x III-3 can be written as *DA*/*d*C x *d*B/*d*B. All three progeny are parental types (IV-1 is *DA*/*d*B, IV-2 is *DA*/*d*B, and IV-3 is *d*C/*d*B). III-13 x III-14 can be written as *d*A/*D*B x *d*C/*d*C. One is a recombinant type (IV-6 is *d*B/*d*C), while three are parental types (IV-4 is *d*A/*d*C, IV-5 is *d*A/*d*C, and IV-7 is *D*B/*d*C).

Thus, for those offspring whose recombinant vs. parental type status can be determined, there are 4/19 recombinants and 15/19 parental types. Is this sufficient evidence to believe the loci are linked?

Apply a χ^2 test with the hypothesis that the two loci are unlinked. One will have expected values of 19/2 = 9.5, a χ^2 value of 6.37, df. = 1, and $0.05 < P < 0.01$. Thus, the hypothesis is rejected as being unlikely, and one can take this data as evidence that the DS12-88 could be linked to the *D*/*d* gene.

15.15 Imagine that you have been able to clone the structural gene for an enzyme in a catecholamine biosynthetic pathway from the adrenal gland of rats. How could you use this cloned DNA as a probe to determine whether this same gene functions in the brain?

Answer: If the same gene functions in the brain, the gene for the enzyme must be transcribed into a precursor mRNA, processed to a mature mRNA and then translated to produce the functional enzyme. Thus, transcripts for the gene should be found in the brain. To address this issue, label the cloned DNA, and use it to probe a Northern blot having mRNA isolated from brain tissue. If the mRNA is rare, it may be prudent to use mRNA isolated from a specific region of the brain, e.g., the hypothalamus. An alternative, quite sensitive approach would be to sequence the cloned DNA, analyze the sequence to identify the coding region, and then design PCR primers that could be used to amplify cDNA made from mRNA isolated from various brain regions. Obtaining a PCR product in such an investigation would provide evidence that the gene is transcribed in the brain. In this alternative method, it would be important to be sure that no genomic DNA was present in the PCR amplification mixture, as the gene for the enzyme would be found in genomic DNA in both tissues.

15.16 Imagine that you find an RFLP in the rat genomic region homologous to your cloned catecholamine synthetic gene from 15.15, and that in a population of rats displaying this polymorphism there is also a behavioral variation. You find that some of the rats are normally calm and placid, but others are hyperactive, nervous, and easily startled. Your hypothesis is that the behavioral difference seen is caused by variations in your gene. How could you use your cloned sequence to test this hypothesis?

Answer: Initially, perform a genomic Southern analysis to see if the RFLP is limited to animals with or without the behavioral phenotype. Isolate DNA from the affected and normal rats, cleave it with the restriction enzyme that was used to identify the RFLP, separate the DNA by size on an agarose gel, blot the gel in the manner of Southern to a membrane, make a probe from DNA in the cloned catecholamine synthetic gene region, hybridize the probe to the Southern blot, and detect the location of the probe signal. If normally behaving rats and abnormally behaving rats share identical RFLPs, evidence against the hypothesis has been gathered. If normally behaving rats and abnormally behaving rats do not share identical RFLPs, evidence against the hypothesis has not been gathered. However, this does not demonstrate that the behavioral difference is caused by the DNA difference.

To gather more supportive evidence, cross the affected and unaffected rats to control strains, and follow the segregation of the RFLP and the behavioral phenotype. If the two traits co-segregate, one would have evidence of linkage. The frequency of recombination will give an estimate of how close the DNA region is to the site of the mutation causing the behavioral phenotype. If the RFLP and the behavioral difference can be separated by recombination, the variations in this genomic region cannot be the molecular basis of the behavioral phenotype.

15.17 The maps of the sites for restriction enzyme R in the wild type and the mutated cystic fibrosis genes are shown schematically in the following figure:

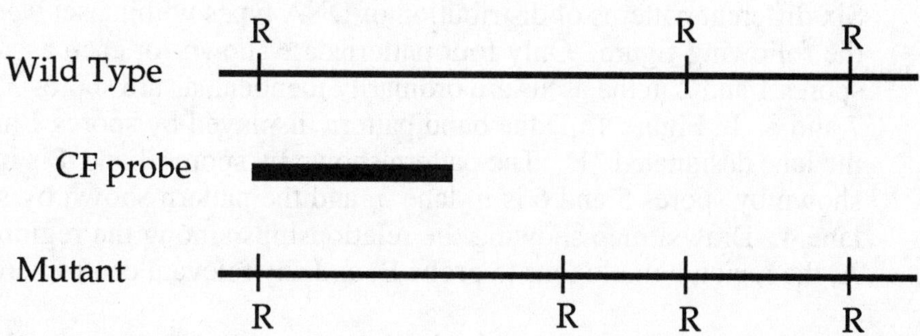

Samples of DNA obtained from a fetus (F) and her parents (M and P) were analyzed by gel electrophoresis followed by the Southern blot technique and hybridization with the radioactively labeled probe designated "CF probe" in the above figure. The autoradiographic results are shown in the following figure:

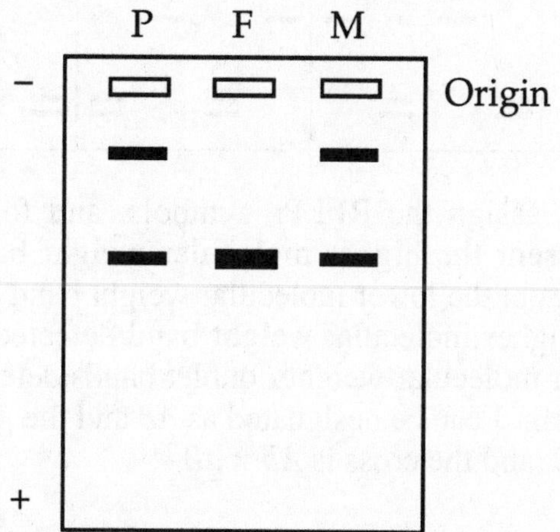

Given that cystic fibrosis is a recessive mutation, will the fetus be affected? Explain.

Answer: Chromosomes bearing CF mutations have a shorter fragment than chromosomes bearing wild-type alleles. Both parents (M and P lanes) have two bands, indicating that each has a normal and mutant chromosome. The fetus (F lane) also has two bands, indicating that it too is heterozygous. Since it has a normal gene, it will be normal.

15.18 The PCR technique was used to amplify two genomic regions (homologous to probes A and B) in DNA from *Neurospora*. Two strains of opposite mating type (strains J and K) were found to differ from each other in both amplified regions, as shown in the following figure, when the amplified DNA was cut with *Eco*RI.

Probe A: Strain J, Strain K — Probe B: Strain J, Strain K

Strain J was crossed to strain K, 100 asci resulting from the cross were dissected, and the PCR reaction was done on DNA from the individual spores in each ascus. Six different patterns of distribution of DNA types within asci were seen as shown in the following figure. Only four patterns are shown for each ascus. Remember that spores 1 and 2 in the ascus are ordinarily identical, as are spores 3 and 4, 5 and 6, and 7 and 8. In Figure 15.D the band pattern displayed by spores 1 and 2 is indicated in the lane designated "1". The pattern shown by spores 3 and 4 is in lane 2, the pattern shown by spores 5 and 6 is in lane 3, and the pattern shown by spores 7 and 8 is in lane 4. Draw a map showing the relationships among the region detected by probe A, the region homologous to probe B, and any relevant centromeres.

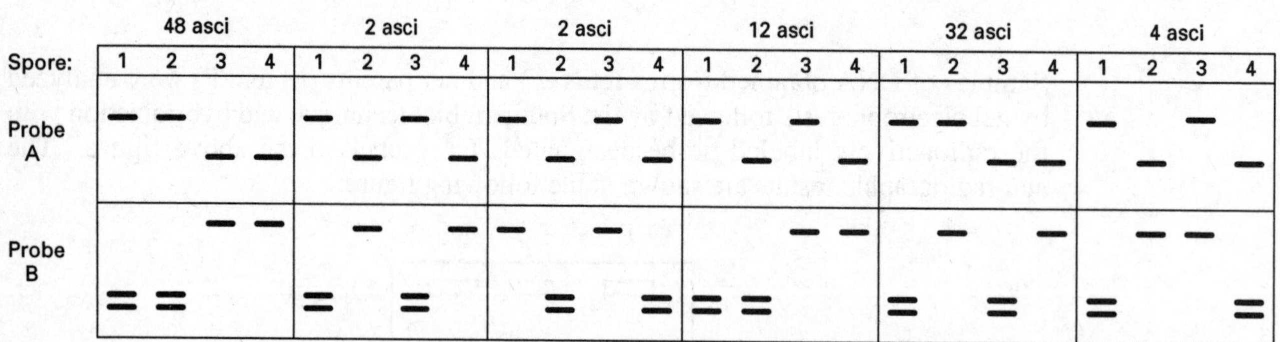

Answer: First, assign the RFLPs symbols, and follow them as genetic traits. Let *A* represent the higher molecular weight band detected by probe A, and let *a* represent the lower molecular weight band detected by probe A. Let *B* represent the higher molecular weight band detected by probe B, and let *b* represent the lower molecular weight doublet bands detected by probe B. Then, the genotype of strain J can be designated as *Ab* and the genotype of strain K can be designated as *aB*, and the cross is *Ab* x *aB*.

Now, rewrite the *Ab* x *aB* cross results using these symbols:

Number:	48	2	2	12	32	4
Tetrad and Spore: 1	*A b*	*A b*	*A B*	*A b*	*A b*	*A b*
2	*A b*	*a B*	*a b*	*a b*	*A B*	*a B*
3	*a B*	*A b*	*A B*	*A B*	*a b*	*A B*
4	*a B*	*a B*	*a b*	*a B*	*a B*	*a b*
Segregation Pattern:	MI MI	MII MII	MII MII	MII MI	MI MII	MII MII
Tetrad Type:	PD	PD	NPD	T	T	T

Percent of MII patterns for *A/a* = (2 + 2 + 12 + 4)/100 x 100% = 20%
Percent of MII patterns for *B/b* = (2 + 2 + 32 + 4)/100 x 100% = 40%.

PD tetrads: 48 + 2 = 50
NPD tetrads: 2
T tetrads: 12 + 32 + 4 = 48

Since PD >> NPD, the genes *A/a* and *B/b* are linked. The map distance between them is given by [(1/2 T + NPD)/Total # asci] x 100% = {[1/2(48) + 2]/100} x 100% = 26 mu. The map distance of each gene to the centromere is given by 1/2 (% MII segregation patterns) x 100%. *A/a* is 10 mu from the centromere, and *B/b* is 20 mu from the centromere. Thus, one can construct the following map:

A←10→ ←20→B
mu mu

15.19 One application of DNA fingerprinting technology has been to identify stolen children and return them to their parents. Bobby Larson was taken from a supermarket parking lot in New Jersey in 1978, when he was 4 years old. In 1990, a 16-year-old boy called Ronald Scott was found in California, living with a couple named Susan and James Scott, who claimed to be his parents. Authorities suspected that Susan and James might be the kidnappers, and that Ronald Scott might be Bobby Larson. DNA samples were obtained from Mr. and Mrs. Larson, and from Ronald, Susan and James Scott. Then DNA fingerprinting was done, using a probe for a particular VNTR family, with the results shown the following figure. From the information in the figure, what can you say about the parentage of Ronald Scott? Explain.

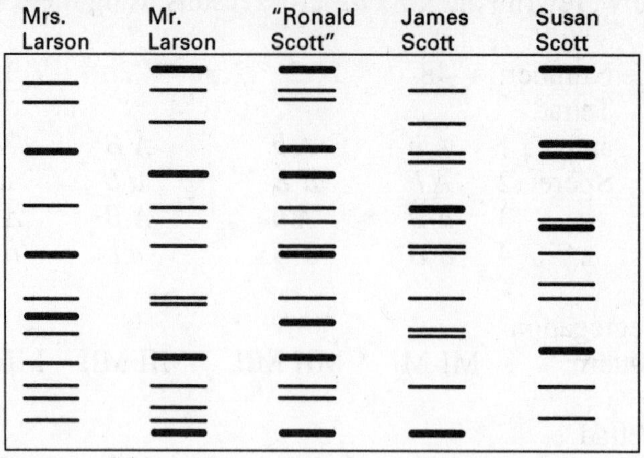

Answer: James and Susan Scott are not the parents of "Ronald Scott." There are several bands in the fingerprint of the boy that are not present in either James or Susan Scott, and thus could not have been inherited from either of them (for example, bands **a** and **b** in the answer figure). In contrast, whenever the boy's DNA exhibits a band that is missing from one member of the Larson couple, the other member of the Larson couple has that band (for example, bands **c** and **d**). Thus, there is no band in the boy's DNA that he could not have inherited from one or the other of the Larsons. These data thus support an argument that the boy is, in fact, Bobby Larson. This data should be used together with other, non-DNA based evidence to support the claim that the boy is Bobby Larson.

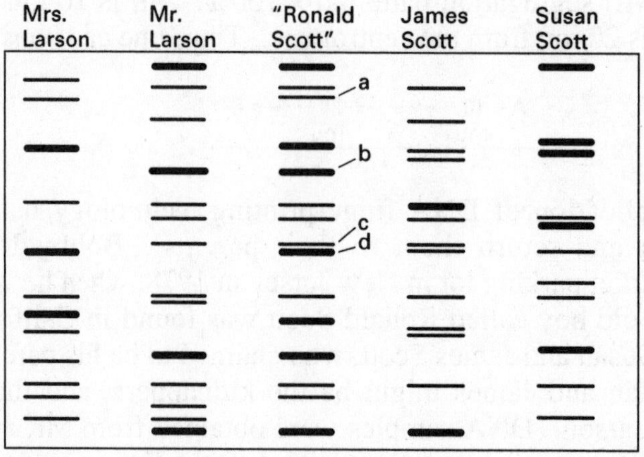

CHAPTER 16

REGULATION OF GENE EXPRESSION IN BACTERIA AND BACTERIOPHAGES

I. CHAPTER OUTLINE

GENE REGULATION OF LACTOSE UTILIZATION IN *E. COLI*
- Lactose as a Carbon Source for *E. coli*
- Experimental Evidence for the Regulation of the *lac* Genes
- Jacob and Monod's Operon Model for the Regulation of the *lac* Genes
- Positive Control of the *lac* Operon
- Molecular Details of *lac* Operon Regulation

TRYPTOPHAN OPERON OF *E. COLI*
- Gene Organization of the Tryptophan Biosynthesis Genes
- Regulation of the *trp* Operon
- Regulation of Other Amino Acid Biosynthesis Operons

SUMMARY OF OPERON FUNCTION

GENE REGULATION IN BACTERIOPHAGES
- Regulation of Gene Expression in Phage Lambda

II. REVIEW OF KEY TERMS, SYMBOLS AND CONCEPTS

Without consulting the text and in your own words, write a brief definition of each term in the groups below. Then, either using a short phrase or a simple diagram, identify the relationship(s) between specific pairs of terms within a set. Finally, consult the text (and perhaps a friend who has also done the exercise) to check your answers.

1	2	3	4
regulated genes	coordinate induction	operon model	*lac* operon
constitutive genes	operon	*lacA, -Z, -Y, -I* genes	negative control
inducible genes	polycistronic mRNA	negative control	allolactose
inducers and effectors	polygenic mRNA	allolactose	positive control
controlling site	effector molecule	partial diploids	catabolite repression
	operator	cis-dominance	glucose effect
	repressor	trans-dominance	CAP-cAMP complex
	promoter	O^c, I^-, I^s, I^{-d}	CAP site
		mutations	

5	6
trp operon	lytic vs. lysogenic pathway
effector	genetic switch
aporepressor	lambda repressor, *cro*
attenuation	antiterminator
attenuator site	P_R, P_{RM}, P_L
pause signal	O_{R1}, O_{R2}, O_{R3}
antitermination	
feedback inhibition	
allosteric shift	

III. THINKING ANALYTICALLY

This chapter provides a thorough presentation of the circuitry used to regulate the expression of bacterial and bacteriophage genes. While the regulation of a particular set of genes can be complex, it is invariably based on a general paradigm. This is that the structural organization of a set of genes is related to both their induction by effector molecules and subsequent expression (as diagrammed in Figure 16.1). As you study the regulation of each set of genes, do not lose site of this general principle. Revisit it as you consider each of the operons discussed in this chapter and analyze how this general principle is used and elaborated upon.

Once you have a solid understanding of the general principle underlying prokaryotic gene expression, the challenge is to understand and retain the complex details of the regulatory circuitry. It will help considerably to first study the text figures and then diagram (from memory) the structure of each operon and the role of regulatory factors within it. This is especially valuable when considering the regulation of genes in λ phage. Note that this will take time, as it requires repeated practice and comparative review!

Strengthen your understanding of the principles underlying prokaryotic gene expression and your detailed knowledge of the regulatory circuitry of operons by examining the consequences of mutations in different operon elements. Initially, consider why mutations in regulatory elements such as the promoter or operator will be cis-dominant, while some mutations in the genes for diffusible regulatory factors *can* be trans-dominant. Then, consider how the properties of different mutations in a specific operon can provide you with a basis to develop a model of its structure. Finally, relate the function of the operon to a biochemical or growth pathway within the cell.

IV. QUESTIONS FOR PRACTICE

A. Multiple Choice Questions

Choose the correct answer or answers for the following questions.

1. Genes that, in general, respond to the needs of a cell or organism in a controlled manner are known as
   a. inducer genes.
   b. effector genes.
   c. regulated genes.
   d. constitutive genes.

2. A mutation that causes a gene to always be expressed, irrespective of what the environmental conditions may be, is known as a
   a. inducer mutation.
   b. effector mutation.
   c. regulator mutation.
   d. constitutive mutation.

3. A gene that is stimulated to undergo transcription in response to a particular molecular event that occurs at a controlling site near that gene is said to be
   a. inducible.
   b. constitutive.
   c. a promoter.
   d. an inducer.

4. An operon which is inducible may be under
   a. positive control only.
   b. negative control only.
   c. constitutive control only.
   d. both positive and negative control.

5. The effector molecule which induces the *lac* protein-coding genes is
   a. lactose.
   b. allolactose.
   c. glucose.
   d. β-galactosidase.

6. Before transcription of the *lac* operon can occur, RNA polymerase must bind strongly to the promoter. This happens when
   a. CAP binds to the CAP site in the promoter.
   b. a CAP-cAMP complex binds to the CAP site in the promoter.
   c. catabolite repression occurs.
   d. *lacI*$^+$ is mutated to *lacI*$^-$.

7. Which of the following is an example of an effector molecule acting via positive control?
   a. lactose inducing the *lac* operon.
   b. glucose causing catabolite repression.
   c. tryptophan attenuating the *trp* operon.
   d. the lambda *cI* gene product, at a certain cellular level, favoring the lysogenic pathway.

8. Bacterial operons of protein-coding genes for the synthesis of amino acids, such as the tryptophan operon, are customarily classified as
   a. negatively controlled.
   b. positively controlled.
   c. repressible operons.
   d. inducible operons.

9. Integrated phage λ can be induced by ultraviolet light to enter the lytic pathway by

a. cleaving repressor monomers.
b. initiating transcription of the *cro* gene.
c. converting the RecA protein to a protease.
d. all of the above, directly or indirectly.

10. Which of the following are critical to the genetic switch controlling the choice between the lysogenic and lytic pathways in phage λ?
   a. the different affinities of the O_{R1}, O_{R2} and O_{R3} sites in the P_{RM} and P_R promoters for dimers of the λ repressor.
   b. the different affinities of the O_{L1}, O_{L2} and O_{L3} sites in the P_L promoter for dimers of the λ repressor..
   c. the concentration of λ repressor protein dimers.
   d. whether RNA polymerase binds to P_{RM} or P_R.

Answers: 1c; 2d; 3a; 4d; 5b; 6c; 7; 8c; 9d; 10a, c & d.

B. Thought Questions

1. What three proteins are synthesized when lactose is the sole carbon source in *E. coli*? What does each do?
2. What are the roles of *lacA, lacI, lacO, lacY,* and *lacZ* in *E. coli* carbohydrate metabolism? In what order are they arranged in the DNA molecule?
3. Describe the sequence of events that occur in the *lac* operon when *E. coli* is grown in the presence of both glucose and lactose.
4. Distinguish between cis-dominant mutations and trans-dominant mutations. Why is *lacO*^{c} cis-dominant but *lacI*^{s} trans-dominant? Would a mutation that led to constitutive expression of the *cI* gene be cis- or trans-dominant?
5. How is attenuation related to the coupling of transcription and translation in prokaryotes?
6. What are the fundamental differences between the *lac* and *trp* operons in *E. coli*?
7. How do structural changes in DNA or RNA conformation play a role in (a) catabolite repression by glucose and CAP-cAMP? (b) antitermination and termination in attenuation of the *trp* operon?
8. Distinguish between feedback inhibition and catabolite repression.
9. In what different or similar ways is allostery used in the regulation of inducible and repressible operons.
10. Speculate as to what phenotype each of the following λ mutations would have: (a) A mutation in the *cI* gene that resulted in repressor molecules being unable to form dimers. (b) A mutation in the *cro* gene that resulted in the *cro* protein being unable to form dimers. (c) A mutation in O_{R1} that resulted in decreased affinity for repressor dimers. (d) A mutation in O_{R3} that resulted in increased affinity for repressor dimers. (e) A mutation in O_{R1} that resulted in increased affinity for *cro* dimers. (f) A mutation in O_{R3} that resulted in decreased affinity for *cro* dimers.

V. SOLUTIONS TO TEXT PROBLEMS

16.1 How does lactose bring about the induction of synthesis of β-galactosidase, permease, and transacetylase? Why does this event not occur when glucose is also in the medium?

Answer: The addition of lactose to *E. coli* cells brings about a rapid synthesis of these three enzymes by the induction of a single promoter that lies upstream of the genes for these three enzymes. The three genes are part of a *lac* operon that is transcribed as a single unit. When lactose is added, it is metabolized (isomerized) to allolactose, which binds to a repressor protein. Without bound allolactose, the repressor protein blocks transcription from the *lac* promoter. Hence, the *lac* operon is normally under a negative control mechanism. When allolactose is bound, the repressor protein is inactivated and is unable to bind to the operator site to block transcription. As a result, RNA polymerase binds to the promoter and initiates transcription of a single mRNA that encodes all three proteins.

One of the enzymes that is synthesized, β-galactosidase, cleaves lactose to produce glucose and galactose (which is converted to glucose in a subsequent enzymatic step). Consequently, if glucose is present in the medium, it is redundant to induce the *lac* operon. Glucose blocks induction of the *lac* operon by utilizing a positive control mechanism. In this catabolite repression, glucose causes a great reduction in the amount of cAMP in the cell. For normal induction of the *lac* operon, cAMP must complex with a CAP (catabolite gene activator protein) that in turn, binds to a CAP site upstream of the *lac* promoter and activates transcription. In the absence of cAMP, the cAMP-CAP complex is absent, and so transcription cannot be activated.

16.2 Operons produce polygenic mRNA when they are active. What is a polygenic mRNA? What advantages, if any, do they confer on a cell in terms of its function?

Answer: Polygenic mRNAs contain coding information for more than one protein. These mRNAs are transcribed from operons that contain several genes encoding related functions such as catalyzing steps of a biosynthetic pathway. One advantage conferred by utilizing such mRNAs is that cells can regulate all of the steps of a pathway coordinately. By using a polygenic mRNA, the synthesis of each of a set of enzymes acting in one pathway can be produced by a single regulatory signal.

16.3 If an *E. coli* mutant strain synthesizes β-galactosidase whether or not the inducer is present, what genetic defect(s) might be responsible for this phenotype?

Answer: One possibility is that the repressor protein bound by the inducer cannot bind to the operator (i.e., the repressor is non-functional and the strain is I^-), so that its presence or absence makes no difference. Another possibility is that there are base pair alterations in the operator region that make it unrecognizable by the repressor protein (i.e., the operon is constitutively expressed and the strain is O^c).

16.4 Distinguish the effects you would expect from (a) a missense mutation and (b) a nonsense mutation in the *lacZ* (β-galactosidase) gene of the *lac* operon.

Answer:

a. A missense mutation results in partial or complete loss of β-galactosidase activity, but no loss of permease and transacetylase activities.
b. A nonsense mutation is likely to have polar effects unless the mutation is very close to the normal chain-terminating codon for β-galactosidase. If the nonsense mutation occurred near the 5'-end of the *lacZ* gene, the ribosome would continue to slide along the polygenic mRNA towards the *lacY* gene. However, it would typically dissociate before reaching the start codon for

permease because of the greater distance. Thus, few ribosomes would be translating the genes lying downstream of the *lacZ* gene, leading to decreased permease and transacetylase production.

16.5 The elucidation of the regulatory mechanisms associated with the enzymes of lactose utilization in *E. coli* was a landmark in our understanding of regulatory processes in microorganisms. In formulating the operon hypothesis as applied to the lactose system, Jacob and Monod found that results from particular partial-diploid strains were invaluable. Specifically, in terms of the operon hypothesis, what information did the partial diploids provide that haploids could not?

Answer: The use of partial diploids allowed observation of the consequences of placing sequences in *trans* and in *cis*. Partial diploids were used to show that some regulatory sequences must lie in *cis* to *lacZ* (the *lacO* region upstream of the *lacZ* gene) to exert a regulatory effect. For example, O^c mutations are cis-acting: they cause constitutive activation of the *lac* promoter that they lie upstream of, but not of any other promoter. Partial diploids also were used to show that the *lacI* gene encoded a *trans*-acting factor (a diffusible product that could bind to the *lacO* region) and that promoter function did not require a diffusible substance. For example, a *lacI*$^+$ gene on a plasmid could function in *trans* to regulate a *lac* promoter in *cis* to a *lacI*$^-$ gene.

16.6 For the *E. coli lac* operon, write the partial-diploid genotype for a strain that will produce β-galactosidase constitutively and permease by induction.

Answer: In order to observe such a phenotype, there are three requirements: (1) a functional *lacZ* and a non-functional *lacY* gene must both lie downstream of a functional operator and promoter, (2) a non-functional *lacZ* and a functional *lacY* gene must lie downstream of a inducible (i.e. O^+) promoter, and (3) a functional repressor gene must be present in the cell (I^+). A genotype that satisfies these requirements is *lacI*$^+$ *lacO*^{c} *lacP*$^+$ *lacZ*$^+$ *lacY*$^-$/*lacI*$^+$ *lacO*$^+$ *lacP*$^+$ *lacZ*$^-$ *lacY*$^+$. Only one *lacI*$^+$ gene is required, so one may be *lacI*$^-$.

16.7 Mutants were instrumental in the elaboration of the model for the regulation of the lactose operon.

a. Discuss why *lacO*^{c} mutants are cis-dominant but not trans-dominant.
b. Explain why *lacI*^{s} mutants are trans-dominant to the wild-type *lacI*$^+$ allele but *lacI*$^-$ mutants are recessive.
c. Discuss the consequences of mutations in the repressor gene promoter as compared with mutations in the structural gene promoter.

Answer: a. *lacO*^{c} mutants are mutants in the operator region that is normally bound by the repressor protein. *lacO*^{c} mutants result from a DNA alteration that precludes the repressor protein from binding the operator. Since the repressor normally blocks transcription initiation from a downstream promoter, *lacO*^{c} mutants result in constitutive transcription at that promoter. Since the operator acts only on an adjacent, and not any other, promoter, mutants in the operator are only cis-dominant. The *lacO*^{c} has no effect on other lactose operons in the same cell because *lacO*^{c} does not code for a product that could diffuse through the cell and affect other DNA sequences.

b. In wild-type strains, *lacI* encodes a repressor that can block transcription at the lac operon by binding to an operator region. By binding the operator, the repressor blocks RNA polymerase from binding to the promoter. This activity of the repressor can be altered if allolactose is present, which binds to the repressor and inhibits it from binding the operator. The super-repressor (*lacI*^{s}) mutation results in a repressor protein that can bind the operator, but cannot bind allolactose. Once it binds to the operator, it cannot leave the operator, and transcription is always blocked. This results in a dominant mutation, since even if normal repressor molecules (made by *lacI*$^+$) are present, the super-repressor molecules do not vacate the operator region, and the operon cannot be induced. On the other hand, lacI$^-$ mutants either do not make repressor protein, or make repressor that is unable to bind to the operator. In a partial diploid that has both *lacI*$^-$ and *lacI*$^+$ genes, repressor proteins are made (by *lacI*$^+$) and are capable of diffusing to any lac operator region to regulate a *lac* operon. Hence, *lacI*$^-$ is recessive to *lacI*$^+$ because the defect caused by the absence of the repressor proteins in *lacI*$^-$ mutants can be overcome by the synthesis of diffusible repressor protein from the *lacI*$^+$ gene.

c. Mutations in the repressor gene promoter result either in an increase, or in a decrease in the level of expression of the repressor gene. Such mutations are likely to have little effect on the control of expression of the lac operon structural genes since the repressor molecule itself will be unaltered. If very few repressor molecules are made, it is possible that there would be increased structural gene expression in the absence of inducer. Promoter mutations for the structural genes can also increase or decrease the level of expression of the induced operon, and thereby alter the level of enzymes present. Most known promoter mutations cause an almost complete loss of expression of the three genes.

16.8 This question involves the lactose operon of *E. coli* where I = *lacI* (the repressor gene), $P = P_{lac}$ (the promoter), O = *lacO* (the operator), Z = *lacZ* (the β-galactosidase gene), and Y = *lacY* (the permease gene). Complete Table 16.A, using + to indicate if the enzyme in question will be synthesized and – to indicate if the enzyme will not be synthesized.

Table 16.A

Genotype	Inducer Absent: β-galactosidase	Inducer Absent: Permease	Inducer Present: β-galactosidase	Inducer Present: Permease
a. $I^+ P^+ O^+ Z^+ Y^+$				
b. $I^+ P^+ O^+ Z^- Y^+$				
c. $I^+ P^+ O^+ Z^+ Y^-$				
d. $I^- P^+ O^+ Z^+ Y^+$				
e. $I^s P^+ O^+ Z^+ Y^+$				
f. $I^+ P^+ O^c Z^+ Y^+$				
g. $I^s P^+ O^c Z^+ Y^+$				
h. $I^+ P^+ O^c Z^+ Y^-$				
i. $I^{-d} P^+ O^+ Z^+ Y^+$				
j. $I^- P^+ O^+ Z^+ Y^+$ / $I^+ P^+ O^+ Z^- Y^-$				
k. $I^- P^+ O^+ Z^+ Y^-$ / $I^+ P^+ O^+ Z^- Y^+$				
l. $I^s P^+ O^+ Z^+ Y^-$ / $I^+ P^+ O^+ Z^- Y^+$				
m. $I^+ P^+ O^c Z^- Y^+$ / $I^+ P^+ O^+ Z^+ Y^-$				
n. $I^- P^+ O^c Z^+ Y^-$ / $I^+ P^+ O^+ Z^- Y^+$				
o. $I^s P^+ O^+ Z^+ Y^+$ / $I^+ P^+ O^c Z^+ Y^+$				
p. $I^{-d} P^+ O^+ Z^+ Y^-$ / $I^+ P^+ O^+ Z^- Y^+$				
q. $I^+ P^- O^c Z^+ Y^-$ / $I^+ P^+ O^+ Z^- Y^+$				
r. $I^+ P^- O^+ Z^+ Y^-$ / $I^+ P^+ O^c Z^- Y^+$				
s. $I^- P^- O^+ Z^+ Y^+$ / $I^+ P^+ O^+ Z^- Y^-$				
t. $I^- P^+ O^+ Z^+ Y^-$ / $I^+ P^- O^+ Z^- Y^+$				

Answer:

	Inducer Absent:		Inducer Present:	
Genotype	β-galactosidase	Permease	β-galactosidase	Permease
a. $I^+ P^+ O^+ Z^+ Y^+$	−	−	+	+
b. $I^+ P^+ O^+ Z^- Y^+$	−	−	−	+
c. $I^+ P^+ O^+ Z^+ Y^-$	−	−	+	−
d. $I^- P^+ O^+ Z^+ Y^+$	+	+	+	+
e. $I^s P^+ O^+ Z^+ Y^+$	−	−	−	−
f. $I^+ P^+ O^c Z^+ Y^+$	+	+	+	+
g. $I^s P^+ O^c Z^+ Y^+$	+	+	+	+
h. $I^+ P^+ O^c Z^+ Y^-$	+	−	+	−
i. $I^{-d} P^+ O^+ Z^+ Y^+$	+	+	+	+
j. $I^- P^+ O^+ Z^+ Y^+$ $I^+ P^+ O^+ Z^- Y^-$	−	−	+	+
k. $I^- P^+ O^+ Z^+ Y^-$ $I^+ P^+ O^+ Z^- Y^+$	−	−	+	+
l. $I^s P^+ O^+ Z^+ Y^-$ $I^+ P^+ O^+ Z^- Y^+$	−	−	−	−
m. $I^+ P^+ O^c Z^- Y^+$ $I^+ P^+ O^+ Z^+ Y^-$	−	+	+	+
n. $I^- P^+ O^c Z^+ Y^-$ $I^+ P^+ O^+ Z^- Y^+$	+	−	+	+
o. $I^s P^+ O^+ Z^+ Y^+$ $I^+ P^+ O^c Z^+ Y^+$	+	+	+	+
p. $I^{-d} P^+ O^+ Z^+ Y^-$ $I^+ P^+ O^+ Z^- Y^+$	+	+	+	+
q. $I^+ P^- O^c Z^+ Y^-$ $I^+ P^+ O^+ Z^- Y^+$	−	−	−	+
r. $I^+ P^- O^+ Z^+ Y^-$ $I^+ P^+ O^c Z^- Y^+$	−	+	−	+
s. $I^- P^- O^+ Z^+ Y^+$ $I^+ P^+ O^+ Z^- Y^-$	−	−	−	−
t. $I^- P^+ O^+ Z^+ Y^-$ $I^+ P^- O^+ Z^- Y^+$	−	−	+	−

16.9 A new sugar, sugarose, induces the synthesis of two enzymes from the *sug* operon of *E. coli*. Some properties of deletion mutations affecting the appearance of these enzymes are as follows (here, + = enzyme induced normally, i.e., synthesized only in the presence of the inducer; C = enzyme synthesized constitutively; 0 = enzyme cannot be detected):

Mutation of	Enzyme 1	Enzyme 2
Gene *A*	+	0
Gene *B*	0	+
Gene *C*	0	0
Gene *D*	C	C

a. The genes are adjacent in the order *ABCD*. Which gene is most likely to be the structural gene for enzyme 1?
b. Complementation studies using partial-diploid (F') strains were made. The episome (F') and chromosome each carried one set of *sug* genes. The results were as follows (symbols are the same as in previous table):

Genotype of F'	Chromosome	Enzyme 1	Enzyme 2
$A^+ B^- C^+ D^+$	$A^- B^+ C^+ D^+$	+	+
$A^+ B^- C^- D^+$	$A^- B^+ C^+ D^+$	+	0
$A^- B^+ C^- D^+$	$A^+ B^- C^+ D^+$	0	+
$A^- B^+ C^+ D^+$	$A^+ B^- C^+ D^-$	+	+

From all the evidence given, determine whether the following statements are true or false:

1. It is possible that gene *D* is a structural gene for one of the two enzymes.
2. It is possible that gene *D* produces a repressor.
3. It is possible that gene *D* produces a cytoplasmic product required to induce genes *A* and *B*.
4. It is possible that gene *D* is an operator locus for the *sug* operon.
5. The evidence is also consistent with the possibility that gene *C* could be a gene that produces a cytoplasmic product required to induce genes *A* and *B*.
6. The evidence is also consistent with the possibility that gene *C* could be the controlling end of the *sug* operon (end from which mRNA synthesis presumably commences).

Answer: First consider the data in general terms. Mutations in genes *A* or *B* result in a loss of one, but not both enzyme activities. These are likely to be structural genes for the enzymes (*B* = enzyme 1, *A* = enzyme 2). The mutation in genes *C* and *D* result in loss of both enzyme activities, suggesting that these genes or regions regulate or affect both genes.

a. Gene *B* is likely to be the gene for enzyme 1, since only a mutation in gene B produced a loss of enzyme 1 activity with no affect on enzyme 2 activity. (By similar logic, gene A codes for enzyme 2.)
b. 1. False. A mutation in *D* leads to the constitutive synthesis of both enzymes 1 and 2, so *D* cannot be a structural gene for either enzyme.
2. True. *D* could encode a repressor. Suppose the repressor acted as the lac repressor does in the *lac* operon. If mutations in *D* inactivated the repressor, then an absence of repressor would lead to constitutive activation of the operon. In this model, D^- mutants would be recessive to D^+ mutants, which is seen in the analysis of partial diploids ($A^- B^+ C^+ D^+/A^+ B^- C^+ D^-$ shows inducible activity of both A^+ and B^+).
3. False. If *D* was needed to induce the sug operon, D^- mutants should produce no enzymes. This is not observed.

4. False. One does see that D^- mutants are constitutive, as would be expected if a D^- operator region could not be bound by a repressor to repress transcription. However, not all of the results support this view. If D was an operator, consider what phenotype would be expected in the partial diploid $A^- B^+ C^+ D^+/A^+ B^- C^+ D^-$. If D^- was a defective operator, this partial diploid would express the A^+ gene (enzyme 2) constitutively, in a cis-dominant manner, and not in an inducible manner. This is not seen. What is seen is a trans-dominant effect where the A^+ gene is inducible. Thus, D is not an operator.

5. False. Since the products of genes A and B are not inducible in C^- mutants, one might speculate that gene C produced a cytoplasmic product that was required to induce genes A and B. However, consider the partial diploid data. In two of the partial diploids, ($A^+ B^- C^- D^+/A^- B^+ C^+ D^+$ and $A^- B^+ C^- D^+/A^+ B^- C^+ D^+$) the wild-type genes on the C^- chromosome are not expressed, even though a C^+ gene is on another chromosome. This indicates that C shows cis-dominance, and not trans-dominance. If C^+ encoded a cytoplasmic factor, it would diffuse and be capable of acting in a trans-dominant fashion. Since it does not, C does not encode a cytoplasmic, trans-acting factor.

6. True. The cis-dominant affects that C^- mutants show in partial diploids could be explained if the mutations were in the controlling end of the sug operon, in a region such as the promoter.

16.10 Four different polar mutations, *1, 2, 3,* and *4*, in the *lacZ* gene of the lactose operon were isolated following mutagenesis of *E. coli*. Each caused total loss of β-galactosidase activity. Two revertant mutants, due to suppressor mutations in genes unlinked to the *lac* operon, were isolated from each of the four strains: suppressor mutations of polar mutation *1* are *1A* and *1B*; those of polar mutation 2 are *2A* and *2B*; and so on. Each of the eight suppressor mutations was then tested, by appropriate crosses, for its ability to suppress each of the four polar mutations; the test involved examining the ability of a strain carrying the polar mutation and the suppressor mutation to grow with lactose as the sole carbon source. The results follow (+ = growth on lactose and – = no growth):

POLAR MUTATION	SUPPRESSOR MUTATION *1A*	*1B*	*2A*	*2B*	*3A*	*3B*	*4A*	*4B*
1	+	+	+	+	+	+	+	+
2	+	–	+	+	+	+	–	–
3	+	–	+	–	+	+	–	–
4	+	+	+	+	+	+	+	+

A mutation to a UAG codon is called an amber nonsense mutation, and a mutation to a UAA codon is called an ochre nonsense mutation. Suppressor mutations allowing reading of UAG and UAA are called amber and ochre suppressors, respectively.

a. Which of the polar mutations are probably amber? Which are probably ochre?
b. Which of the suppressor mutations are probably amber suppressors? Which are probably ochre suppressors?
c. How would you explain the anomalous failure of suppressor *2B* to permit growth with polar mutation *3*? How could you test your explanation most easily?

d. Explain precisely why ochre suppressors suppress amber mutants but amber suppressors do not suppress ochre mutants.

Answer: a. An ochre suppressor mutation will allow reading of the UAA codon, and have a 3'-AUU-5' anticodon. Because the 5'-U in the anticodon can wobble base pair with either a 3'-A or 3'-G in a codon, the ochre suppressor will also allow reading of the UAG codon, and be an amber mutant suppressor. An amber suppressor mutation will allow reading of a UAG codon, and have will have a 3'-AUC-5' anticodon. It will be able to suppress only amber mutants. Thus, amber mutants will be those suppressible by all given suppressors, while ochre mutants will be those suppressible by only a subset of suppressors. Mutants *1* and *4* appear to be amber mutants, while mutants *2* and *3* appear to ochre mutants.

b. An ochre suppressor will allow growth of all four polar mutations, while an amber suppressor will allow growth of only *1* and *4*. Therefore, *1B*, *4A*, and *4B* are amber suppressors, while *1A*, *2A*, *3A* and *3B* are ochre suppressors. *2B* is unlike either of these two groups, and will be discussed in part **c**.

c. Since *2B* suppresses *1*, *2*, and *4*, it is probably an ochre suppressor. *3* might not be suppressed because of the mechanism of suppression. Any of a number of tRNA molecules could have been mutated to have an anticodon that will base pair with an ochre triplet (UAA). It is likely that this tRNA will bring an amino acid to the ochre triplet that is different from that coded by the wild-type message. In some cases, this could result in a non-functional β-galactosidase enzyme. In turn, this would lead to the cell being unable to grow on lactose as a sole carbon source, so that suppression would not be seen. The hypothesis that non-functional β-galactosidase protein would be made can be tested. One can generate antibodies to wild-type β-galactosidase protein, and use them to immunoprecipitate β-galactosidase protein from mutant cells. The hypothesis would gain support if protein can be precipitated, even though no enzyme activity can be detected.

d. The fact that ochre suppressors suppress amber mutants but amber suppressors do not suppress ochre mutants can be explained by the wobble hypothesis. A 5'-U in the tRNA anticodon can pair with either a 3'-A or 3'-G in the codon. Thus, a 5'-UAG-3' amber mutant codon and a 5'-UAA-3' ochre mutant codon can both be read by a 3'-AUU-5' ochre suppressor anticodon. This allows ochre suppressors to suppress both ochre and amber nonsense mutants. On the other hand, 5'-C in the tRNA anticodon can only pair with a 3'-G in the codon. Thus, a 3'-AUC-5' amber suppressor anticodon can only pair with a 5'-UAG-3' amber mutant codon.

16.11 What consequences would a mutation in the catabolite activator protein (CAP) gene of *E. coli* have for the expression of a wild-type *lac* operon?

Answer: The CAP, in a complex with cAMP, is required to facilitate RNA polymerase binding to the *lac* promoter. The RNA polymerase binding occurs only in the absence of glucose, and only if the operator is not occupied by repressor (i.e., lactose is also absent). Suppose the CAP were mutated so that it could no longer bind to the CAP site. In this case, RNA polymerase would not be able to recognize and bind to the promoter, and the operon would not be expressed. Such a mutation could be distinguished from other mutations that eliminate operon expression (e.g., I^s, P^-) by several means. First, it should be recessive in partial diploids with a normal allele, and not show cis- or trans-dominance. Second, it should map to a location other than the *lac* operon.

16.12 The lactose operon is an inducible operon, whereas the tryptophan operon is a repressible operon. Discuss the differences between these two types of operons.

Answer: Both inducible and repressible operons allow sensitive control of transcription. They differ from each other in the details of how such control is achieved. In the *lac* operon, a repressor protein bound to an operator blocks transcription unless lactose is present. Thus, if lactose is absent, the system is OFF. When lactose is added, it is converted to allolactose and acts as an effector molecule to release the repressor from the operator, so that RNA polymerase can transcribe the operon. In the tryptophan operon, the control strategy is the opposite. When tryptophan is abundant in the medium, the operon is turned off, as it is unnecessary to synthesize the enzymes needed to build tryptophan. Tryptophan also acts as an effector molecule. It binds to an aporepressor protein and converts it into an active repressor. This repressor is capable of binding the *trp* operator to reduce transcription of the *trp* operon protein-coding genes by RNA polymerase. Transcription of the *trp* operon is reduced by about 70 percent in the presence of tryptophan, while the aporepressor has no affinity for the operator in the absence of tryptophan.

The *trp* operon also can be regulated by attenuation, a mechanism that controls the ratio of the transcripts that include the five structural genes to those that are terminated before the structural genes. Under conditions where some tryptophan is present in the medium, short 140 bp transcripts are produced, and transcription is attenuated. Under conditions of tryptophan starvation or limitation, full length transcripts are produced. A model for how attenuation occurs is described in text Figures 16.15, 16.16 and 16.17.

16.13 In the presence of high intracellular concentrations of tryptophan, only short transcripts of the *trp* operon are synthesized because of attenuation of transcription 5' to the structural genes. This is mediated by the recognition of two Trp codons in the leader sequence. If these codons were mutated to be amber (UAG) nonsense codons, what effect would this have on the regulation of the operon in the presence or absence of tryptophan? Explain.

Answer: Mutation of the Trp codons to amber nonsense codons would result in the ribosome stalling at these codons whether or not tryptophan was present. As shown in Figure 16.16a, stalling in this position would allow the pairing of regions 2 and 3 and prevent the pairing of regions 3 and 4. This would allow transcription to continue, and result in anti-termination. Attenuation would not occur, the complete operon would be transcribed and all of the structural genes would be translated. As tryptophan will be synthesized, one would expect the aporepressor to be activated by its presence and transcription initiation to be blocked by about 70 percent.

Attenuation could be restored in the presence of an amber suppressor mutation, although it would be under the control of a different amino acid. Such a mutation results when a tRNA has a mutant anticodon that recognizes the amber (UAG) nonsense codon. The operon would be attenuated under the control of the amino acid that bound the suppressor tRNA.

16.14 In the bacterium *Salmonella typhimurium* seven of the genes coding for histidine biosynthetic enzymes are located adjacent to one another in the chromosome. If excess histidine is present in the medium, the synthesis in all seven enzymes is coordinately repressed, whereas in the absence of histidine all seven genes are coordinately expressed. Most mutations in this region of the chromosome result in the loss of activity of only one of the enzymes. However, mutations mapping to one end of the gene cluster result in the loss of all seven enzymes, even though none of the structural genes have been lost. What is the counterpart of these mutations in the *lac* operon system?

Answer: *lacP*⁻ (promoter mutations)

16.15 Upon infecting an *E. coli* cell, bacteriophage λ has a choice between the lytic and lysogenic pathways. Discuss the molecular events that determine which pathway is taken.

Answer: See text, pp. 529-534, particularly Figures 16.20, 16.23 and 16.24.

16.16 How do the lambda repressor protein and the Cro protein regulate their own synthesis?

Answer: Both the repressor and Cro proteins are capable of binding to operator regions in a concentration-dependent manner. They can therefore block transcription of their own genes at high concentration. See text Figures 16.22, 16.23 and 16.24. The lambda repressor protein can exist either as a monomer or a dimer, depending its concentration. As the concentration of repressor increases in the cell, repressor dimers bind (in order) to each of three operator sites near the *cI* (and *cro*) gene, O_{R1}, O_{R2} and O_{R3}. When concentrations of the repressor are such that both O_{R1} and O_{R2} are bound, RNA polymerase is prevented from binding to P_R, and further transcribing the *cro* gene. At this concentration of repressor, RNA polymerase can still bind to P_{RM} and transcribe *cI* to produce more repressor. As more repressor is produced however, O_{R3} is bound by repressor, and RNA polymerase can no longer bind to P_{RM} and transcribe *cI*. Thus, *cI*'s transcription is dependent on the concentration of repressor. The Cro protein controls its own production similarly. At high Cro protein concentration, all three O_R binding sites become occupied, and transcription initiation at the P_R promoter is blocked. At lower Cro protein concentrations, only O_{R3} is bound, so that P_{RM} and repressor synthesis is blocked but P_R is not blocked and *cro* is transcribed. See text pp. 530-534.

16.17 If a mutation in the phage lambda *cI* gene results in a non functional *cI* gene product,, what phenotype would you expect the phage to exhibit?

Answer: The *cI* gene product is a repressor protein that functions to keep the lytic functions of the phage repressed when lambda is in the lysogenic state. A *cI* mutant strain would be unable to repress lysis, so that the phage would always follow a lytic pathway.

16.18 Bacteriophage λ can form a stable association with the bacterial chromosome because the virus manufactures a repressor. This repressor prevents the virus from replicating its DNA, making lysozyme and all the other tools used to destroy the bacterium. When you induce the virus with UV light, you destroy the repressor, and the virus goes through its normal lytic cycle. This repressor is the product of a gene called the *cI* gene and is a part of the wild-type viral genome. A bacterium that is lysogenic for λ^+ is full of repressor substance, which confirms immunity against any λ virus added to these bacteria. These added viruses can inject their DNA, but the repressor from the resident virus prevents replication, presumably by binding to an operator on the incoming virus. Thus this system has many analogous elements to the lactose operon. We could diagram a virus as shown in the figure. Several mutations of the *cI* gene are known. The c_i mutation results in an inactive repressor.

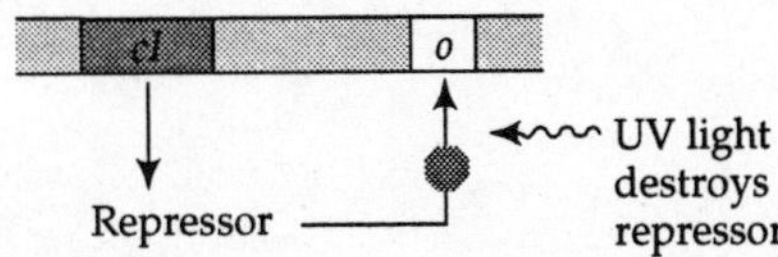

a. If you mix λ containing a c_i mutation, can it lysogenize (form a stable association with the bacterial chromosome)? Why?
b. If you infect a bacterium simultaneously with a wild-type c^+ and a c_i mutant of λ, can you obtain stable lysogeny? Why?
c. Another class of mutants called c^{IN} makes a repressor that is insensitive to UV destruction. Will you be able to induce a bacterium lysogenic for c^{IN} with UV light? Why?

Answer:

a. No. The repressor is necessary to keep the lytic function of the phage repressed, and allow the phage to enter lysogeny. In the presence of a c_i mutation, only the lytic pathway can be taken.
b. Yes. A normal repressor will be made from the wild-type c^+ gene. This repressor is diffusible, and will act trans-dominantly to repress lytic growth.
c. No. UV irradiation of lysogenic bacteria destroys repressor function, which in turn leads to induction, including excision of lambda and lytic growth. In a c^{IN} mutant, the repressor would not be destroyed following UV irradiation, so that the prophage would not be excised and lysogeny would be retained.

a. If you have a cI-containing ts mutation, can it lysogenize (form) a stable lysogen with the prophage [illegible]? Why?

CHAPTER 17

REGULATION OF GENE EXPRESSION AND DEVELOPMENT IN EUKARYOTES

I. CHAPTER OUTLINE

LEVELS OF CONTROL OF GENE EXPRESSION IN EUKARYOTES

Transcriptional Control
RNA Processing Control
Transport Control
mRNA Translation Control
mRNA Degradation Control
Protein Degradation Control

GENE REGULATION IN DEVELOPMENT AND DIFFERENTIATION

Gene Expression in Higher Eukaryotes
Constancy of DNA in the Genome During Development
Differential Gene Activity Among Tissues and During Development
Immunogenetics and Chromosome Rearrangements During Development

GENETIC REGULATION OF DEVELOPMENT IN *DROSOPHILA*

Drosophila Developmental Stages
Embryonic Development
Imaginal Discs
Homeotic Genes

II. REVIEW OF KEY TERMS, SYMBOLS AND CONCEPTS

Without consulting the text and in your own words, write a brief definition of each term in the groups below. Then, either using a short phrase or a simple diagram, identify the relationship(s) between specific pairs of terms within a set. Finally, consult the text (and perhaps a friend who has also done the exercise) to check your answers.

1	2	3	4
transcriptional control positive regulatory element/protein negative regulatory element/protein promoter enhancer silencer combinatorial gene regulation	DNA binding protein DNA binding domain helix-turn-helix zinc finger leucine zipper helix-loop-helix	DNase I sensitive region DNase hypersensitive site erythroblast globin, ovalbumin genes histones, nucleosomes DNA binding proteins DNA methylation	short term gene regulation UAS steroid hormone polypeptide hormone steroid response element (RE) second messenger hormone receptor heat shock response, gene, element plant hormones gibberellins, auxins, cytokinins ethylene, abscisic acid

5	6	7	8
RNA processing control alternative poly(A) site alternative splicing regulation cascade transport control spliceosome retention snRNPs translation control adenylation control element mRNA degradation control protein degradation control proteolysis ubiquitin N-end rule	immune system antigen antibody immunoglobulin clonal selection H-, L-chains C-, V-, J-, D-segments somatic recombination	development differentiation totipotency differential gene activity α, β, δ, ε, γ globin genes polytene chromosomes endoreduplication chromomeres polytene puffs ecdysone	*Drosophila* development polar cytoplasm syncytium syncytial, cellular blastoderm molecular gradient parasegment, segment maternal genes segmentation genes homeotic genes

9

Drosophila development
imaginal discs
serial transplantation
transdetermination
determined state
homeotic mutation
bithorax complex
Antennapedia complex
homeobox
homeodomain
helix-turn-helix motif

III. THINKING ANALYTICALLY

The complexity of eukaryotic gene regulation has several roots. First, eukaryotic genes and chromosomes are differently organized than their prokaryotic counterparts. Second, the transfer of information from DNA to protein is more complicated in eukaryotes than in prokaryotes, as mRNAs and proteins are processed in additional ways. Third, eukaryotic organisms are capable of substantially more development and differentiation, with different cell types in a multicellular organism having very diverse patterns of gene expression. As you approach this material, keep these roots in mind to help organize your thoughts.

While eukaryotic gene regulation is more complex than in prokaryotes, many of the fundamental principles of gene regulation in bacteria and viruses are retained. For example, the principles of gene regulation used by λ phage to choose between a lytic and lysogenic pathway are similar to many used in the control of transcription in eukaryotic genes. It will help you to think about how a principle is retained between prokaryotes and eukaryotes even while confronting and analyzing the differences.

The multiple levels in which eukaryotic gene expression is controlled require you to pay very close attention to detail. Consider each level separately, and make frequent reference to the text figures and diagrams, redrawing them to solidify your understanding. In many instances, it will help you to relate the level of control to the organization of the gene or the chromosome. In others, it will help you to relate the level of control to the stage where information is being transferred from DNA into protein.

Once you have understood the levels at which eukaryotic gene expression can be controlled, you will have a foundation to consider how gene expression can be controlled in a temporal and spatial manner during development. When considering the regulation of gene expression during development, it is important to have a thorough understanding of the "biology" of the system that is developing. First understand how the organism (e.g., in the case of *Drosophila*) or population of cells (e.g., in the immune system) develops, and then consider how genes are regulated in this context. It will again help to construct diagrams and illustrate how genes are regulated along a timeline or developmental pathway.

IV. QUESTIONS FOR PRACTICE

A. Multiple Choice Questions

1. In eukaryotes, coordinated gene expression in a differentiated cell is achieved via
   a. the use of operons.
   b. combinatorial gene regulation.
   c. DNase I sensitivity.
   d. selective deletion of genes not active in differentiated cells.

2. Transcription in eukaryotes is activated if positive regulatory proteins are bound at
   a. the enhancer element.
   b. the promoter element.
   c. the operator.
   d. both a and b.

3. Histones act in eukaryotic gene regulation primarily as
   a. enhancers of gene expression.
   b. repressors of gene expression.
   c. promoters.
   d. proteins preventing DNase I digestion.

4. A gene lying in chromatin actively transcribed by RNA polymerase II
   a. is likely to have DNase I hypersensitive sites in regions upstream of the protein coding region.
   b. is likely to have DNase I hypersensitive sites within the protein coding region.
   c. is likely to be relatively insensitive to DNase I.
   d. is likely to heavily methylated.

5. The receptors for steroid hormones
   a. lie on the cell surface and act via second messengers.
   b. lie on the cell surface and when bound by hormone, are transported into the nucleus and bind response elements to activate transcription.
   c. lie in the cytoplasm and act via second messengers.
   d. lie in the cytoplasm and when bound by hormone, are transported into the nucleus and bind response elements to activate transcription.

6. Sex-type in *Drosophila* is controlled by
   a. the ratio of X chromosomes to autosomes.
   b. a cascade of regulated, alternative RNA splicing.
   c. sex-specific selection of polyadenylation sites.
   d. sex-specific transport control.
   e. all of the above.
   f. a and b only.

7. A wide diversity of immunoglobulin molecules is obtained via
   a. clonal selection.
   b. combinatorial gene regulation.
   c. alternative RNA splicing.
   d. somatic recombination of DNA.

8. Evidence for the totipotency of nuclei in eukaryotic cells is provided by
   a. the existence of homeotic mutations.
   b. Gurdon's nuclear transplantation experiments.
   c. transdetermination of imaginal discs after serial transplantation.
   d. the transcription of fetal globin genes (γ-globin) only during fetal development and not during adult life.
   e. the finding that nuclei in different adult tissues have the same amount of DNA.

9. In *Drosophila*, maternal genes are genes that
   a. are expressed in the mother during oogenesis, and whose products will specify spatial organization in the developing embryo
   b. cause females to lay eggs.
   c. affect the number or polarity of body segments.
   d. affect the identity of a segment.

10. The homeotic mutations of *Drosophila*
   a. cause transformation of one segment to another.
   b. affect the function of proteins that are regulators of transcription.
   c. affect highly conserved functions found in nearly all organisms.
   d. all of the above.

Answers: 1b; 2d; 3b; 4a; 5d; 6f; 7d; 8b; 9a; 10d.

B. Thought Questions

1. Suppose you have cloned gene *X*. (a) Design an experimental procedure for determining whether or not gene *X is available for transcription* in liver and in brain tissue. (b) Now design an experiment to determine if gene *X is transcribed* in liver and/or in brain tissue. How do these experiments differ?

2. Present evidence that argues for or against the following statement: "Histone and non-histone proteins function as repressors of gene expression in eukaryotic cells."

3. Cite two different lines of evidence that methylation plays a role in transcriptional control in some eukaryotes. What evidence is there that methylation does not entirely determine the transcriptional activity of some genes?

4. Consider the cascade of alternative splicing events that are used to regulate sex-type in *Drosophila*. From an evolutionary perspective, what advantages might there be to having a cascade of events to regulate such an important process? (Hint: Consider the nature of the initial signal for sex-type, the X:A ratio, and how a splicing cascade "amplifies" this signal.)

5. Consider how hormones act. (a) Since hormones are diffusible, why are certain cells, but not others, targets of a particular hormonal signal? (b) How is the response of a cell to a peptide hormone fundamentally different from the response of a cell to a steroid hormone? (c) Propose an hypothesis to explain why it is that steroid hormone receptors are located inside the target cells, whereas the receptors for peptide hormones are located on the surfaces of target cells.

6. Suppose Gurdon's experiments on totipotency were repeated using differentiated B lymphocytes as the source of nuclei for transplantation. To what extent would you expect these nuclei to be totipotent? If mature organisms developed from cells with transplanted B-cell nuclei, what would be the nature of their immune system?

7. In *Drosophila*, the polar cytoplasm contains factors determining that the nuclei migrating into this region will become germ-line cells. (a) Design an experiment whose results could

provide evidence for this conclusion. (b) What phenotype would be associated with mutants lacking these factors? (c) Would the mutants described in part b identify maternal genes?

8. What are homeoboxes and what is their role in the regulation of development in eukaryotes?

9. In vertebrates, including mammals such as humans, homeotic gene complexes have been identified using homologous probes. Curiously, the gene complexes show similar structural organization, even though vertebrates are not segmented as are invertebrates. What significance might this finding have? How might having mutations in these genes in vertebrates help demonstrate the function of these genes, and test this significance?

10. A number of organisms have been studied intensely with the aim of understanding the genetic control of development. Describe the features of *Drosophila* and *Caenorhabditis* that make them useful for such analysis. Why is it important to study more than one model organism?

V. SOLUTIONS TO TEXT PROBLEMS

17.1 Eukaryotic organisms have a large number of copies (usually more than a hundred) of the genes that code for ribosomal RNA, yet they have only one copy of each gene that codes for each ribosomal protein. Explain why.

Answer: Eukaryotic organisms need to make enormous numbers of ribosomes. Since the "final product" of rDNA genes is an rRNA molecule, one means to make a large quantity of rRNA is to transcribe many rDNA templates. Given a fixed transcription rate, an increase by a factor of *n* rDNA templates will increase the number of final products by a factor of *n*. Unlike the rRNAs, the ribosomal proteins are made by translating mRNAs. Since the ribosomal protein genes can be transcribed into *stable* mRNAs that can be translated over and over again, it is unnecessary to have multiple copies of the genes. Synthesis of multiple stable mRNA copies of a ribosomal protein gene allows production of enough template to ensure adequate ribosomal protein synthesis. Put another way, since the final product is a protein, it is sufficient to amplify the mRNA template, and translate it repeatedly.

17.2 The human α, β, γ, δ, ε and ζ globin genes are transcriptionally active at various stages of development. Fill in the following table, indicating whether the globin gene in question is sensitive (S) or resistant (R) to DNase I digestion at the developmental stages listed.

	Tissue		
Globin Gene	Embryonic Yolk Sac	Fetal Tissue	Adult Bone Marrow
α			
β			
γ			
δ			
ζ			
ε			

Answer: Transcriptionally active chromatin has a looser structure than that found in an area of a transcriptionally inactive gene. Thus, DNase I sensitivity can be used to indicate genes that are transcriptionally active. Considering when and where the protein products appear, one would expect to find the following patterns of DNase I sensitivity.

	Tissue		
Globin Gene	Embryonic Yolk Sac	Fetal Tissue	Adult Bone Marrow
α	R	S	S
β	R	R	S
γ	R	S	R
δ	R	R	S
ζ	S	R	R
ε	S	R	R

17.3 A cloned DNA sequence was used to probe a Southern blot. There were two DNA samples on the blot, one from white blood cells and the other from a liver biopsy of the same individual. Both samples had been digested with *Hpa*II. The probe bound to a single 2.2 kb band in the white blood cell DNA, but bound to two bands (1.5 and 0.7 kb) in the liver DNA.

a. Is this difference likely to be due to a somatic mutation in a *Hpa*II site? Explain.
b. How would it affect your answer if you knew that white blood cell and liver DNA from this individual both showed the 2 band pattern when digested with *Msp*I?

Answer:

a. A somatic mutation is rather unlikely. Since humans are diploid, the single 2.2 kb band in the white blood cell DNA indicates that both chromosomes have the same restriction sites. The liver cells have two bands whose sizes add up to the size of the band in white blood cells. Thus, it would appear that the liver has an additional site on *both* chromosomes, not just one chromosome (heterozygous cells would have three bands). For this to occur by a mutational mechanism, two different somatic mutations would have had to occur in the same site, an unlikely event.
b. *Msp*I recognizes the same CCGG sequence as *Hpa*II, but unlike *Hpa*II, it will cleave the methylated sequence $C^{m}CGG$ as well as the unmethylated CCGG. That both liver cells and white blood cells show the same 2 band pattern when digested with *Msp*I, but not *Hpa*II, indicates that there is a central methylated site in white blood cells, but not in liver. Increased DNA methylation has, in some instances, been correlated with a decrease in transcriptional activity. However, not all methylated DNA is transcriptionally silent.

17.4 What is a hormone?

Answer: A hormone is typically a low molecular weight chemical messenger synthesized in low concentrations in one tissue or cell and transmitted in body fluids to another part of the organism. It acts as an effector molecule to produce a specific effect on target cells bearing a receptor for the hormone that maybe remote from its point of origin. Hormones function to regulate gene activity, physiology, growth, differentiation, or behavior.

17.5 How do hormones participate in the regulation of gene expression in eukaryotes?

Answer: Hormones can regulate gene expression by a number of means. Two fundamentally different mechanisms are illustrated by the actions of steroid hormones and polypeptide hormones. Steroid hormones traverse the cell membrane and bind to a cytoplasmic receptor. The complex of the hormone and receptor protein can then act directly to regulate gene expression by binding to specific response elements (REs) in the genome and regulating transcription. The stability of mRNAs and the processing of precursor mRNAs can be affected as well.

Polypeptide hormones, such as insulin and certain growth factors, can bind to receptors that reside on cell surfaces. The activated, bound receptor then transduces a signal via a second-messenger system inside the cell. Some activated receptors increase the activity of the membrane-bound enzyme adenylate cyclase, which results in an increase in cAMP levels. Increased intracellular levels of cAMP serve as an intracellular signal (a second messenger) to activate other cellular processes that ultimately lead to an alteration in patterns of gene expression. See text pp. 552-557.

17.6 The following figure shows the effect of the hormone estrogen on ovalbumin synthesis in the oviduct of 4-day-old chicks. Chicks were given daily injections of estrogen ("Primary Stimulation") and then after 10 days the injections were stopped. Two weeks after withdrawal (25 days), the injections were resumed ("Secondary Stimulation").

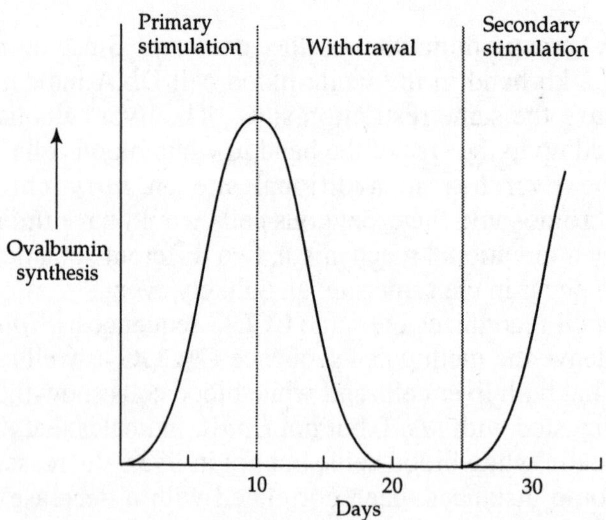

Provide possible explanations of these data.

Answer: The data indicate that the synthesis of ovalbumin is dependent upon the presence of the hormone estrogen. These data do not address the mechanism by which estrogen achieves its effects. Theoretically, it could act (1) to increase transcription of the ovalbumin gene by binding to an intracellular receptor that, as an activated complex, stimulates transcription at the ovalbumin gene; (2) to stabilize the ovalbumin precursor mRNA; (3) to increase the processing of the precursor ovalbumin mRNA; (4) to increase the transport of the processed ovalbumin mRNA out of the nucleus; (5) to stabilize the mature ovalbumin mRNA once it has been transported into the cytoplasm; (6) to stimulate translation of the ovalbumin mRNA in the cytoplasm; and/or (7) to stabilize (or process) the newly synthesized ovalbumin protein. Experiments in which the levels of ovalbumin mRNA were measured have shown that the production of ovalbumin mRNA is primarily regulated at the level of transcription.

17.7 Distinguish between the terms *development* and *differentiation.*

Answer: Development is a process of regulated growth and cellular change. It results from the interaction of the genome with the cytoplasm and external environment. It involves a programmed sequence of phenotypic events that are typically irreversible. Differentiation refers to an aspect of cellular change in development and involves the formation of distinctly different types of cells, tissues and organs through the processes of specific regulation of gene expression. Differentiation is thus a part of development and leads to cells that have characteristic structural and functional properties.

17.8 What is totipotency? Give an example of the evidence for the existence of this phenomenon.

Answer: Totipotency refers to the capacity of a nucleus to direct a cell through all the stages of development. In other words, a cell taken from a differentiated tissue is totipotent if it can be isolated and if a complete functional organism can develop from it. The implication is that the cell contains all the genetic information present in a zygote so that the developmental program for the complete organism can be executed. The classic demonstration of nuclear totipotency was done by Gurdon with *Xenopus laevis* and is described in detail in the text on pp. 569-570.

17.9 Discuss some of the evidence for differential gene activity during development.

Answer: The evidence for differential gene activity during development is vast. Classic lines of evidence stem from (1) studies on the differential expression of the α, β, γ, δ, ζ, and ε classes of globin genes during development, (2) differential puffing patterns in the polytene chromosomes in Dipteran insects and (3) the studies on genes that are expressed in a temporal and spatially specific manner during the development of *Drosophila.* These lines of evidence are presented in the text on pp. 571-583.

17.10 The enzyme lactate dehydrogenase (LDH) consists of four polypeptides (a tetramer). Two genes are known to specify two polypeptides, A and B, which combine in all possible ways (A_4, A_3B, A_2B_2, AB_3, and B_4) to produce five LDH isozymes. If, instead, LDH consisted of three polypeptides (i.e., it was a trimer), how many possible isozymes would be produced by various combinations of polypeptides A and B?

Answer: If LDH were a trimer that could be composed of two different polypeptides, one would have four kinds of molecules: A_3, A_2B, AB_2, and B_3.

17.11 Discuss the expression of human hemoglobin genes during development.

Answer: Distinct genes code for α-like and β-like globin polypeptides which form different types of hemoglobin at different times during human development. In the embryo, hemoglobin is initially made in the yolk sac and consists of two ζ-polypeptides and two ε-polypeptides. ζ-polypeptides are α-like, while ε-polypeptides are β-like. At about three months of gestation, hemoglobin synthesis switches to the fetal liver and spleen. Here, hemoglobin is made that consists of two α-polypeptides and two β-like polypeptides, either two γA-polypeptides or two γG-polypeptides. Just before birth, hemoglobin synthesis switches to the bone marrow, were α-polypeptides and β-polypeptides are predominantly made, along with some β-like, δ-polypeptides.

17.12 Discuss the organization of the hemoglobin genes in the human genome. Is there any correlation with the temporal expression of the genes during development?

Answer: All of the α-like genes (the ζ-, α-genes) are located in a gene cluster on chromosome 16, while all of the β-like (ε-, γG-, γA-, δ-, β-genes) genes are located in a gene cluster on chromosome 11. At a very general level, the pattern of organization is somewhat similar, but it is by no means identical. The sets of genes are arranged in the chromosome in an order that exactly parallels the timing in which genes are transcribed during human development. In both gene clusters, the genes transcribed in the embryo are at the left end of the cluster, the genes transcribed in the fetus are to the right of these, and the genes transcribed in the adult are the most rightwardly positioned. This is exceptionally intriguing, especially in light of the fact that the genes are also transcribed in different tissues.

17.13 In humans, β-thalassemia is a disease caused by failure to produce sufficient β-globin chains. In many cases, the mutation causing the disease is a deletion of all or part of the β-globin structural gene. Individuals homozygous for certain of the β-thalassemia mutations are able to survive because their bone marrow cells produce γ-globin chains. The γ-globin chains combine with α-globin chains to produce fetal hemoglobin. In these people, fetal hemoglobin is produced by the bone marrow cells throughout life, whereas normally it is produced in the fetal liver. Use your knowledge about gene regulation during development to suggest a mechanism by which this expression of γ-globin might occur in β-thalassemia.

Answer: There are a number of possibilities here. One is that the γ-globin genes in bone marrow are under negative regulation by β-globin (or some metabolite of it). When β-globin is not formed, the γ-globin gene is derepressed.

17.14 What are the polytene chromosomes? Discuss the molecular nature of the puffs that occur in polytene chromosomes during development.

Answer: Polytene chromosomes occur in Dipteran insects, such as *Drosophila.* Polytene chromosomes are formed by endoreduplication, in which repeated cycles of chromosome duplication occur without nuclear division or chromosome segregation. Since they can be 1,000 times as thick as corresponding chromosomes in meiosis or in the nuclei of normal cells, they can be stained and viewed at the light microscope level. Distinct bands, or chromomeres, are visible, and genes are located both in bands and in interband regions. A puff results when a gene in a band or interband region is expressed at very high levels in a particular developmental stage. Puffs are accompanied by a loosening of the chromatin structure that allows for efficient transcription of a particular DNA region. When increased transcriptional activity at the gene ceases at a later developmental stage, the puff disappears, and the chromosome resumes its compact configuration. In this way, the appearance and disappearance of puffs provides a visual representation of differential gene activity.

17.15 Puffs of regions of the polytene chromosomes in salivary glands of *Drosophila* are surrounded by RNA molecules. How would you show that this RNA is single-stranded and not double-stranded?

Answer: That RNA molecules are present in puffs has been demonstrated by feeding or injecting developing flies with radioactive uridine, an RNA precursor. After such a treatment, salivary glands were dissected from larvae and a spread of the polytene chromosomes was subjected to autoradiography to detect the location of the incorporated uridine. Silver grains were evident over puffed regions, indicating that they contained RNA. Evidence that the RNA is single-stranded might be obtained by treating such spreads with RNase. If such spreads were treated with an RNase that was capable of digesting only single-stranded RNA, one would expect the radiolabel to be recovered in the solution, and not remain bound to the salivary gland chromosome. A lack of signal on puffs following autoradiographic detection (compared to controls, of course) would provide evidence that the RNA in puffs was single-stranded.

17.16 In experiment A, ^{3}H-thymidine (a radioactive precursor of DNA) is injected into larvae of *Chironomus*, and the polytene chromosomes of the salivary glands are later examined by autoradiography. The radioactivity is seen to be distributed evenly throughout the polytene chromosomes. In experiment B, ^{3}H-uridine (a radioactive precursor of RNA) is injected into the larvae, and the polytene chromosomes are examined. The radioactivity is first found only around puffs; later, radioactivity is also found in the cytoplasm. In experiment C, actinomycin D (an inhibitor of transcription) is injected into larvae and then ^{3}H-uridine is injected. No radioactivity is found associated with the polytene chromosomes, and few puffs are seen. Those puffs that are present are much smaller than the puffs found in experiments A and B. Interpret these results.

Answer: Experiment A results in all of the DNA becoming radioactively labeled. The distribution of radioactive label throughout the polytene chromosomes indicates that DNA is a fundamental, and major component of these chromosomes. The even distribution of label suggests that each region of the chromosome has been replicated to the same extent. This provides support for the contention that band and interband regions are the result of different types of packaging, and not different amounts of DNA replication.

Experiment B results in radioactive labeling of RNA molecules. The finding that label is found first in puffs indicates that these are sites of transcriptional activity, and arises from molecules that are in the process of being synthesized. The later appearance of label in the cytoplasm reflects the completed RNA molecules that have been processed and transported into the cytoplasm where they will be translated.

Experiment C provides additional support for the hypothesis that transcriptional activity is associated with puffs. The inhibition of RNA transcription by actinomycin D blocks the appearance of signal over puffs, indicating that it blocks the incorporation of ^{3}H-uridine into RNA in puffed regions. The fact that the puffs are much smaller indicates that the puffing process itself is associated with the onset of transcriptional activity for the gene(s) in a specific region of the chromosome.

17.17 The following figure shows the percentage of ribosomes found in polysomes in unfertilized sea urchin oocytes (0 h) and at various times after fertilization:

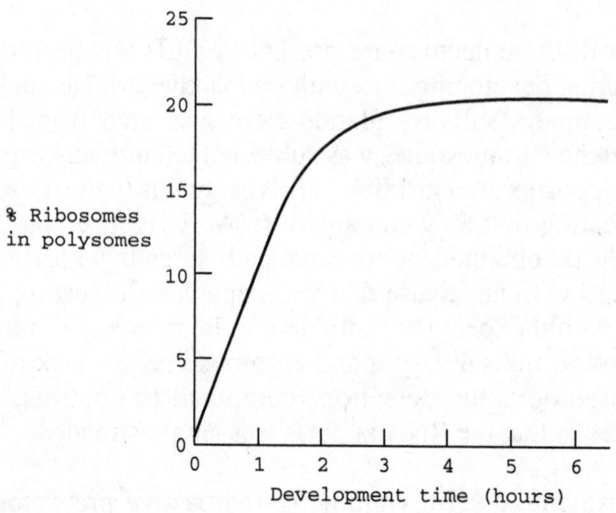

In the unfertilized egg, less than 1 percent of ribosomes are present in polysomes, while at 2 h post-fertilization, about 20 percent of ribosomes are present in polysomes. It is known that no new mRNA is made during the time period shown. How may the data be interpreted?

Answer: Since no new mRNA is synthesized during this time period, pre-existing, maternally packaged mRNAs that have been stored in the oocyte must be recruited into polysomes as development begins following fertilization.

17.18 The mammalian genome contains about 10^5 genes. Mammals can produce about 10^6 to 10^8 different antibodies. Explain how it is possible for both of the above sentences to be true.

Answer: The ability to make 10^6 to 10^8 different antibodies arises from the combinatorial way in which antibody genes are generated in different antibody-producing cells during their development, and not the existence of this many separate antibody genes in each and every mammalian cell. A template that exists in germ-line cells is used differently by different developing antibody producing cells to generate antibody diversity.

Antibody molecules consist of two light (L) chains and two heavy (H) chains. The amino acid sequence of one domain of each type of chain is variable, and generates antibody diversity. In the germ line DNA of mammals, coding regions for these immunoglobulin chains exist in tandem arrays of gene segments. For light chains, there are many variable (V) region gene segments, a few joining (J) segments and one constant (C) gene segment. Somatic recombination during development results in the production of a recombinant V-J-C DNA molecule which, when transcribed, produces a unique, functional L chain. From a particular gene in one cell, only one L chain is produced. A large number of L chains are obtained by recombining the gene segments in many different ways. Diversity in these L chains results from variability in the sequences of the multiple V segments, variability in the sequences of the four J segments and variability in the number of nucleotide pairs deleted at the V-J joints. H chains are similar, except that several D (diversity) segments can be used between the V and J segments, increasing the possible diversity of recombinant H chain genes.

17.19 Ashley and Connie are identical twins. Ashley has blood type A, and therefore must have anti-B antibodies in her serum. (For the purpose of this problem, ignore the effects of possible new mutations.)

a. What is Connie's blood type?
b. If Ashley's genotype is I^A/I^A, what is Connie's blood type genotype?
c. If Ashley has anti-B antibodies, what type of antibodies would Connie have?
d. Would Ashley's and Connie's anti-B antibodies have identical polypeptide sequences? Explain.
e. Would Ashley's and Connie's β-globin chain have the identical polypeptide sequence? Explain.

Answer: If Ashley and Connie are identical twins, they were formed as separate organisms after the first mitotic cleavage of a single embryo. They had identical genes when they became distinct organisms after the first mitotic cleavage. After this point, they would retain identical genes except for those genes that undergo somatic recombination during development, such as immunoglobulin genes.

a. A (blood type genes do not undergo somatic recombination).
b. I^A/I^A.
c. anti-B antibodies.
d. No. Although the antibodies in each twin are able to recognize the B-blood-type-antigen, they need not be identical in primary structure. Antibodies recognize antigens using structural interactions based on the folding of the antibody, that is, its tertiary structure. Two different antibodies might be capable of recognizing the same antigen. In this case, the antibodies that recognize the B-antigen arose during the development of the immune system in each twin. In each twin, a different set of V-J-(D)-C somatic recombination events could lead to the production of an antibody that recognizes the B-antigen. Hence the antibodies would have a different amino acid sequence.
e. Yes. The β-globin genes do not undergo somatic recombination during development, so the twins will have an identical set of β-globin genes. These genes in each twin will therefore produce identical polypeptides.

17.20 Recall that antibody molecules (Ig) are composed of four polypeptide chains (one light chain type and one heavy chain type) held together by disulfide bonds.

a. If for the light chain there were 300 different V_K and four J_K segments, how many different light chain combinations would be possible?
b. If for the heavy chain there were 200 V_H segments, 12 D segments, and 4 J_H segments, how many heavy chain combinations would be possible?

c. Given the information in parts (a) and (b), what would be the number of possible types of IgG molecules (L + H chain combinations)?

Answer: a. If one does not take into account the variability in the number of nucleotide pairs deleted at the V_K-J_K joint, one would have 300 x 4 = 1,200 different light chain combinations. This number is a lower bound for an estimate, since significantly more variability can be obtained by using an imprecise V-J joint.

b. If one does not take unto account the imprecise joining of gene segments that comprise the chains' variable region, one would have 200 x 12 x 4 = 9,600 different heavy chain combinations. As in part a, this is a lower bound for an estimate of heavy chain variability.

c. An estimate of the minimal number of combinations would be 9,600 x 1,200 = 1.152×10^7. (About 11.5 million combinations).

17.21 Define *imaginal disc*, *homeotic mutant*, and *transdetermination*.

Answer: An imaginal disc is a larval structure in *Drosophila* and other insects that undergo a metamorphic transformation from a larva to an adult. During metamorphosis in the pupal period, it will differentiate into an adult structure. It is essentially a sac of cells, a disc, that is set aside early in embryonic development, and although it increases in cell number by mitosis, it remains undifferentiated during larval growth. Although the disc remains undifferentiated up until the time hormonal signals activate the differentiation of adult structures during the pupal period, its fate is *determined* very early in development. Thus, separate imaginal discs give rise to different adult structures, such as legs, eyes, antennae, and wings.

Homeotic mutations are mutations that alter the identity of particular segments, transforming them into copies of other segments. Homeotic mutations thus affect the determination, or fate of a disc. Mutations in the *Antennapedia* gene complex can, for example, transform the fate of cells of the antenna disc so that a leg is formed where an antenna should be. The general inference from such observations is that homeotic genes normally play a pivotal role in establishing the segmental identify of an undifferentiated cell.

Transdetermination is a phenomenon seen in disc transplantation studies. When a disc is transplanted from a larva to an adult abdomen, it can grow and be 'incubated' without differentiating. It can be repeatedly transplanted to new abdomens for incubation, and remain undifferentiated. If it is later transplanted back to a larva that is allowed to undergo metamorphosis, it will differentiate into an adult structure. In rare instances, the determined fate of the disc is altered so that the disc differentiates to another fate. The disc does not totally de-differentiate, and its fate is not totally random. Rather, it takes on another determined fate. For example, a leg disc may become an antennal structure, or a wing structure, but not a genital structure. (See Figure 17.32, p. 579). This process of selective switching of fate is referred to as transdetermination.

17.22 Imagine that you observed the following mutants (*a–e*) in *Drosophila*. Based on the characteristics given, assign each of the mutants to one of the following categories: maternal gene, segmentation gene, or homeotic gene.

a. Mutant *a*: In homozygotes phenotype is normal, except wings are oriented backwards.

b. Mutant *b*: Homozygous females are normal but produce larvae that have a head at each end and no distal ends. Homozygous males produce normal offspring (assuming the mate is not a homozygous female).
c. Mutant *c*: Homozygotes have very short abdomens, which are missing segments A2 through A4.
d. Mutant *d* Affected flies have wings growing out of their heads in place of eyes.
e. Mutant *e*: Homozygotes have shortened thoracic regions and lack the second and third pair of legs.

Answer: a. As the *polarity* of an individual, differentiated structure is altered, *a* is likely to be a segmentation gene. In particular, it is likely to be a member of the subclass of segmentation genes known as segment polarity genes.
b. The abnormal appearance of progeny of homozygous mothers, but not homozygous fathers, suggests that homozygous mothers produce abnormal oocytes. This view is bolstered by the phenotype of the abnormal progeny. Progeny with defect(s) at one end could arise if a maternally produced gradient in the oocyte was abnormal. Thus, *b* is likely to be a maternal gene.
c. The absence of a set of segments is characteristic of the phenotypes of *gap* gene mutants, a subclass of segmentation gene mutants.
d. It would appear that the fate of the eye disc has been changed to that of a wing disc. The segmental *identity* of cells is controlled by homeotic genes, and so *d* is likely to be a homeotic mutation.
e. In mutant *e*, a set of segments is missing (T2 and T3). Thus, like *c*, *e* is likely to be a mutant in a *gap* gene, a subclass of segmentation gene mutants.

17.23 If actinomycin D, an antibiotic that inhibits RNA synthesis, is added to newly fertilized frog eggs, there is no significant effect on protein synthesis in the eggs. Similar experiments have shown that actinomycin D has little effect on protein synthesis in embryos up until the gastrula stage. After the gastrula stage, however, protein synthesis is significantly inhibited by actinomycin D, and the embryo does not develop any further. Interpret these results.

Answer: Pre-existing mRNA that was made by the mother and packaged into the oocyte prior to fertilization is translated up to the gastrula stage. After gastrulation, new mRNA synthesis is necessary for the production of proteins needed for subsequent embryonic development.

17.24 It is possible to excise small pieces of early embryos of the frog, transplant them to older embryos, and follow the course of development of the transplanted material as the older embryo develops. A piece of tissue is excised from a region of the late blastula or early gastrula that would later develop into an eye and is transplanted to three different regions of an older embryo host (see part a in the figure below). If the tissue is transplanted to the head region of the host, it will form eye, brain, and other material characteristic of the head region. If the tissue is transplanted to other regions of the host, it will form organs and tissues characteristic of those regions in normal development (e.g., ear, kidney, etc.). In contrast, if tissue destined to be an eye is excised from a neurula and transplanted into an older embryo host to exactly the same places as used for the blastula/gastrula transplants, in every case the transplanted tissue differentiates into an eye (see b in the figure below). Explain these results.

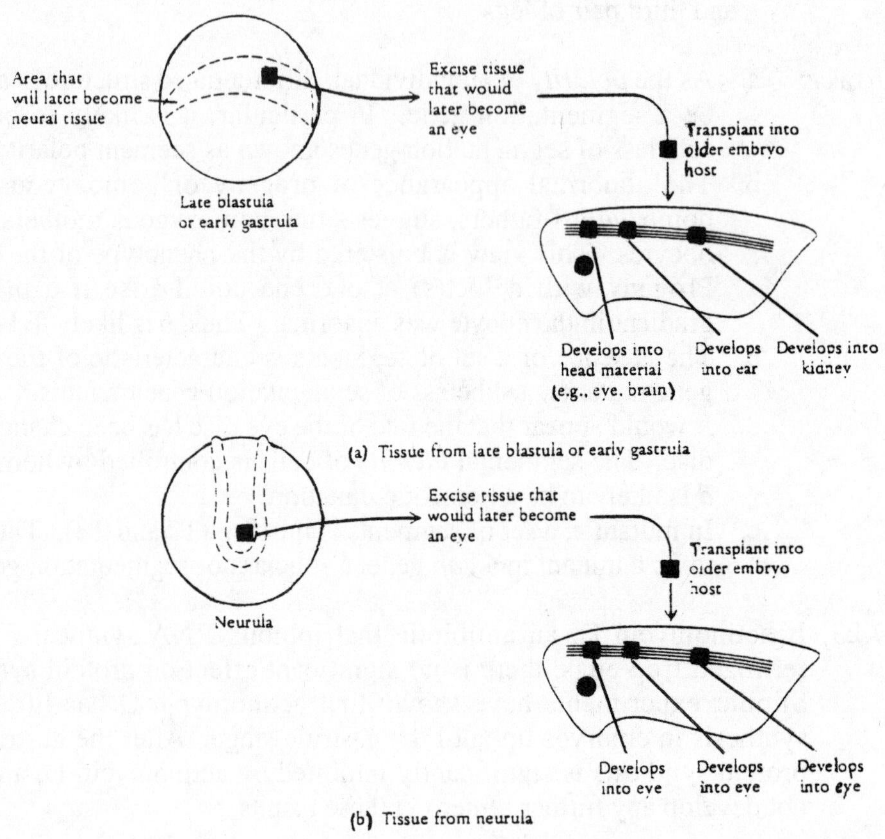

(a) Tissue from late blastula or early gastrula

(b) Tissue from neurula

Answer: This experiment demonstrates the phenomenon of *determination* and when it occurs during development. The tissue taken from the blastula/gastrula has not yet been committed to its final differentiated state in terms of its genetic programming; that is, it has not yet been *determined.* Thus, when the tissue is transplanted into the host, it adopts the fate of nearby tissues, and becomes determined in the same way as they are. Presumably, cues from the tissue surrounding the transplant determine its fate.

In contrast, tissues in the neurula stage are stably determined. By the time the neurula developmental stage has been reached, a developmental program has been set. In other words, the fate of neurula tissue transplants is *determined.* Upon transplantation, they will differentiate according to their own set genetic program. Tissue transplanted from a neurula to an older embryo cannot be influenced by the determined surrounding tissues. It will develop into the tissue type for which it is determined, in this case, an eye.

CHAPTER 18

GENE MUTATION

I. CHAPTER OUTLINE

ADAPTATION VERSUS MUTATION
MUTATIONS DEFINED
- Types of Mutations
- Reverse Mutations and Suppressor Mutations

CAUSES OF MUTATION
- Spontaneous Mutations
- Induced Mutations

THE AMES TEST: A SCREEN FOR POTENTIAL MUTAGENS
DNA REPAIR MECHANISMS
- Direct Correction of Mutational Lesions
- Repair Involving Excision of Base Pairs
- Human Genetic Diseases Resulting From DNA Replication and Repair Errors

SCREENING PROCEDURES FOR THE ISOLATION OF MUTANTS
- Visible Mutations
- Nutritional Mutations
- Conditional Mutations

II. REVIEW OF KEY TERMS, SYMBOLS AND CONCEPTS

Without consulting the text and in your own words, write a brief definition of each term in the groups below. Then, either using a short phrase or a simple diagram, identify the relationship(s) between specific pairs of terms within a set. Finally, consult the text (and perhaps a friend who has also done the exercise) to check your answers.

1	2	3	4
base pair mutation	mutation	mutagen	forward mutation
gene mutation	somatic mutation	base-pair substitution	reverse mutation (reversion)
adaptation	germline mutation	transition mutation	true reversion
fluctuation test	chromosomal mutation	transversion mutation	suppressor mutation
random mutation	gene mutation	missense mutation	second-site mutation
	point mutation	nonsense mutation	intragenic suppressor
		neutral mutation	intergenic suppressor
		silent mutation	suppressor gene
		frameshift mutation	

5	6	7	8
spontaneous mutation mutation rate mutation frequency tautomer tautomeric shift depurination deamination mutational hot spot	induced mutation thymine dimer SOS system base analog 5BU, 2AP base-modifying agent nitrous acid, hydroxylamine MMS intercalating agent acridine orange, proflavin site-specific mutagenesis reverse genetics	Ames test mutagen carcinogen mitogenesis	DNA repair proofreading mutator mutation photoreactivation light repair excision (dark) repair alkylating repair glycosylase

9	10
DNA repair AP site, endonuclease nick translation mismatch repair *dam* methylase hemimethylation SOS response error-prone repair error-free (prone)-bypass synthesis	mutant screen visible mutation replica plating nutritional mutation auxotrophic mutation conditional mutation

III. THINKING ANALYTICALLY

Mutation is a fundamental process important to all of genetics, and for that matter, to the evolution of life itself. In this chapter, the conceptual framework used by geneticists to consider the nature of mutation is presented. This is followed by a detailed discussion of the types and causes of mutation, strategies to identify mutants and mutagens, and the biological mechanisms used to repair mutations. Consequently, the material presented here is both highly conceptual and quite detailed. Focus first on the concepts, and as in previous chapters, learn to use the terms and definitions to explain the concepts. In order to follow some of the complicated processes described in this chapter, it is essential that one have the precise meaning of each term well in mind.

As you consider the molecular basis of mutation, it will help to review the chemical structure of the bases, so that you can envision how each of the mutagens acts. Employ diagrams as you consider the effects of specific mutagens and repair processes on the structure of DNA. It will also be invaluable to use diagrams to follow how a mutant DNA sequence can result in specific alterations in codon usage in an mRNA. It is essential to proceed methodically (and slowly – so as not to overlook a subtle point such as changes in polarity) through any analysis involving base pair changes. Use terms carefully, remembering that mutations affect the bases in DNA, and then, indirectly, the bases in transcribed RNA and the amino acids in proteins.

Two topics in this chapter, suppressor mutations and screening procedures used in the isolation of new mutations, warrant particularly careful attention. Proceed by carefully and methodically organizing the information that is presented. Then, supplement your reading with diagrams and flowcharts.

IV. QUESTIONS FOR PRACTICE

A. Multiple Choice Questions

Match the best choice below to each of the statements for Questions 1-8.

a. missense mutation	e. nonsense mutation
b. transversion mutation	f. transition mutation
c. neutral mutation	g. frameshift mutation
d. forward mutation	h. suppressor mutation

1. A purine-pyrimidine base-pair is mutated to a different purine-pyrimidine base pair.

2. A point mutation in the DNA that changes a codon in the mRNA and causes an amino acid substitution that does not alter the function of the translated protein.

3. A DNA change in which a GC base pair is replaced by a TA base pair.

4. A mutation that results in the addition or deletion of a base-pair within the coding region of a gene.

5. Any point mutation that is expressed as a change in phenotype from wild type to mutant.

6. A base-pair change in the DNA that results in the change of a mRNA codon to either UAG, UAA, or UGA.

7. A mutation in the DNA that changes a codon in the mRNA and causes an amino acid substitution that may or may not produce a change in the function of a translated protein.

8. Any new mutation that restores some or all of the wild type phenotype to a previously isolated mutation.

9. The Ames test is used to screen for
   a. potential mutagens and carcinogens.
   b. the presence of enol forms of thymine and guanine.
   c. frameshift mutations.
   d. spontaneous mutations.

10. In the Ames test, rat liver extracts are used
    a. to chemically alter and detoxify potential mutagens.
    b. to chemically alter and toxify potential mutagens.
    c. to determine if an environmental chemical that itself is not mutagenic may become mutagenic when processed in the liver.
    d. all of the above.

11. A population of cells is subjected to UV-irradiation and then placed in the dark. There will be significant
   a. production of base tautomers.
   b. breakage of phosphodiester bonds.
   c. formation of pyrimidine dimers.
   d. photolyase activity.

12. Single bacterial cells sensitive to the infection of a phage are inoculated into a large number of separate culture dishes, and allowed to grow in parallel. After many generations, a sample is taken from each culture, inoculated with the phage and separately plated. What result is expected to be seen?
   a. All plates will have similar numbers of bacterial colonies.
   b. No bacterial colonies will be seen on any of the plates.
   c. Only one plate in 10^6 will have bacterial colonies.
   d. Some plates will have no bacterial colonies, some will have a few, and some will have many.

Answers: 1f; 2c; 3b; 4g; 5d; 6e; 7a; 8h; 9a; 10d; 11c; 12

B. Thought Questions

1. What is the evidence that mutations occur spontaneously and at a low frequency, regardless of selection for or against them?

2. The proofreading ability of DNA polymerase is extremely good, but not quite perfect. What error rates are associated with DNA replication, from whence do they arise, and how might this be important in evolutionary terms?

3. Distinguish between a reversion mutation, an intragenic suppressor mutation and a intergenic suppressor mutation.

4. Distinguish between a missense and a nonsense mutation, and between a missense, a silent and a neutral mutation.

5. How would you classify the following mutations? (a) A single base change in a promoter region that effects the transcription of a gene. (b) A single base change in a promoter region that has no affect. (c) An RFLP difference between two strains that has no apparent phenotypic affect?

6. What is the basis for the Ames test? What are its specific merits? Are there any kinds of mutagens that it might not detect? (e.g., "kinds of mutagens" refers to those that specifically cause point mutations, frameshifts, small deletions, chromosome rearrangements, etc.)

7. In the last several years, significant insight into how DNA is repaired has come from studies in bacteria and yeast, as well as from the molecular cloning of disease genes from humans with defective DNA repair. (a) Speculate as to the phenotype(s) of bacterial and yeast mutations that are defective in DNA repair. (b) How might comparison of these three systems be beneficial?

8. In *Drosophila*, the first mutations isolated (e.g., the first *white-eyed* mutant) were spontaneous. Currently, most new mutations are isolated by using mutagens. Why are mutagens currently employed, and what benefit do mutagens have for the study of genetics and development? How is site-specific *in vitro* mutagenesis used to study the function of cloned genes?

9. What spontaneous mutation rates are found in different organisms? Are these rates relatively similar or quite different? Why do you think this is so?

10. Dyes are routinely used to stain nuclei in bacteriological and histological preparations as well as to visualize DNA and RNA molecules separated by size using agarose gel electrophoresis. Some of these dyes bind to the backbone of nucleic acids while others intercalate between bases.

How would you specifically test whether a particular dye is mutagenic? How would you use the results of the test to design general precautions for laboratory workers using these dyes?

11. If a substance is mutagenic, is it necessarily carcinogenic? Why or why not? If a substance is carcinogenic, is it necessarily mutagenic? Why or why not?

V. SOLUTIONS TO TEXT PROBLEMS

18.1 Mutations are (choose the correct answer):
  a. Caused by genetic recombination.
  b. Heritable changes in genetic information.
  c. Caused by faulty transcription of the genetic code.
  d. Usually but not always beneficial to the development of the individuals in which they occur.

Answer: b. Heritable changes in genetic information.

18.2 Answer true or false: Mutations occur more frequently if there is a need for them.

Answer: False. Mutations occur spontaneously at a more or less constant frequency, regardless of selective pressure. Once they occur however, they can be selected for or against, depending on the advantage or disadvantage they confer. It is important not to confuse the frequency of mutations with selection.

18.3 Which of the following is *not* a class of mutation?
  a. Frameshift
  b. Missense
  c. Transition
  d. Transversion
  e. None of the above (i.e., all are classes of mutation)

Answer: e. None of the above (i.e., all are classes of mutation)

18.4 Ultraviolet light usually causes mutations by a mechanism involving (choose the correct answer):
  a. One-strand breakage in DNA
  b. Light-induced change of thymine to alkylated guanine
  c. Induction of thymine dimers and their persistence or imperfect repair
  d. Inversion of DNA segments
  e. Deletion of DNA segments
  f. All of the above

Answer: c. Induction of thymine dimers and their persistence or imperfect repair. The key to this answer is the word "usually". The other choices might apply rarely, but not usually.

18.5 For the middle region of a particular polypeptide chain, the normal amino acid sequence and the amino acid sequence of several mutants were determined, as shown below (... indicates additional, unspecified amino acids). For each mutant, say what DNA level change has occurred, whether the change is a base pair substitution mutation (transversion or transition, missense or nonsense) or a frameshift mutation, and in which codon the mutation occurred. (Refer to the codon dictionary in Figure 14.9, p. 435.)

		CODON								
		1	2	3	4	5	6	7	8	9
a.	Normal:	... Phe	Leu	Pro	Thr	Val	Thr	Thr	Arg	Trp
b.	Mutant 1:	... Phe	Leu	His	His	Gly	Asp	Asp	Thr	Val
c.	Mutant 2:	... Phe	Leu	Pro	Thr	Met	Thr	Thr	Arg	Trp
d.	Mutant 3:	... Phe	Leu	Pro	Thr	Val	Thr	Thr	Arg	
e.	Mutant 4:	... Phe	Pro	Pro	Arg					
f.	Mutant 5:	... Phe	Leu	Pro	Ser	Val	Thr	Thr	Arg	Trp

Answer: Start methodically and write out the potential codons for the normal protein. Let N represent any nucleotide, R a purine (A or G) and Y a pyrimidine (C or U). Then the codons can be written as:

	1	2	3	4	5	6	7	8	9
Normal:	Phe	Leu	Pro	Thr	Val	Thr	Thr	Arg	Trp
CODON:	UUY	UUR or CUN	CCN	ACN	GUN	ACN	ACN	CGN or AGR	UGG

Mutant 1: The alteration of all of the amino acids after codon 2 suggests that this mutation is a frameshift, either a deletion or an addition of a base pair in the DNA region that codes for the mRNA near codons 2 and 3. His is CAY, which can be generated from CCN by the insertion of a single A between the two C's. If this is the case, one has:

	1	2	3	4	5	6	7	8	9
Mutant 1:	Phe	Leu	His	His	Gly	Asp	Asp	Thr	Val
Frameshift:	UUY	UUR or CUN	CAC	NAC	NGU	NAC	NAC	NCG or NGR	NUG or RUG

Now identify the unknown bases, given the amino acid sequence:

To code for His, the new fourth codon must be CAC.
To code for Gly, the new fifth codon must be GGU.
To code for Asp, the new sixth codon must be GAC.
To code for Asp, the new seventh codon must be GAC.
To code for Thr, the new eighth codon must be ACG.
To code for Val, the new ninth codon must be GUG.

Thus, one has:

	1	2	3	4	5	6	7	8	9
Mutant 1:	Phe	Leu	His	His	Gly	Asp	Asp	Thr	Val
CODON:	UUY	UUR or CUN	CAC	CAC	GGU	GAC	GAC	ACG	GUG

This analysis indicates that the codons used in the normal protein are:

	1	2	3	4	5	6	7	8	9
Normal:	Phe	Leu	Pro	Thr	Val	Thr	Thr	Arg	Trp
CODON:	UUY	UUR or CUN	CCC	ACG	GUG	ACG	ACA	CGG	UGG

Mutant 2: Codon 5 in mutant 2 encodes Met, instead of Val. This single change could be caused by a point mutation, where the G in codon GUG is changed to an A leading to an AUG codon. This would occur by a CG → TA transition in the DNA.

Mutant 3: Mutant 3 has a normal amino acid sequence, but is prematurely terminated, indicating that a nonsense mutation occurred. The nonsense mutation occurred in the ninth codon. If an A base is substituted (for one of the G's) or inserted (before or after the first G), one could have either UAG or UGA. Thus, either a frameshift or base substitution (CG → TA transition) occurred.

Mutant 4: Mutant 4 shows premature termination at codon 5 (indicating a nonsense codon there), and has a missense mutation at codons 2 and 4. Compare the possible sequences of the mutant with the normal sequence to see if one mutational event can account for all of these phenomena.

	1	2	3	4	5	6	7	8	9
Normal:	Phe	Leu	Pro	Thr	Val	Thr	Thr	Arg	Trp
Normal Codon:	UUY	UUR or <u>CUN</u>	CCC	A<u>CG</u>	<u>GUG</u>	<u>A</u>CG	ACA	CGG	UGG
Mutant 4:	Phe	Pro	Pro	Arg	Stop				
Possible Mutant Codons:	UUY	CCN (CCC)	CCN (CCA)	AGR or <u>CGN</u>	UAA UAG <u>UGA</u>				

Consider codon 2 carefully. In order for Pro to be encoded by CCN in the mutant protein, and be obtained by a single change from the second codon, Leu (in the normal protein) must be encoded by CUN. If the U were deleted (by a AT deletion in the DNA) and the N were a C, a CCC Pro codon and a frameshift would result. Codon 3 would become CCA and still code for Pro, codon 4 would become CGG and code for Arg, and codon 5 would become AGA, a nonsense codon.

Mutant 5: Mutant 5 shows an alteration of only the fourth amino acid (Thr to Ser), suggesting a point mutation. If the fourth codon were changed from **A**CG to **U**CG (by a TA to an AT transversion in the DNA), this missense mutation would occur.

18.6 In mutant strain *X* of *E. coli*, a leucine tRNA which recognizes the codon 5'-CUG-3' in normal cells has been altered so that it now recognizes the codon 5'-GUG-3'. A missense mutation, which affects amino acid 10 of a particular protein, is suppressed in mutant *X* cells.

a. What are the anticodons of the two Leu tRNAs, and what mutational event has occurred in mutant *X* cells?
b. What amino acid would normally be present at position 10 of the protein (without the missense mutation)?
c. What amino acid would be put in at position 10 if the missense mutation is not suppressed (i.e., in normal cells)?
d. What amino acid is inserted at position 10 if the missense mutation is suppressed (i.e., in mutant *X* cells)?

Answer: a. If the normal codon is 5'-CUG-3', the anticodon of the normal tRNA is 5'-CA<u>G</u>-3'. If a mutant tRNA recognizes 5'-GUG-3', it must have an anticodon that is 5'-CA<u>C</u>-3'. The mutational event was a CG to GC transversion.
b. Since a leucine bearing (mutant) tRNA can suppress the mutation, one presumes that leucine is normally present at position 10.
c. The mutant tRNA recognizes the codon 5'-GUG-3', which codes for Val. In normal cells, a Val.tRNA.Valine would recognize the codon.
d. Leu

18.7 In any kind of chemotherapy, the object is to find a means to kill the invading pathogen or cancer cell without killing the cells of the host. To do this successfully, one must find and exploit biological differences between target organisms and host cells. Explain the nature of the biological difference between host cells and HIV-I virus that permits the use of azidothymidine (AZT) for chemotherapy.

Answer: The reverse transcriptase of HIV-1 recognizes AZT as a substrate (as if it was thymidine), and incorporates it into viral cDNA. The viral cDNA is normally incorporated into the host cell's genome using host cell DNA polymerases. Once incorporated, the viral cDNA directs new viral synthesis. The presence of AZT in viral cDNA blocks its incorporation into the host cell's DNA (and consequently, the subsequent viral replication) because AZT is not a good substrate for cellular DNA polymerases.

18.8 The mutant *lacZ-1* was induced by treating *E. coli* cells with acridine, while *lacZ-2* was induced with 5BU. What kinds of mutants are these likely to be? Explain. How could you confirm your predictions by studying the structure of the β-galactosidase in these cells?

Answer: Acridine is an intercalating agent, and so can be expected to induce frameshift mutations. 5BU is incorporated into DNA in place of T. During DNA replication, it is likely to be read as C by DNA polymerase because of a keto- to an enol-shift. This results in point mutations, usually TA to CG transitions. Considering these expectations, *lacZ-1* would probably result in a completely altered amino acid sequence after some point, although it might be truncated (due to the introduction–out of frame–of a nonsense codon). *lacZ-2* is likely to contain a single amino acid difference due to a missense mutation, although it too could contain a nonsense codon.

18.9

a. The sequence of nucleotides in an mRNA is:

5'-AUGACCCAUUGGUCUCGUUAG-3'

Assuming that ribosomes could translate this mRNA, how many amino acids long would you expect the polypeptide chain made with this messenger to be?

b. Hydroxylamine is a mutagen that results in the replacement of an AT base pair for a GC base pair in the DNA; that is, it induces a transition mutation. When applied to the organism that made the mRNA molecule shown in part a, a strain was isolated in which a mutation occurred at the 11th position of the DNA that coded for the mRNA. How many amino acids long would you expect the polypeptide made by this mutant to be? Why?

Answer: a. One would have codons read as:

5' - AUG - ACC - CAU - UGG - UCU - CGU - UAG - 3'

The last codon is a nonsense (chain termination) codon, while the others are sense codons. The chain would be six amino acids long.

b. The new sequence would be

5' - AUG - ACC - CAU - UAG - ...

Since UAG is a nonsense (chain termination) codon, the new chain would only be three amino acids long.

18.10 In a series of 94,075 babies born in a particular hospital in Copenhagen, 10 were achondroplastic dwarfs (this is an autosomal dominant condition). Two of these 10 had an achondroplastic parent. The other 8 achondroplastic babies each had two normal parents. What is the apparent mutation rate at the achondroplasia locus?

Answer: There were eight new mutations in 94,075 normal couples. Since the phenotype is dominant, the phenotype is seen when just one of the parental genes is mutated. There were 2 x 94,075 copies of the gene that could have undergone mutation. Therefore, the apparent mutation rate at this locus is

[8/(2 x 94,075)] = 8/188,150 = 4 x 10^{-5} mutations per locus per generation.

18.11 Three of the codons in the genetic code are chain-terminating codons for which no naturally occurring tRNAs exist. Just like any other codons in the DNA, though, these codons can change as a result of base-pair changes in the DNA. Confining yourself to single base-pair changes at a time, determine which amino acids could be inserted in a polypeptide by mutation of these chain-terminating codons: (a) UAG; (b) UAA; (c) UGA. (The genetic code is listed in Figure 14.9.)

Answer:

NUCLEOTIDE ALTERED	UAG	CODE	UAA	CODE	UGA	CODE
FIRST	AAG	Lys	AAA	Lys	AGA	Arg
	CAG	Gln	CAA	Gln	CGA	Arg
	GAG	Glu	GAA	Glu	GGA	Gly
SECOND	UUG	Leu	UUA	Leu	UUA	Leu
	UCG	Ser	UCA	Ser	UCA	Ser
	UGG	Trp	UGA	STOP	UAA	STOP
THIRD	UAC	Tyr	UAC	Tyr	UGC	Cys
	UAU	Tyr	UAU	Tyr	UGU	Cys
	UAA	STOP	UAG	STOP	UGG	Trp

18.12 The amino acid substitutions in the following figure occur in the α and β chains of human hemoglobin. Those amino acids connected by lines are related by single nucleotide changes. Propose the most likely codon or codons for each of the numbered amino acids. (Refer to the genetic code listed in Figure 14.9, p. 435.)

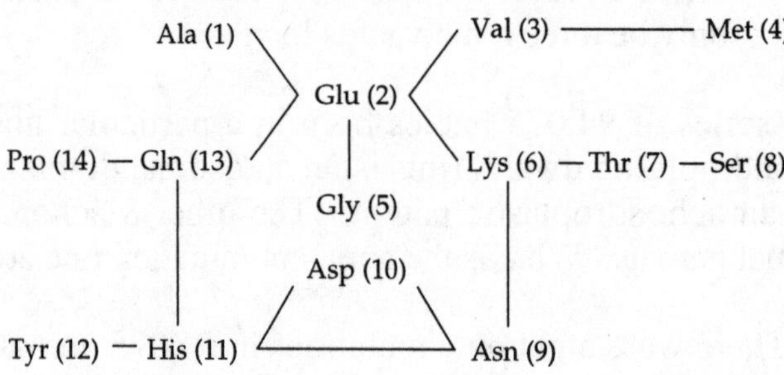

Answer:

Ala (1) GCG
Val (3) GUG — Met (4) AUG
Glu (2) GAG
Pro (14) CCG — Gln (13) CAG
Lys (6) AAG — Thr (7) ACG — Ser (8) UCG
Gly (5) GGG
Asp (10) GAU or GAC
Tyr (12) UAU or UAC — His (11) CAU or CAC
Asn (9) AAU or AAC

18.13 Yanofsky studied the tryptophan synthetase of *E. coli* in an attempt to identify the base sequence specifying this protein. The wild type gave a protein with a glycine in position 38. Yanofsky isolated two *trp* mutants, *A23* and *A46*. Mutant *A23* had Arg instead of Gly at position 38, and mutant *A46* had Glu at position 38. Mutant *A23* was plated on minimal medium, and four spontaneous revertants to prototrophy were obtained. The tryptophan synthetase from each of four revertants was isolated, and the amino acids at position 38 were identified. Revertant 1 had Ile, revertant 2 had Thr, revertant 3 had Ser, and revertant 4 had Gly. In a similar fashion, three revertants from *A46* were recovered, and the tryptophan synthetase from each was isolated and studied. At position 38 revertant 1 had Gly, revertant 2 had Ala, and revertant 3 had Val. A summary of these data is given in the figure. Using the genetic code in Figure 14.9 (p. 435), deduce the codons for the wild type, for the mutants *A23* and *A46*, and for the revertants, and place each designation in the space provided in the following figure.

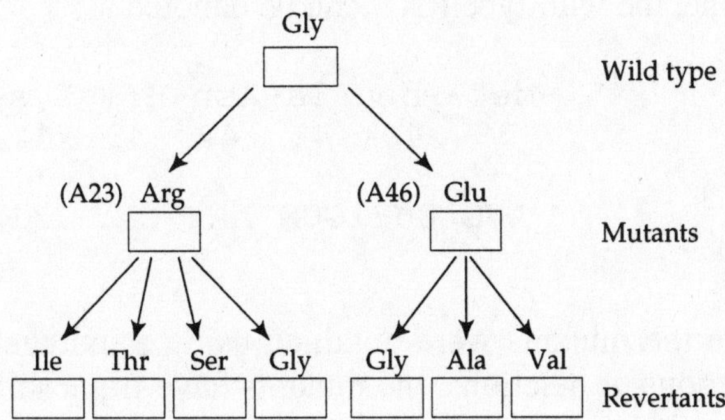

Answer:

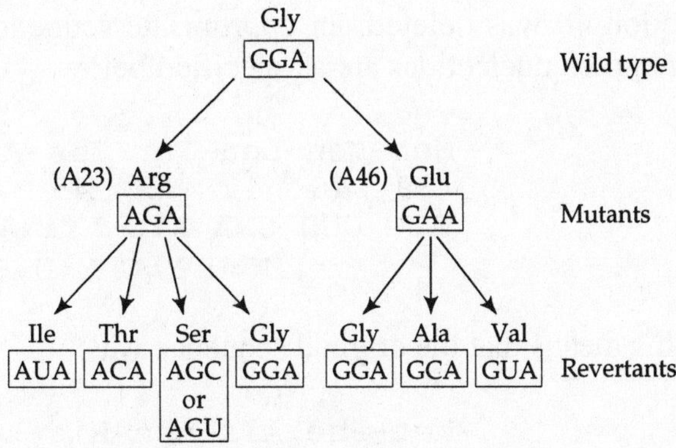

18.14 Consider an enzyme chewase from a theoretical microorganism. In the wild-type cell the chewase has the following sequence of amino acids at positions 39 to 47 (reading from the amino end) in the polypeptide chain:

```
-Met-Phe-Ala-Asn-His-Lys-Ser-Val-Gly-
  39  40  41  42  43  44  45  46  47
```

A mutant of the organism was obtained; it lacks chewase activity. The mutant was induced by a mutagen known to cause single base-pair insertions or deletions. Instead of making the complete chewase chain, the mutant makes a short polypeptide chain only 45 amino acids long. The first 38 amino acids are in the same sequence as the first 38 of the normal chewase, but the last 7 amino acids are as follows:

```
-Met-Leu-Leu-Thr-Ile-Arg-Val-
  39  40  41  42  43  44  45
```

A partial revertant of the mutant was induced by treating it with the same mutagen. The revertant makes a partly active chewase, which differs from the wild-type enzyme only in the following region:

```
-Met-Leu-Leu-Thr-Ile-Arg-Gly-Val-Gly-
 39  40  41  42  43  44  45  46  47
```

Using the genetic code given in Figure 14.9, deduce the nucleotide sequences for the mRNA molecules that specify this region of the protein in each of the three strains.

Answer: Using the genetic code, and letting N represent any nucleotide, R a purine, and Y a pyrimidine, the wild-type RNA can be denoted as:

```
-Met-Phe-Ala-Asn-His-Lys-Ser-Val-Gly-
 39  40  41  42  43  44  45  46  47
                         UCN
AUG UUY GCN AAY CAY AAR  or GUN GGN
                         AGY
```

Since the mutants were obtained using a mutagen that causes single base-pair insertions or deletions, the mutants have frameshift mutations. Considering this fact and comparing the mutant amino acid sequence to that of the wild-type, one expects a base-pair to be missing or added in codon 40. Try various options to obtain an appropriate amino acid sequence after such a frameshift. If the Y of codon 40 was deleted, an appropriate sequence can be obtained providing the unknown nucleotides are as specified below:

```
-Met-Leu-Leu-Thr-Ile-Arg-Val-
 39  40  41  42  43  44  45
AUG UUG CNA AYC AYA ARA GYG UNG
        N=U Y=C Y=U R=G Y=U N=A
```

This means that the original sequence was:

```
-Met-Phe-Ala-Asn-His-Lys-Ser-Val-Gly-
 39  40  41  42  43  44  45  46  47
AUG UUY GCU AAC CAU AAG AGU GUA GGN
```

In the revertant, the reading frame is restored at codon 46, the Val (GUG) at codon 45 is altered to a Gly (GGN), and the chain termination codon at position 46 (UAG) is altered to a Val (GUN). This would have occurred if a G was inserted before or after the first G in codon 45. One would have:

```
-Met-Leu-Leu-Thr-Ile-Arg-Gly-Val-Gly-
 39  40  41  42  43  44  45  46  47
AUG UUG CUA ACC AUA AGA GGU GUA GGN
```

18.15 Two mechanisms in *E. coli* were described for the repair of DNA damage (thymine dimer formation) after exposure to ultraviolet light: photoreactivation and excision (dark) repair. Compare and contrast these mechanisms, indicating how each achieves repair.

Answer: Photoreactivation requires the enzyme photolyase, and light (photons) with wavelengths between 320 and 370 nm. Dark repair does not require light, but requires several different enzymes. First, an endonuclease makes a single-stranded nick on the 5'-side of the dimer. Then, an exonuclease trims away part of one strand, including the dimer. Next, DNA polymerase fills in the single-stranded region in the 5'-to-3' direction. Finally, the gap is sealed by DNA ligase.

18.16 After a culture of *E. coli* cells was treated with the chemical 5-bromouracil, it was noted that the frequency of mutants was much higher than normal. Mutant colonies were then isolated, grown, and treated with nitrous acid; some of the mutant strains reverted to wild type.

a. In terms of the Watson-Crick model, diagram a series of steps by which 5BU may have produced the mutants.
b. Assuming the revertants were not caused by suppressor mutations, indicate the steps by which nitrous acid may have produced the back mutations.

Answer:

a. 5-bromouracil in its normal state is a T analog, base pairing with A. In its rare state, it resembles C and can base pair with G. It will induce an AT → GC transition as diagrammed below.

A–T → A–5BU → G–5BU → G–C

b. Nitrous acid can deaminate C to U, resulting in a CG → TA transition.

18.17 A single, very hypothetical strand of DNA is composed of the base sequence indicated in this figure.

5'-T–HX–U–A–G–BU-enol–2AP–C–BU–X–2AP-imino-3'

In the sequence above, A indicates adenine, T indicates thymine, G indicates guanine, C denotes cytosine, U denotes uracil, BU is 5-bromouracil, 2AP is 2-aminopurine, BU-enol is a tautomer of 5BU, 2AP-imino is a rare tautomer of 2AP, HX is hypoxanthine, and X is xanthine; 5' and 3' are the numbers of the free, OH-containing carbons on the deoxyribose part of the terminal nucleotides.

a. Opposite the bases of the hypothetical strand, and using the shorthand of the figure, indicate the sequence of bases on a complementary strand of DNA.
b. Indicate the direction of replication of the new strand by drawing an arrow next to the new strand of DNA from part a.
c. When postmeiotic germ cells of a higher organism are exposed to a chemical mutagen before fertilization, the resulting offspring expressing an induced mutation are almost always mosaics for wild-type and mutant tissue. Give at least one reason that in the progenies of treated individuals these mosaics are found and not the so-called complete or whole-body mutants.

Answer: a and b.

```
5'—T—HX—U—A—G—BU-enol—2AP—C—BU—X—2AP-imino—3'
3'—A— C—A—T—C—    G    —T —G— A—C—    C     —5'
 ←
```

c. If postmeiotic germ cells are treated, a single site in one strand of the double helix affected by the mutagen will not be repaired until mitotic divisions begin during embryogenesis. Given the semi-conservative replication of DNA, the normal strand will be replicated into two normal strands. If the mutated site on the other strand is not repaired, when the strand is replicated, it will give rise to a double helix with two mutant strands. In this scenario, the products of the first mitotic division consist of one normal daughter and one mutant daughter helix. This will produce two different cell types, resulting in a mosaic individual.

For some mutagens, such as BU or 2AP, there can be a variation on this method of mosaic production. If the BU (or 2AP) remains in the DNA strand it was initially incorporated into, the first mitotic division may result in a normal double helix (the new strand is copied from the old normal strand), as well as a double helix in which the BU (or 2AP) persists on one strand, but the "wrong" base has paired with it (due to the rare form of BU) on the opposite strand. In this case, a mutant site will be produced on one of the daughter double helices that result following replication, that is, after the second mitotic division (at the four-cell stage). In principle, a mutation-bearing cell could be introduced at the mitotic division subsequent to the pairing of a mutant base with the rare form in any round of DNA synthesis. In each case, a mosaic individual would be produced, albeit a mosaic with relatively fewer mutation-bearing cells.

The following information applies to Problems 18.18 through 18.22. A solution of single-stranded DNA is used as the template in a series of reaction mixtures. It has the following base sequence:

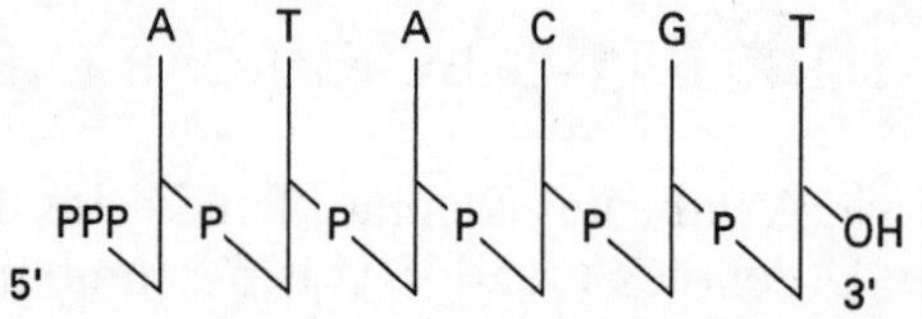

A T A C G T

5' PPP P P P P P OH 3'

where A = adenine, G = guanine, C = cytosine, T = thymine, H hypoxanthine, and HNO_2 = nitrous acid. For Problems 18.18 through 18.22, use this shorthand system and draw the products expected from the reaction mixtures. Assume that a primer is available in each case.

18.18 The DNA template + DNA polymerase + dATP + dGTP + dCTP + dTTP + Mg^{2+}.

Answer:

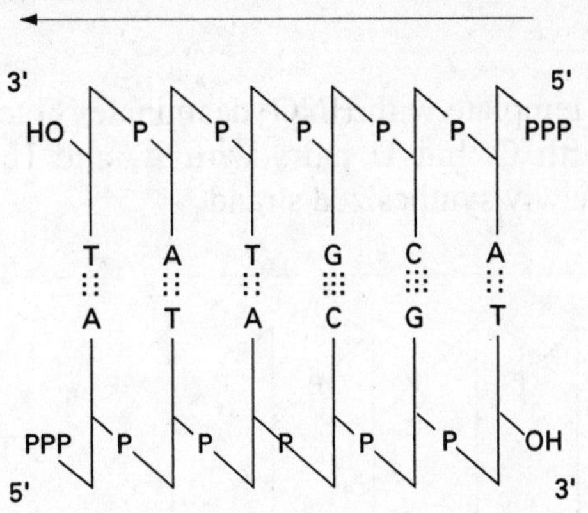

3' HO P P P P P PPP 5'
T A T G C A
A T A C G T
5' PPP P P P P P OH 3'

18.19 The DNA template + DNA polymerase + dATP + dGMP + dCTP + dTTP + Mg^{2+}.

Answer: The absence of dG<u>T</u>P leads to a block in polymerization after the first two bases.

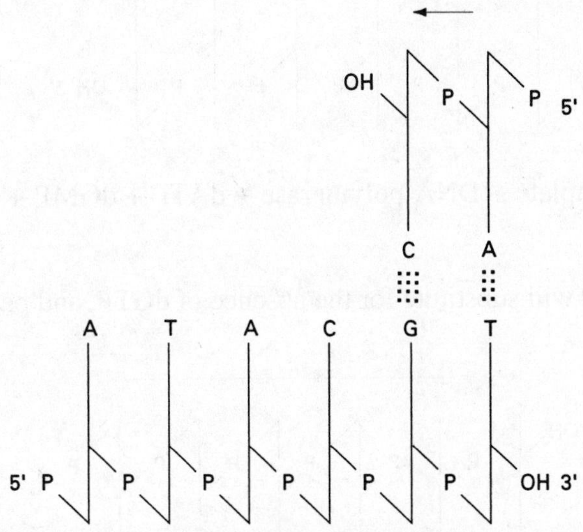

OH P P 5'
C A
A T A C G T
5' P P P P P P OH 3'

18.20 The DNA template + DNA polymerase + dATP + dHTP + dGMP + dTTP + Mg^{2+}.

Answer: While the dHTP can substitute for dGTP, there is no dCTP, so polymerization cannot continue past the first base.

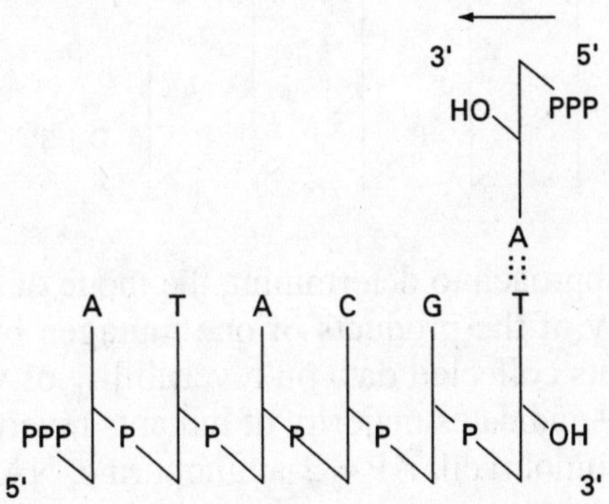

3' 5'
HO PPP
A
A T A C G T
PPP P P P P P OH
5' 3'

18.21 The DNA template is pretreated with HNO_2 + DNA polymerase + dATP + dGTP + dCTP + dTTP + Mg^{2+}.

Answer: Pretreatment of the template with HNO_2 deaminates G to X, C to U, and A to H. X will still pair with C, but U pairs with A, and H pairs with C, causing "mutations" in the newly synthesized strand.

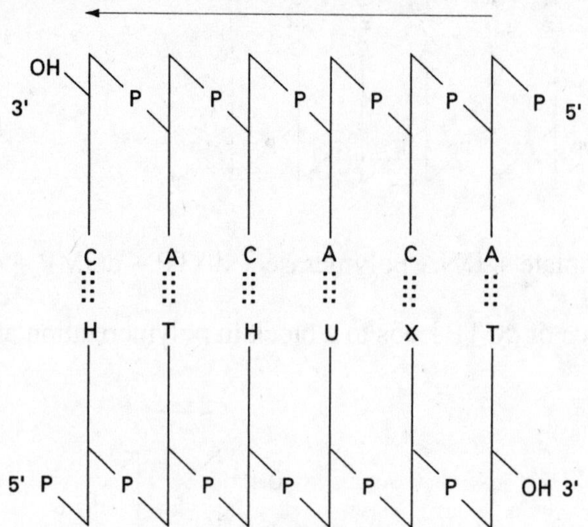

18.22 The DNA template + DNA polymerase + dATP + dGMP + dHTP + dCTP + dTTP + Mg^{2+}.

Answer: The dHTP will substitute for the absence of dGTP, and pair with C.

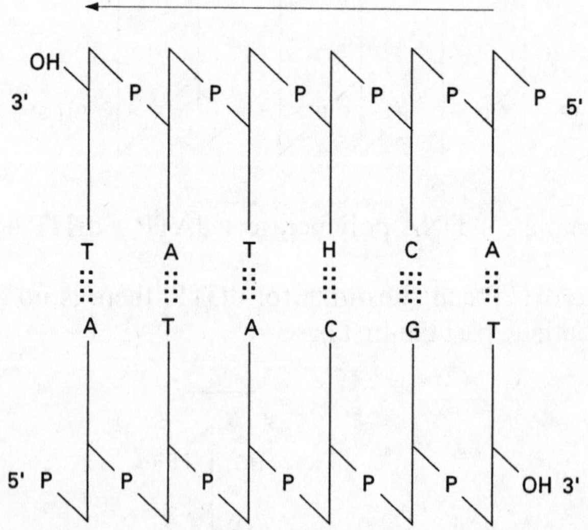

18.23 A strong experimental approach to determining the mode of action of mutagens is to examine the revertibility of the products of one mutagen by other mutagens. The following table represents collected data on revertibility of various mutagens on *rII* mutations in phage T2; + indicates majority of mutants reverted, – indicates virtually no reversion; BU = 5-bromouracil, AP = 2-aminopurine, NA = nitrous acid, and HA = hydroxylamine. Fill in the empty spaces.

MUTATION INDUCED BY	PROPORTION OF MUTATIONS REVERTED BY				BASE-PAIR SUBSTITUTION INFERRED
	BU	AP	NA	HA	
BU	+			–	
AP		–	+		
NA	+	+		+	
HA			+	–	GC → AT

Answer: First consider what is known about HA. We are told that HA causes a GC → AT transition, and cannot revert mutations it induces. Therefore, HA cannot cause AT → GC transitions.

Now consider the relationship between NA and HA. Since NA can revert HA mutations, NA can cause AT → GC transitions. Since HA can revert NA mutations, NA also can cause GC → AT transitions. Thus, NA should be able to revert mutations it induces.

Now consider the relationship between NA and AP, and HA and AP. NA can revert AP mutations, so AP must cause GC → AT and/or AT → GC transitions. Since AP can also revert NA mutations, which we have established are GC → AT and AT → GC transitions, it must cause both AT → GC and GC → AT transitions. One would expect that AP would be able to revert HA mutations (since they are GC → AT transitions). However, one would not expect HA to revert all AP mutations, since some are AT → GC transitions. It is curious that AP cannot revert its own mutations, given that it causes both AT → GC and GC → AT transitions.

Now consider the relationships of AP, NA and HA to BU. BU can revert mutations caused by NA, indicating that it can cause both AT → GC and GC → AT transitions. It can revert its own mutations, consistent with its ability to cause both AT → GC and GC → AT transitions. Mutations induced by it are unable to be reverted by HA (as HA cannot cause AT → GC transitions), but should be able to be reverted by AP. One would expect BU to be able to revert mutations induced by HA, as it can cause AT → GC transitions.

MUTATION INDUCED BY	PROPORTION OF MUTATIONS REVERTED BY				BASE-PAIR SUBSTITUTION INFERRED
	BU	AP	NA	HA	
BU	+	+	+	–	GC ↔ AT
AP	+	– (?)	+	–	GC ↔ AT
NA	+	+	+	+	GC ↔ AT
HA	+	+	+	–	GC → AT

18.24 a. Nitrous acid deaminates adenine to form hypoxanthine which forms two hydrogen bonds with cytosine during DNA replication. Following treatment with nitrous acid, a mutant is recovered which contains a protein with an amino acid substitution: the

amino acid valine (Val) is located in a position occupied by methionine (Met) in nonmutant organisms. What is the simplest explanation for this observation?
b. Hydroxylamine adds a hydroxyl (OH) group to cytosine, causing it to pair with adenine. Could mutant organisms like those in part (a) be back mutated (returned to normal) using hydroxylamine? Explain.

Answer: a. The codon for Met is 5'-AUG-3'. This would be encoded by 3'-TAC-5' in the template DNA strand, which pairs with 5'-ATG-3' in the nontemplate DNA strand. In this case, the A in the nontemplate strand was deaminated to form hypoxanthine, and paired with C in the template strand. The new template was 3'-CAC-5' and the new codon was 5'-GUG-3', coding for Val.

b. In the mutant in a, the template strand was 3'-CAC-5', and the nontemplate strand was 5'-GTG-3'. Since hydroxylamine acts on cytosine bases, it would act on the template strand. If it acted at the 3'-C, DNA replication would result in the nontemplate strand becoming 5'-ATG-3'. After an additional round of replication, one of the daughter cells would have a template strand that is 3'-TAC-5'. Thus, hydroxylamine could be used to obtain revertants.

18.25 A protein contains the amino acid proline (Pro) at one site. Treatment with nitrous acid, which deaminates C to make it U, produces two different mutants. One mutant has a substitution of serine (Ser) and the other has a substitution of leucine (Leu) at the site.

Treatment of the two mutants produces new mutant strains, each with phenylalanine (Phe) at the site. Treatment of these new Phe-carrying mutants produces no change. The results are summarized in the following figure:

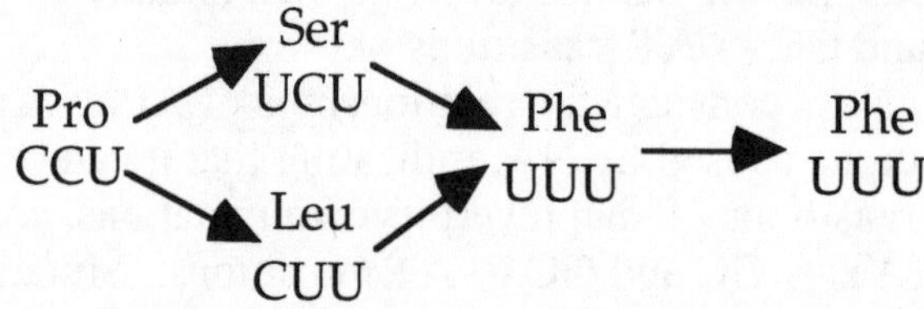

Using the appropriate codons, show how it is possible for nitrous acid to produce these changes, and why further treatment has no influence. (Assume only single nucleotide changes occur at each step.)

Answer: Nitrous acid deaminates C to make it U. U will pair with A, so that treatment with nitrous acid leads to CG → TA transitions. Analyze how this treatment would affect the codons of this protein. Use N to represent any nucleotide, and Y to represent a pyrimidine (U or C). Then, the codons for Pro are CCN, the (relevant) codons for Ser are UCN, the codons for Leu are CUN, and the codons for Phe are UUY. (Nucleotides are written in the 5'→3' direction unless specifically noted)

The codon CCN for Pro would be represented by CCN in the nontemplate DNA strand. Deamination of the 5' C would lead to a nontemplate strand of UCN, and a template strand of 3'-AGN-5'. This would produce a UCN codon, encoding Ser. Deamination of the middle C would lead to a nontemplate strand of CUN, and a template strand of 3'-GAN-5'. This would produce a CUN codon, encoding Leu.

Further treatment of either mutant would result in deamination of the remaining C, and a template strand of 3'-AAN-5'. This would result in a UUN codon. Since we are told Phe is obtained, N must be C or U, and the template strand must be 3'-AAA-5' or 3'-AAG-5'.

To explain why further treatment with nitrous acid has no effect, consider that nitrous acid acts via deamination and that T has no amine group. If the template strand was 3'-AAA-5', the nontemplate strand would have been TTT. Since T cannot be deaminated, nitrous acid will not have any effect on the non-template strand.

18.26 Three *ara* mutants of *E. coli* were induced by mutagen X. The ability of other mutagens to cause the reverse change (*ara* to *ara*$^+$) was tested, with the results shown in the following table.

FREQUENCY OF *ara*$^+$ CELLS AMONG TOTAL CELLS AFTER TREATMENT

	MUTAGEN				
MUTANT	NONE	BU	AP	HA	FRAMESHIFT
ara-1	1.5×10^{-8}	5×10^{-5}	1.3×10^{-4}	1.3×10^{-8}	1.6×10^{-8}
ara–2	2×10^{-7}	2×10^{-4}	6×10^{-5}	3×10^{-5}	1.6×10^{-7}
ara–3	6×10^{-7}	10^{-5}	9×10^{-6}	5×10^{-6}	6.5×10^{-7}

Assume all *ara*$^+$ cells are true revertants. What base changes were probably involved in forming the three original mutations? What kind(s) of mutations are caused by mutagen X?

Answer: Use the revertant frequencies under "none" to estimate the spontaneous reversion frequency.

ara-1: BU and AP, but not HA or a frameshift, can revert *ara-1*. Both BU and AP cause CG→TA and TA→CG transitions, while HA only causes CG→TA transitions. If HA cannot revert *ara-1*, it must require a TA→CG transition to be reverted and be caused by a CG→TA transition.

ara-2: BU, AP and HA, but not a frameshift, can revert *ara-2*. Since HA only causes CG→TA transitions, *ara-2* must have been caused by a TA→CG transition.

ara-3: By the same logic as for *ara-2*, *ara-3* must have been caused by a TA→CG transition.

Providing that this is a representative sample, mutagen *X* appears to cause both TA→CG and CG→TA transitions. It does not appear to cause frameshift mutations.

[illegible] treatment [illegible] would result in the [illegible] of the [illegible] and a template strand of 5'-AGAA-3'. [illegible] would result in a [illegible] codon. Since we [illegible] the [illegible] mutation [illegible], and the template strand [illegible] 5'-AAG-3' or 5'-AGG-3'.

To explain why [illegible] mutations [illegible], consider that [illegible] and that [illegible]. If the template strand was 5'-AAA-3', the [illegible] strand would [illegible] [illegible] not [illegible] any effect on the [illegible].

	MUTAGEN				
MUTANT	NONE	BU	AP	HA	PROFLAVIN
mut-1	[illegible]	[illegible]	[illegible]	[illegible]	[illegible]
mut-2	[illegible]	[illegible]	[illegible]	[illegible]	[illegible]
mut-3	[illegible]	[illegible]	[illegible]	[illegible]	[illegible]

CHAPTER 19

TRANSPOSABLE ELEMENTS, TUMOR VIRUSES, AND ONCOGENES

I. CHAPTER OUTLINE

TRANSPOSABLE ELEMENTS IN PROKARYOTES
- Insertion Sequences
- Transposons
- IS Elements and Transposons in Plasmids
- Bacteriophage Mu

TRANSPOSABLE ELEMENTS IN EUKARYOTES
- Transposons in Plants
- *Ty* Elements in Yeast
- *Drosophila* Transposons
- Human Retrotransposons

TUMOR VIRUSES AND ONCOGENES
- Tumor Viruses
- Retroviruses
- HIV: The AIDS Virus

II. REVIEW OF KEY TERMS, SYMBOLS AND CONCEPTS

Without consulting the text and in your own words, write a brief definition of each term in the groups below. Then, either using a short phrase or a simple diagram, identify the relationship(s) between specific pairs of terms within a set. Finally, consult the text (and perhaps a friend who has also done the exercise) to check your answers.

1	2	3	4
transposable element transposition transposition event	insertion (IS) element IS module inverted terminal repeat transposase target site target site duplication transposon (Tn)	composite transposition (e.g., Tn*10*) non-composite transposition (e.g., Tn*3*) replicative transposition conservative transposition simple insertion cointegrate cointegration model transposase, resolvase, β-lactamase	IS element *F* factor plasmid episome *R* plasmid *RTF* region

5	6	7	8
bacteriophage Mu mutator temperate bacteriophage replicative transposition	mutator gene *Mu1, Tam, Tgm* elements *Ac*, *Ds* controlling elements null mutation autonomous, nonautonomous element unstable (mutable) vs. stable allele donor site conservative transposition	*Ty* element long terminal (direct) repeat delta repeat target site duplication retrovirus reverse transcriptase retrotransposon mobile element *copia* element	mobile element hybrid dysgenesis *P* element *P* cytotype *M* cytotype

9	10	11	12
retrotransposon LINES SINES *Alu* elements *L1* elements	dedifferentiation tumor neoplasm transformation cancer oncogenesis	tumor virus oncogene DNA vs. RNA tumor virus retrovirus *RSV, FLV, MMTV, HIV-1* *gag, pol, env, src* genes	retrovirus replication-competent transformation-competent transducing retrovirus viral-oncogene proto-oncogene cellular oncogene growth factor protein kinase

13

retrovirus
HIV
AIDS
gag, pol, env, tat genes
CD4 receptor
helper T cell
gp120 glycoprotein

III. THINKING ANALYTICALLY

As the material in this chapter is broadly descriptive, it will be useful initially to organize and categorize the information that is presented. Distinguish between the various categories of transposable elements by considering transposable elements from four perspectives: (1) structure, (2) mode of transposition, (3) the effect of a transposed element on the structure of a gene, and (4) the effect of a transposed element on the function of a gene. Then, as you study tumor viruses, use this framework to examine the similarities between retrotransposons and retroviruses. As you go further and study oncogenesis, continue to take an organizational approach, and analyze the differences and the similarities between viral and cellular processes. As in earlier chapters, it will help to relate verbal descriptions of structures and processes to diagrams of them.

IV. QUESTIONS FOR PRACTICE

A. Multiple Choice Questions

Choose the correct answer or answers for the following questions.

1. How are transposition and recombination fundamentally different?
   a. Recombination requires DNA homology but transposition does not.
   b. Transposition requires DNA homology but recombination does not.
   c. Only during transposition can a piece of DNA be moved from a virus to a cell.
   d. Only during recombination can a deletion occur.

2. Which of the following can result from the insertion of a transposon in bacteria?
   a. Gene inactivation.
   b. An increase or a decrease in transcriptional activity of a gene.
   c. Deletions or insertions.
   d. All of the above.

3. Which of the following can be consequences of transposition in eukaryotes?
   a. Conversion of a heterozygote from the dominant to the recessive phenotype.
   b. Production of a null mutation.
   c. Increased or decreased efficiency of promoter regions.
   d. All of the above.

4. In what ways are bacterial IS and Tn elements alike?
   a. Both have inverted repeat sequences at their ends.

b. Both integrate into target sites and cause a target site duplication.
c. Both contain a transposase gene.
d. Both contain antibiotic resistance genes.
e. Both always use replicative transposition.
f. a, b, and c.
g. All of the above.

5. How might plasmids *X* and *Y* differ? *X* is an *F* factor in an F^+ bacterium sensitive to the antibiotics ampicillin, streptomycin and tetracycline. *Y* is an *R* plasmid in a bacterium resistant to these antibiotics.
   a. *Y* is likely to be smaller in size.
   b. *X* is likely to lack the *tra* genes.
   c. *Y* is likely to have additional genes between flanking IS modules.
   d. *Y* is likely to lack the *RTF* genes.

6. Which of the following statements is not true about Mu?
   a. It is a temperate bacteriophage.
   b. Like λ, it is excised and replicated during lytic infection.
   c. It is a transposon that integrates by conservative transposition.
   d. Homologous recombination between two Mu elements can generate inversions.

7. Why does the insertion of an autonomous element result in the production of an unstable (mutable) allele?
   a. Although an autonomous element cannot excise, it becomes a mutable hot-spot.
   b. An autonomous element can excise, thereby causing a mutation.
   c. Autonomous alleles will be unstable when in the company of a nonautonomous element.
   d. Autonomous alleles generally insert into promoter regions.

8. *Ty* elements in yeast are more similar to retroviruses than to autonomous *Ac* elements in corn because (choose the correct answer):
   a. Both *Ty* elements and retroviruses utilize an RNA transposition intermediate, whereas *Ac* elements do not.
   b. Although all three kinds of transposons synthesize a reverse transcriptase, the enzyme in corn never becomes functional.
   c. Introns are usually found in retroviruses and *Ty* elements, but not in *Ac* elements.
   d. More than one of the preceding.

9. Which two of the following are true about hybrid dysgenesis in *Drosophila*?
   a. It occurs in the progeny of M females after they are mated to P males.
   b. It occurs in the progeny of P females after they are mated to M males.
   c. It is a result of the mobilization of P elements in an M cytotype.
   d. It is a result of the mobilization of M elements in a P cytotype.

10. What is the relationship between a proto-oncogene and its corresponding viral oncogene?
   a. Only the viral oncogene contains introns.
   b. If the viral oncogene encodes a transcription factor, the corresponding proto-oncogene will encode a growth factor.

c. The proto-oncogene encodes a product essential to the normal development and function of the organism. The viral oncogene encodes an altered product that has aberrant function.
d. The proto-oncogene product is usually produced in greater amounts. The viral oncogene, being under the control of the retroviral promoter, enhancer, and poly(A) signals, produces lesser amounts of a product.

Answers: 1a; 2d; 3d; 4f; 5c; 6b; 7b; 8a; 9a & c; 10c.

B. Thought Questions

1. Contrast conservative and replicative transposition. How can conservative transposition of the *Ac* element result in gene duplication? Can particular IS sequences behave in both ways?

2. Consider how the distinction between direct and inverted repeats is central to the recognition of insertion sequences. The insertion sequence itself contains inverted terminal repeats. Thus, inverted repeats are present as a part of the transposable element both before and after the insertion. On the other hand, the act of insertion results in a staggered cut in the host DNA. When the gaps are filled in after insertion, direct repeats will be generated flanking the inverted terminals of the transposable element. Indeed, insertion sequences have been identified by the presence of a pair of inverted repeats flanked by direct repeats. Using this logic, find the insertion sequence in the following single-stranded length of DNA.

CGTAGCCATTTGCGATATGCATCCGAATATCGCAAGCCATGCCA

3. What are the similarities and differences between antibiotic resistance factors in *Shigella* and fertility factors in *E. coli?*

4. In *Drosophila*, hybrid dysgenesis results in the mobilization of transposable elements. This in turn results in progeny flies that have normal somatic cells but are sterile. Why is the somatic tissue unaffected and only the germline affected? Males from three *Drosophila* strains recently isolated from wild populations in different geographical areas are crossed with females from the same M cytotype laboratory strain. In the first cross, 12 percent of the progeny are sterile. In the second, 64 percent of the progeny are sterile. In the third, 97 percent of the progeny are sterile. Generate an hypothesis to address why different degrees of sterility are seen in these three strains.

5. In what situations can a nonautonomous element behave as an autonomous element? Give specific examples in corn and *Drosophila*.

6. Following the mobilization of *Ty* elements in yeast, two mutants were recovered that affected the activity of enzyme *X*. One of these mutants lacked enzyme activity, and was a null mutation. The other mutant had three times as much enzyme activity. Both mutants had a *Ty* element inserted into the *X* gene. Generate an hypothesis to explain how this is possible.

7. What differences and similarities are there between transposable elements that utilize an RNA intermediate and those that do not?

8. What kinds of repetitive elements are SINES and LINES? Support for a hypothesis that a particular repetitive element is a transposon often comes analyzing its DNA sequence for open reading frames that encode transposase-like products. In humans, *Alu* sequences do not encode any of the enzymes that are presumably needed for retrotransposition. What aspects of their structure suggest that they might be capable of retrotransposition? What evidence is there that *Alu* sequences are retrotransposons capable of moving via an RNA intermediate, much like the yeast *Ty* element?

9. Distinguish between a viral oncogene, a cellular oncogene, and a proto-oncogene. Is each of these always associated with a transducing retrovirus?

10. There are at least 7 functional classes of oncogene products. What are these functional classes? How can alteration in such diverse functions lead to a similar oncogenic phenotype?

11. Describe the structure of the HIV virus and viral genome. How does the virus specifically target the immune system? Does the HIV virus itself have viral oncogenes? Why are individuals infected with HIV more susceptible to cancer and infection?

V. SOLUTIONS TO TEXT PROBLEMS

19.1 Compare and contrast the types of transposable elements in bacteria.

Answer: Four kinds of transposable elements exist in bacteria: insertion sequences (IS elements), transposable elements (Tn elements), plasmids and certain temperate bacteriophages, such as Mu.

IS elements are generally the simplest in structure, having a transposase gene flanked by perfect or nearly perfect inverted repeat sequences. They integrate into target sites (with which they have no homology), where they cause target site duplications. They can cause mutations by a number of mechanisms, including disruption of a gene's coding sequence or regulatory region. Their transposition requires the use of the host cell's replicative enzymes, as well as an element encoded transposase.

Tn elements are more complex than IS elements, and exist in two forms, composite transposons and non-composite transposons. Composite transposons have a central gene-bearing region flanked by IS elements. Transposition of such elements, and the genes (e.g., for antibiotic resistance) they contain, occurs because of the function of their IS elements. Noncomposite transposons also contain genes, such as those for antibiotic resistance, but do not terminate with IS elements. They do have repeated sequences required for transposition, however. Tn elements can transpose by one of several means. Tn*3*-like noncomposite transposons utilize a replicative transposition mechanism employing a cointegrate – a fusion between a transposable element and the recipient DNA. Tn*10*-like composite transposons can move by conservative (nonreplicative) transposition.

The level of complexity is increased when one considers plasmids, *F* factors, episomes and *R* plasmids. Analysis of their structure has indicated that they contain IS and/or Tn elements. Indeed, the resistance genes in *R* plasmids are contained in transposons. Chromosome integration can occur when recombination occurs between homologous sequences of the insertion elements.

Mu is an example of a temperate bacteriophage that is also a transposon. In addition to containing genes related to phage function, the Mu genome has a G segment that can invert, as well as left and right inverted repeats. In this sense, it is similar to other prokaryotic transposable elements. It is also similar in the mode of transposition. Mu integrates into the host using a conservative transposition mechanism. This results in a prophage flanked by a 5 bp direct repeat of the host target-site sequence. Mu can cause insertions, and, by homologous recombination between two copies of Mu, deletions, inversions, and translocations.

19.2 What are the properties in common between bacterial and eukaryotic transposable elements?

Answer: The structure and, at a general level, the function of eukaryotic and prokaryotic transposable elements is very similar. For example, both Tn and *Ac* elements have genes within them, and have inverted repeats at their ends. Both prokaryotic and eukaryotic elements may affect gene function in a variety of different ways, depending on the element involved and how it integrates into or nearby a gene. The integration events of eukaryotic elements, like those of most prokaryotic transposable elements, involve nonhomologous recombination. Some eukaryotic elements, such as *Ty* elements in yeast, and retrotransposons, move via an RNA intermediate, unlike the IS and Tn prokaryotic elements. For additional details, see text pp. 630-634 and pp. 637-643.

19.3 An IS element became inserted into the *lacZ* gene of *E. coli*. Later, a small deletion occurred in this gene, which removed forty base pairs, starting to the left of the IS element. Ten *lacZ* base pairs were removed, including the left copy of the target site, and the thirty leftmost base pairs of the IS element were removed. What will be the consequence of this deletion?

Answer: The left inverted repeat of the IS element has been removed, so that the two ends of this IS element are no longer homologous. The element will not be able to move out of this location and insert into another site.

19.4 A geneticist was studying glucose metabolism in yeast, and had deduced both the normal structure of the enzyme glucose-6-phosphatase (G6Pase) and the DNA sequence of its coding region. She had been using a wild-type strain called *A* to study another enzyme for many generations, when she noticed a morphologically peculiar mutant had arisen from one of the strain A cultures. She grew the mutant up into a large stock and found that the defect in this mutant involved a markedly reduced G6Pase activity. She isolated the G6Pase protein from these mutant cells and found it was present in normal amounts, but had an abnormal structure. The N-terminal 70 percent of the protein was normal. The C-terminal 30 percent was present but altered in sequence by a frame shift reflecting the insertion of 1 base pair, and the N-terminal 70 percent and the C-terminal 30 percent were separated by 111 new amino acids unrelated to normal G6Pase. These amino acids represented predominantly the AT rich codons (Phe, Leu, Asn, Lys, Ile, Tyr). There were also two extra amino acids added at the C-terminal end. Explain these results.

Answer: The extra 111 amino acids plus the one base-pair shift indicates that 334 base pairs were inserted into the G6Pase structural gene. This is the size of a "delta dropping." That is, this is consistent with an initial *Ty* transposition into the G6Pase gene that was followed by recombination between the two deltas. Recombination between the two deltas would excise the *Ty* element but leave a delta dropping behind in the G6Pase gene. The delta elements are 334 base-pairs long, and are 70 percent AT, the characteristics of the inserted sequence. If the delta element were positioned so that it would be translated and not generate a stop codon, it would yield 111 amino acids and one extra base pair, which would cause the frame shift. The two extra amino acids at the C-terminal end of G6Pase were added presumably because the frameshift did not allow the normal termination codon to be read.

19.5 Consider two theoretical yeast transposons, A and B. Each contains an intron. Each transposes to a new location in the yeast genome and then is examined for the presence of the intron. In the new locations, you find that A has no intron, while B does. What can you conclude about the mechanisms of transposon movement for A and B from these facts?

Answer: Since introns are spliced out only at the RNA level, a transposition event that results in the loss of an intron (such as that used by *Ty* elements) indicates that the transposition occurred via an RNA intermediate. Thus, A is likely to move via an RNA intermediate. The lack of intron removal during B transposition suggests that it uses a DNA→DNA transposition mechanism (either conservative or replicative or some other mechanism).

19.6 An investigator has found a retrovirus capable of infecting human nerve cells. This is a complete virus, capable of reproducing itself, and it contains no oncogenes. People who are infected suffer a debilitating encephalitis. The investigator has shown that when he infects nerve cells in culture with the complete virus, the nerve cells are killed as the virus reproduces, but if he infects cultured nerve cells with a virus in which he has created deletions in the *env* or *gag* genes, no cell death occurs. The investigator is interested in finding ways to bring about nerve cell growth or regeneration in people who have suffered nerve damage. For example, in a patient with a severed spinal cord, nerve regeneration might relieve paralysis. The investigator has cloned the human nerve growth factor gene, and wants to insert it into the genome of his retrovirus from which he has deleted parts of the *env* and *gag* genes. He would then use the engineered retrovirus to infect cultured nerve cells. Adult nerve cells do not normally produce large amounts of nerve growth factor. If he is successful in inducing growth in them without causing any cell death, he would like to move on to clinical trials on injured patients. When the investigator applied for grant support to do this work, his application was denied on grounds that there were inadequate safeguards in the plan. Why might this work be dangerous? What comparisons can you draw between the virus the investigator wants to create and, for example, Avian myeloblastosis virus?

Answer: The investigator is very close to engineering a new cancer virus, in which the cloned nerve growth factor gene would be the viral oncogene. While the engineered virus would, by itself, not be able to reproduce, this is not sufficient to insure its safe use. Many "wild" cancer viruses are also defective in their ability to reproduce, and reproduce with the help of other, helper viruses. If the engineered virus were to infect that would could supply the *env* and *gag* functions, the new virus would reproduce and spread. Infection of normal nerve cells *in vivo* by the engineered retrovirus could result in abnormally high levels of nerve growth factor, and perhaps lead to the production of nervous system cancers.

In Avian myeloblastosis virus (AMV) the *pol* and *env* genes are partially deleted. Thus, like the investigator's virus, AMV needs a helper virus to reproduce. In AMV, the *myb* oncogene has been inserted, which encodes a nuclear protein involved in the control of gene expression. In the new virus, the oncogene would be the cloned nerve growth factor gene. Given the complexity of the regulation of eukaryotic gene regulation, there are many routes via which oncogenes can act (See Table 19.5).

CHAPTER 20

EXTRANUCLEAR GENETICS

I. CHAPTER OUTLINE

ORGANIZATION OF EXTRANUCLEAR GENOMES
- Mitochondrial Genome
- Chloroplast Genome
- RNA Editing
- The Origin of Mitochondria and Chloroplasts

RULES OF EXTRANUCLEAR INHERITANCE

EXAMPLES OF EXTRANUCLEAR INHERITANCE
- Leaf Variegation in the Higher Plant *Mirabilis jalapa*
- The [*poky*] Mutant of *Neurospora*
- Yeast *petite* Mutants
- Extranuclear Genetics of *Chlamydomonas*
- Human Genetic Diseases and Mitochondrial DNA Defects
- Exceptions to Maternal Inheritance
- Infectious Heredity—Killer Yeast

MATERNAL EFFECT

GENOMIC IMPRINTING

II. REVIEW OF KEY TERMS, SYMBOLS AND CONCEPTS

Without consulting the text and in your own words, write a brief definition of each term in the groups below. Then, either using a short phrase or a simple diagram, identify the relationship(s) between specific pairs of terms within a set. Finally, consult the text (and perhaps a friend who has also done the exercise) to check your answers.

1	2	3	4
extranuclear inheritance/gene	mitochondria	chloroplast	extranuclear inheritance
non-Mendelian inheritance/gene	H, L strands	SSC, LSC, IRA, IRB regions	variegation in *Mirabilis jalapa*
uniparental inheritance	D-loop	23S, 5S, 4.5S, 16S rRNAs	[*poky*] in *Neurospora*
biparental inheritance	continuous replication	RNA editing	*petite, grande* in yeast
maternal effect	ORF, URF	guide RNA (gRNA)	*nuclear petite*
nuclear genotype	12S, 16S rRNAs	endosymbiont hypothesis	*neutral petite*
maternal gene			*suppressive petite*
genomic (parental) imprinting			

5	6
mitochondrial DNA	killer yeast
heteroplasmy	L, M viruses
Leber's hereditary optic neuropathy	infectious inheritance
Kearns-Sayre syndrome	
myoclonic epilepsy and ragged-red fiber disease	

III. THINKING ANALYTICALLY

After developing skills to analyze nuclear inheritance patterns, non-Mendelian patterns of inheritance are quite a twist at first. Always keep in mind that the patterns seen in extranuclear inheritance are quite well defined. In general, they follow from the cytoplasmic contributions of one (female) parent to the offspring.

Consideration of the variation in inheritance patterns, and in the structure and function of the mitochondrial and chloroplast genomes provides a valuable perspective on nuclear inheritance patterns and nuclear genome structure and function. After organizing the factual information that is present in this chapter, take time to compare and contrast the organization, replication and expression of mitochondrial and chloroplast genes with nuclear genes. The salient features of an important biological process can often be better understood by examining variations in it.

Unlike extranuclear inheritance patterns that are controlled by extranuclear genes, maternal effects and genomic imprinting are controlled by nuclear genes. Keep this distinction clear. In considering maternal effects, it will help to diagram the generations of a cross, and illustrate how contributions from the parental generation are important for the development of the progeny. Carefully read the material in the chapter and do the problems to discover how to distinguish a maternal effect from an extranuclear, maternally inherited trait.

IV. QUESTIONS FOR PRACTICE

A. Multiple Choice Questions

1. DNA is found in
   a. the nucleus of eukaryotes.
   b. mitochondria.
   c. chloroplasts.
   d. all subcellular organelles.
   e. all of the above except d.

2. In general, extranuclear genes show
   a. maternal inheritance.
   b. Mendelian inheritance.
   c. maternal effects.
   d. dominance.

3. Mitochondrial DNA is
   a. typically single-stranded and coiled in nucleosomes.
   b. uniform in size in all species.
   c. uniformly double-stranded, supercoiled, and circular.
   d. found only in animals and fungi.

4. In the D-loop model of replication of mitochondrial DNA, replication
   a. is synchronous with nuclear DNA replication.
   b. originates at distinct locations on the H and L strands and is continuous.
   c. originates on both strands in the D-loop and is semi-discontinuous.
   d. only occurs in the maternal cytoplasm.

5. The ribosomal proteins of mitochondrial ribosomes are
   a. coded for in the nucleoid region.
   b. coded for in the nuclear genome, and are the same as those used in cytoplasmic ribosomes.
   c. coded for in the nuclear genome, and are different from those used in cytoplasmic ribosomes.
   d. encoded on genes in between mitochondrial rRNA genes.

6. Which of the following is not a consequence of the differences in the genetic code used by most mitochondrial and nuclear genes?
   a. Fewer tRNAs are needed in the mitochondria.
   b. Fewer amino acids are used in mitochondrial proteins.
   c. Wobble is used more extensively.
   d. Some nonsense codons in mitochondrial genes are sense codons in nuclear genes, and *visa versa.*

7. Which of the following is true about both mitochondrial and chloroplast genomes?
   a. Both have double-stranded, circular, supercoiled genomes that typically exist in multiple copies per organelle.
   b. Both are small in size, typically only 16-18 kb.
   c. Both have only one copy of rRNA genes.
   d. Both have a similar density in a CsCl density gradient.

8. Which of the following is not a typical characteristic of extranuclear traits?
    a. They show uniparental inheritance.
    b. They cannot be mapped relative to nuclear genes.
    c. They show non-Mendelian segregation in crosses.
    d. They are affected by substitution of a nucleus with a different genotype.

9. Two brothers have a hereditary disease associated with a particular lesion in mitochondrial DNA. One brother is more severely affected than the other. Which of the following is not a plausible explanation for this observation?
    a. The brothers have different degrees of heteroplasmy.
    b. The brothers have different proportions of two mitochondrial types.
    c. The brothers do not have identical nuclear genomes.
    d. Different mitochondrial genes are affected in the two brothers.

10. Which of the following is not true about genomic imprinting?
    a. It is an example of extranuclear inheritance.
    b. Expression patterns of imprinted genes can vary depending on their parental source.
    c. Expression patterns of imprinted genes are correlated with their methylation states.
    d. Genomic imprinting can affect the severity of disease symptoms.

Answers: 1e; 2a; 3c; 4b; 5c; 6b; 7a; 8d; 9d; 10a.

B. Thought Questions

1. What are the tenets of the endosymbiont hypothesis? What arguments can you marshal to support, or detract from, this hypothesis?

2. Speculate on the evolutionary origin of plant chloroplasts, compared to the presumed origin of mitochondria.

3. Is there any evidence that there are paternal contributions to mtDNA inheritance? What, if any, significance might these contributions have?

4. How might molecular methods be employed to analyze of a maternal line of descent? How might this be useful to trace movements of populations over time? Do any significant complications arise because of paternal contributions to mtDNA inheritance?

4. Distinguish between bi-parental and uni-parental inheritance. Which organisms have organelles that show bi-parental inheritance?

5. One often thinks of extranuclear inheritance being associated with genes in mitochondrial or chloroplast DNA. What other examples of extranuclear inheritance exist? What is the significance of the examples you cite?

6. In what ways does cpDNA differ from nuclear and mitochondrial DNA?

7. The mtDNAs of most organisms contain similar coding information. Given this, speculate why animal mtDNAs are always relatively small (< 20 kb), while mtDNAs of some other organisms can be quite large? (Put another way, what advantages do small extranuclear mtDNA genomes confer?) Why should genes be retained in mtDNA at all, given that at least some proteins (e.g., ribosomal proteins) are able to be imported into the mitochondria?

8. Recall that environmental factors, as well as other genetic factors, were used to explain variable expressivity. What role might genomic imprinting play?

9. What is RNA editing? Where does it occur, how extensively is it used, and what regulatory features does it confer?

10. How do you account for the maternal affect in shell-coiling in the snail, *Limnaea peregra*? What features of this phenomenon indicate that it is controlled via a maternal affect, and not via extranuclear inheritance?

V. SOLUTIONS TO TEXT PROBLEMS

20.1 Compare and contrast the structure of the nuclear genome, the mitochondrial genome, and the chloroplast genome.

Answer: The nuclear genome is organized into linear chromosomes that are long double-stranded DNA molecules packaged by histone and non-histone chromosomal proteins. Both mitochondrial and chloroplast genomes are circular, supercoiled, double-stranded DNA molecules and are not complexed with packaging proteins. While there is some variation in the size of chloroplast genomes (80 to 600 kb) and mitochondrial genomes (<20 kb in animals, 80 kb in yeast, 100 to 2,000 kb in plants), these extracellular genomes are not nearly as large as most eukaryotic chromosomes. Unlike nuclear chromosomes that contain diverse genes and non-coding sequences, mitochondria contain similar, unique genes, with larger mitochondrial genomes containing more non-coding sequences. Diploid, mitotically active eukaryotic cells have two copies of each chromosomal homologue that divide in an exquisitely organized mitosis. On the other hand, extranuclear genomes can exist in a variable number of copies per cell and can replicate independently in many phases of the cell cycle. A single yeast mitochondrion, for example, can have 10-30 nucleoid regions, each containing four to five mitochondrial DNA genomes.

20.2 Imagine you have discovered a new genus of yeast. In the course of your studies on this organism, you isolate DNA and subject it to CsCl density gradient centrifugation. You observe a major peak at a density of 1.75 g/cm^3 and a minor peak at a density of 1.70 g/cm^3. How could you determine whether the minor peak represents organellar (presumably mitochondrial) DNA, as opposed to a relatively AT-rich repeated sequence in the nuclear genome?

Answer: In addition to being relatively AT-rich, mitochondrial DNA is circular, supercoiled, and has a fixed size. To characterize the DNA that is associated with mitochondria, first separate the mitochondria from the nuclei by differential centrifugation or by using sucrose density gradients. Then isolate and characterize the DNA from these two organellar fractions. Assess the density of the DNAs in CsCl density gradients and visualize the DNAs using electron microscopy. DNA associated with the mitochondrial fraction should be circular with a density of 1.70 g/cm^3. Nuclear DNA should have neither of these properties.

20.3 How do mitochondria reproduce? What is the evidence for the method you describe?

Answer: Mitochondria arise from growth and division of pre-existing mitochondria, not from *de novo* synthesis. This view is supported by following the reproduction of density labeled *Neurospora crassa* mitochondria. The mitochondria were initially made dense by growing *Neurospora* in a medium low in protein, and high in choline. Then, the cells were shifted to a medium high in protein and low in choline. After one generation, mitochondria were found to have intermediate density, consistent with a model of growth and division, and not *de novo* synthesis. See text p. 664.

20.4 What genes are present in the human mitochondrial genome?

Answer: Human mitochondrial DNA contains the genes for the 12S and 16S mitochondrial rRNAs, 22 tRNAs, and some of the polypeptide subunits of cytochrome oxidase (COI, COII, COIII), NADH-dehydrogenase (ND1-6), and ATPase (ATPase 6 and 8). See text pp. 664-6, and Figure 20.6.

20.5 What conclusions can you draw from the fact that most nuclear-encoded mRNAs and all mitochondrial mRNAs have a poly(A) tail at the 3' end?

Answer: That most nuclear and mitochondrial mRNAs are polyadenylated suggests that polyadenylation serves some basic, and very likely, important function. However, the features unique to polyadenylation of mitochondrial transcripts give little insight into a general function. For example, polyadenylation is necessary in some mitochondrial mRNAs to complete a missing part of a UAA stop codon, but this function is not required for nuclear mRNAs. Since mitochondrial mRNAs do not exit the mitochondrion, the 3' poly(A) tail is presumably not needed for transport between cellular compartments. Experiments in which *Xenopus* oocytes were injected with globin mRNA with and without a poly(A) tail have suggested that one more general function of the tail is to confer stability on eukaryotic mRNAs.

20.6 Discuss the differences between the universal genetic code of the nuclear genes of most eukaryotes and the code found in human mitochondria. Is there any advantage to the mitochondrial code?

Answer: While plant mitochondria use the universal nuclear genetic code, mitochondria from other organisms, including humans, utilize a code that differs in codon designation. In addition, the mitochondrial code typically has more extensive wobble, so that many fewer tRNAs are needed to read all possible sense codons. In humans, only 22 mitochondrial tRNAs are needed, as opposed to the 32 required for nuclear mRNAs. [There is some variability between mitochondrial codes of different organisms, as well as between the nuclear codes of different organisms (e.g., ciliated protozoa do not use the universal nuclear code).] Fewer mitochondrial tRNAs confer the advantage that fewer tRNA genes are needed in the mitochondria, allowing for a smaller mitochondrial genome.

20.7 When the DNA sequences for most of the mRNAs in human mitochondria are examined, no nonsense codons are found at their termini. Instead, either U or UA is found. Explain this result.

Answer: Chain-termination codons are found only after polyadenylation, when the addition of the poly(A) tail completes the missing part of the UAA stop codon.

20.8 Compare and contrast the cytoplasmic and mitochondrial protein-synthesizing systems.

Answer: Translation initiation in mitochondria (except for plant and yeast mitochondria) is quite different from initiation in the cytoplasm. Animal mitochondrial mRNAs lack a 5' cap and have virtually no 5' leader sequence. Therefore, mitochondrial ribosomes bind to mitochondrial mRNAs and orient translation initiation differently than cytoplasmic ribosomes. In some respects, mitochondrial translation is similar to bacterial translation. A special mitochondrial tRNA.fMet is used in initiation, and special IF's, EFs and RFs that are distinct from cytoplasmic factors (but similar to those in bacteria) are used in initiation, elongation and termination. There are considerable differences between mitochondrial and cytoplasmic tRNAs as well. As discussed in problem 20.6, mitochondrial have fewer tRNAs, and considerably more wobble is employed. Finally, there are considerable differences between the structure of mitochondrial and cytoplasmic ribosomes. While there is some structural diversity among mitochondrial ribosomes, a common feature is that mitochondrial ribosomes lack the 5S and 5.8S rRNA components of cytoplasmic ribosomes. In human mitochondria, only 12S (associated with the small, 35S subunit) and 16S (associated with the large 45S subunit) rRNAs are found. The number of ribosomal proteins found in the cytoplasmic and mitochondrial ribosomes are also different, although the precise content in mitochondrial ribosomes is not well defined. With the exceptions of *Neurospora* and yeast, the proteins of mitochondrial ribosomes are distinct from those in cytoplasmic ribosomes.

20.9 Compare and contrast the organization of the ribosomal RNA genes in mitochondria and in chloroplasts.

Answer: In mitochondria, the two rRNA genes are typically found in distinct positions, and in one copy. While animal mitochondrial rRNA genes are closely linked, and separated only by a spacer with one tRNA gene, in fungi they are widely spaced. It is unusual to find more than one of either the large or small rRNA gene. In chloroplasts, a quite different situation is found. In addition to having four (16S, 23S, 4.5S and 5S), and not just two rRNA genes, chloroplasts have two copies of each gene. One complete set of four chloroplast rRNA genes is clustered within 10-25 kb, and positioned in an inverted orientation to the second set. The inverted repeats are designated IR_A and IR_B, and include other duplicated genes. Recombination between these repeats can lead to inversions of the intervening, unique sequences. See text pp. 665-666 and pp. 670-671.

20.10 What features of extranuclear inheritance distinguish it from the inheritance of nuclear genes?

Answer: Extranuclear inheritance can be distinguished from nuclear inheritance by four characteristics. First, different reciprocal cross results are seen, not related to sex. With extranuclear inheritance, uniparental inheritance is typically seen. Second, extranuclear genes cannot be mapped to the chromosomes in the nucleus. New extranuclear mutations do not show linkage to any known nuclear linkage group. Third, ratios typical of Mendelian segregation are not found. Fourth, extranuclear inheritance is indifferent to nuclear substitution.

20.11 Distinguish between maternal effect and extranuclear inheritance.

Answer: Maternal effect is the determination of gene-controlled characters by the maternal genotype prior to the fertilization of the egg cell. A maternal effect is seen when nuclear genes of the mother function to specify some characteristic of the zygote. For example, a maternal effect could result from a maternally transcribed gene whose product was localized at one pole of the embryo and was responsible for the polarity of the developing embryo.

In contrast to nuclear genes that can function in the mother, the genes involved in extranuclear inheritance are extranuclear, being found in mitochondria and chloroplasts. If the mitochondria are inherited from the mother's cytoplasm, they will be maternally inherited. However, genes of the mitochondria will still show extranuclear inheritance.

20.12 Reciprocal crosses between two types of the evening primrose, *Oenothera hookeri* and *Oenothera muricata*, produce the following effects on the plastids:

0. hookeri female x *0. muricata* male → yellow plastids
0. muricata female x *0. hookeri* male → green plastids

Explain the difference between these results, noting that the chromosome constitution is the same in both types.

Answer: Both offspring have the same nuclear constitution, but have different maternally inherited cytoplasm. Since the plastids are primarily derived from the egg cytoplasm, they are derived from the mother. If *hookeri* has green plastids and is the maternal parent, the maternally inherited *hookeri* plastids become yellow in the *hookeri/muricata* nuclear genetic background. If *muricata* has green plastids and is the maternal parent, the maternally inherited *muricata* plastids remain green in the *hookeri/muricata* nuclear genetic background. It would appear that there is a difference in the effect of the hybrid *hookeri/muricata* nuclear gene combination on the different, maternally derived plastids.

20.13 A form of male sterility in corn is maternally inherited. Plants of a male-sterile line crossed with normal pollen give male-sterile plants. Some lines of corn carry a dominant, so-called restorer (*Rf*) gene, which restores pollen fertility in male-sterile lines.

a. If a male-sterile plant is crossed with pollen from a plant homozygous for gene *Rf*, what will be the genotype and phenotype of the F_1?
b. If the F_1 plants of part a are used as females in a testcross with pollen from a normal plant (*rf/rf*), what would be the result? Give genotypes and phenotypes, and designate the type of cytoplasm.

Answer: a. Let the normal cytoplasm be denoted by [*N*], and the male-sterile cytoplasm be denoted by [*Ms*]. The cross can be written as [*Ms*] *rf/rf* female x [*N*] *Rf/Rf* male. The F_1 would be [*Ms*] *Rf/rf*, and be male-fertile.
b. The cross is [*Ms*] *Rf/rf* female x [*N*] *rf/rf* male. Half of the progeny would be [*Ms*] *Rf/rf* and half would be [*Ms*] *rf/rf*. Thus, half of the progeny will be male fertile and half will be male sterile.

20.14 In *Neurospora* a chromosomal gene *F* suppresses the slow-growth characteristic of the [*poky*] phenotype and makes a [*poky*] culture into a fast-[*poky*] culture, which still has abnormal cytochromes. Gene *F* in combination with normal cytoplasm has no detectable effect. (Hint: Since both nuclear and extranuclear genes have to be considered, it will be convenient to use symbols to distinguish the two. Thus cytoplasmic genes will be designated in square brackets; e.g., [*N*] for normal cytoplasm, [*poky*] for poky.)

a. A cross in which *fast*-[*poky*] is used as the female (protoperithecial) parent and a normal wild-type strain is used as the male parent gives half [*poky*] and half *fast*-[*poky*] progeny ascospores. What is the genetic interpretation of these results?

b. What would be the result of the reciprocal cross of the cross described in part a, that is, normal female x *fast*-[*poky*] male?

Answer: a. The phenotypic results indicate that half of the progeny are *poky* and half are *fast-poky*. This means that all of the progeny have the *poky* mitochondrial phenotype (by maternal inheritance) and that the *F* gene must be a nuclear gene segregating according to Mendelian principles. Using the symbolism above, the cross can be written as *F* [*poky*] female x + [*N*] male, and is expected to give half *F* [*poky*] and half + [*poky*] progeny. Considering the phenotypes, the *poky* progeny are + [*poky*] and the *fast-poky* progeny are *F* [*poky*].

b. The cross could be written as + [*N*] female x *F* [*poky*] male. The progeny will be half + [*N*] and half *F* [*N*]. These two genotypes are both normal, and have indistinguishable phenotypes.

20.15 Distinguish between nuclear (segregational), neutral, and suppressive *petite* mutants of yeast.

Answer: *petites* are a class of mutations that affect mitochondrial function. *Neutral petites* are able to grow on a medium that will support fermentation, but cannot grow on a medium that supports only anaerobic respiration because they lack nearly all of their mitochondrial DNA. Nuclear *petites* result from a nuclear mutation that affects mitochondrial function (e.g., in a nuclear gene that encodes a subunit of a mitochondrial protein). These mutations, if crossed to a wild-type (*grande)* strain, will show a typical 2:2 Mendelian segregation pattern. *Neutral* and *suppressive petites* result from mutations in the mitochondrial genome. They show uniparental inheritance (but not maternal inheritance): when crossed to a normal cell, and (*grande*) diploids go through meiosis, a 0:4 ratio of *petite:grande* is seen. *Suppressive petites*, unlike *neutral petites*, do have an effect on the wild-type. A diploid formed from a *suppressive petite* and a normal cell will have respiratory properties intermediate between the *petite* and *normal*. Mitosis in this diploid results in mostly petites (up to 99%) with a respiratory deficient phenotype, and meiosis results in a 0:4 ratio of *petite:grande*. At a molecular level, the suppressive petite mutations result from partial deletions of the mitochondrial DNA.

20.16 In yeast a haploid nuclear (segregational) *petite* is crossed with a neutral *petite*. Assuming that both strains have no other abnormal phenotypes, what proportion of the progeny ascospores are expected to be *petite* in phenotype if the diploid zygote undergoes meiosis?

Answer: Nuclear genes will show Mendelian segregation, so that +/*petite* should show 2:2 segregation. One should see 1/2 *petite* and 1/2 wild-type (*grande*) progeny.

20.17 When grown on a medium containing acriflavin, a yeast culture produces a large number of very small (*tiny*) cells that grow very slowly. How would you determine whether the slow-growth phenotype was the result of a cytoplasmic factor or a nuclear gene?

Answer: This problem is formally very similar to determining the mode of inheritance for yeast *petite* mutants. Acriflavin intercalates between base pairs, and might be used to introduce *petite*-like mutations. If the *tiny* phenotypes are due to a nuclear gene, then meiosis in cells formed from a cross of *tiny* x *normal* should result in a 2:2 segregation of *tiny:normal* ascospores. If, in contrast, an extranuclear gene is involved, one would expect a 0:4 or 4:0 ratio, and see only *normal* or *tiny* progeny.

20.18 *Drosophila melanogaster* has a sex-linked, recessive, mutant gene called *maroon-like* (*ma-l*). Homozygous *ma-l* females or hemizygous *ma-l* males have light-colored eyes, owing to the absence of the active enzyme xanthine dehydrogenase, which is involved in the synthesis of eye pigments. When heterozygous *ma-l*⁺/*ma-l* females are crossed with *ma-l* males, all the offspring are phenotypically wild type. However, half the female offspring from this cross, when crossed back to *ma-l* males, give all *ma-l* progeny. The other half of the females, when crossed to *ma-l* males, give all phenotypically wild-type progeny. What is the explanation for these results?

Answer: Since *ma-l* is sex-linked, we know it is nuclearly inherited. One can therefore diagram the crosses and the results. In the first cross, $ma\text{-}l^+/ma\text{-}l$ x $ma\text{-}l$/Y gives 1/4 *ma-l*/*ma-l* females, 1/4 $ma\text{-}l^+/ma\text{-}l$ females, 1/4 $ma\text{-}l^+$/Y males and 1/4 *ma-l*/Y males. To explain why the half of the total F_1 progeny that are *ma-l* hemizygotes or homozygotes do not have light-colored eyes, consider the possibility of a maternal effect. The parental female was $ma\text{-}l^+/ma\text{-}l$, and so could have provided normal xanthine dehydrogenase (or stable, normal xanthine dehydrogenase mRNA, or the biochemical product of the xanthine dehydrogenase reaction) to the embryo. The results of the crosses indicate that this maternally transmitted product is sufficient for the F_1 progeny to have a normal phenotype. When the F_1 females are backcrossed to the *ma-l* males, two different crosses are possible. One is $ma\text{-}l^+/ma\text{-}l$ x *ma-l*/Y. Just like the initial cross, this will give all wild-type progeny. The other cross is *ma-l*/*ma-l* x *ma-l*/Y, which will give all *maroon-like* progeny (as the mother can no longer supply the $ma\text{-}l^+$ product to her progeny).

20.19 When females of a particular mutant strain of *Drosophila melanogaster* are crossed to wild-type males, all the viable progeny flies are females. Hypothetically, this result could be the consequence of either a sex-linked, male-specific lethal mutation or a maternally inherited factor that is lethal to males. What crosses would you perform in order to distinguish between these alternatives?

Answer: If the mutation is a sex-specific, male-lethal mutation, the first cross could be diagrammed as *L*/*L* x +/Y, giving *L*/+ (females) and *L*/Y (dead males). If one crossed the F_1 females to normal males, one would have *L*/+ x +/Y. The progeny would be 1/4 *L*/+ (normal females), 1/4 +/+ (normal females), 1/4 +/Y (normal males) and 1/4 *L*/Y (dead males). Thus, one would expect a 2:1 ratio of females to males. If the male-lethality was associated with a maternally inherited factor lethal to males, the F_1 females should possess this factor (as they received cytoplasm from their mother). Consequently, just like their mother, the F_1 females should have no male offspring when mated to wild-type males.

20.20 Reciprocal crosses between two *Drosophila* species, *D. melanogaster* and *D. simulans*, produce the following results:

melanogaster female x *simulans* male → females only
simulans female x *melanogaster* male → males, with few or no females

Propose a possible explanation for these results.

Answer: When *melanogaster* females are crossed with *simulans* males, *melanogaster* cytoplasm is given to the offspring. Female progeny will have an X from the *melanogaster* as well as the *simulans* parent, while male progeny will have an X from the *melanogaster* parent but only a Y (with few structural genes) from the *simulans* parent. Female progeny survive selectively because of nuclear gene products provided on the *simulans* X needed for hybrid survival in *melanogaster* cytoplasm. When a *simulans* female is crossed with a *melanogaster* male, *simulans* cytoplasm is given to the offspring. Female progeny will have a *melanogaster* as well as a *simulans* X, while male progeny will have a *simulans* X and a *melanogaster* Y. Since few or no females are recovered, it appears that the *melanogaster* X encodes products that (generally) cause lethality in the *simulans* cytoplasm.

20.21 Some *Drosophila* flies are very sensitive to carbon dioxide—they become anesthetized when it is administered to them. The sensitive flies have a cytoplasmic particle called *sigma* that has many properties of a virus. Resistant flies lack *sigma*. The sensitivity to carbon dioxide shows strictly maternal inheritance. What would be the outcome of the following two crosses: (a) sensitive female x resistant male and (b) sensitive male x resistant female?

Answer: In part (a), all of the progeny would inherit the *sigma* factor from the (sensitive) female parent. Consequently, all the progeny will be sensitive. In part (b), the resistant female parent lacks the *sigma* factor. All of the progeny will also lack the factor, and so be resistant.

20.22 A few years ago the political situation in Chile was such that very many young adults were kidnapped, tortured, and killed by government agents. When abducted young women had young children or were pregnant, those children were often taken and given to government supporters to raise as their own. Now that the political situation has changed, grandparents of stolen children are trying to locate and reclaim their grandchildren. Imagine that you are a judge in a trial centering on the custody of a child. Mr. and Mrs. Escobar believe Carlos Mendoza is the son of their abducted, murdered daughter. If this is true, then Mr. and Mrs. Sanchez are the paternal grandparents of the child, as their son (also abducted and murdered) was the husband of the Escobars' daughter. Mr. and Mrs. Mendoza claim Carlos is their natural child. The attorney for the Escobar and Sanchez couples informs you that scientists have discovered a series of RFLPs in human mitochondrial DNA. He tells you his clients are eager to be tested, and ask that you order that Mr. and Mrs. Mendoza and Carlos be tested also.

a. Can mitochondrial RFLP data be helpful in this case? In what way?
b. Do all seven parties need to be tested? If not, who actually needs to be tested in this case? Explain your choices.
c. Assume the critical people have been tested, and you have received the results. How would the results determine your decision?

Answer: a. Mitochondrial RFLP data can be helpful to trace the maternal line of descent. If Carlos is the son of Mr. and Mrs. Mendoza, then Carlos and Mrs. Mendoza should have identical (or at least highly similar) RFLPs. If, on the other hand, Carlos is the son of Mrs. Escobar's murdered daughter, then Carlos and Mrs. Escobar should have identical (or at least highly similar) mitochondrial RFLPs. If Mrs. Escobar and Mrs. Mendoza have different mitochondrial RFLPs, it can be determined which of them contributed mitochondria to Carlos.

b. Only Carlos, Mrs. Escobar and Mrs. Mendoza need to be tested. The potential grandfathers need not be tested, as they will not have given any of their mitochondria to Carlos. In addition, Mrs. Sanchez need not be tested, as her son, even if he was Carlos' father, would not have given him mitochondria (see however p. 686–exceptions to maternal inheritance). Only individuals who might have contributed mitochondria maternally need to be tested.

c. If Mrs. Mendoza and Mrs. Escobar do not differ in mitochondria RFLPs, the data will not be helpful. If they do differ, and Carlos matches Mrs. Mendoza, the case should be dismissed. If Carlos matches Mrs. Escobar, then the Escobar and Sanchez couples are indeed the grandparents, and the Mendozas have claimed a stolen child.

20.23 The pedigree in the figure below shows a family in which an inherited disease called Leber's optic atrophy is segregating. This condition causes blindness in adulthood. Studies have recently shown that the mutant gene causing Leber's optic atrophy is located in the mitochondrial genome.

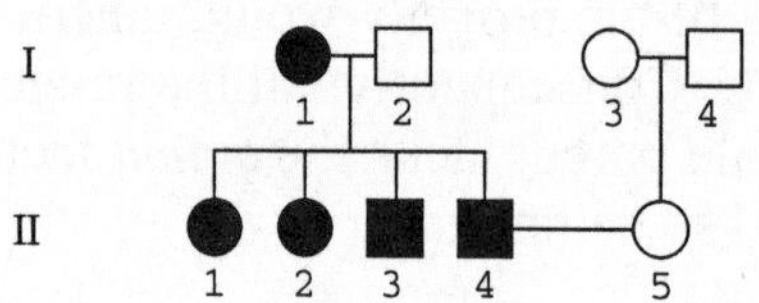

a. Assuming II-4 marries a normal person, what proportion of his offspring should inherit Leber's optic atrophy?
b. What proportion of the sons of II-2 should be affected?
c. What proportion of the daughters of II-2 should be affected?

Answer: a. Since II-4 is a male, he will not contribute any of his mutant mitochondria to his offspring. All of his offspring will have normal mitochondria from their mother, and be normal.

b. Since II-2 is a female, all of her progeny will obtain her mutant mitochondria, and hence all will be affected. All of her sons will be affected.

c. All of the daughters will be affected, as all will have mutant mitochondria from their mother.

20.24 The inheritance of the direction of shell coiling in the snail *Limnaea peregra* has been studied extensively. A snail produced by a cross between two individuals has a shell with a right-hand twist (dextral-coiling). This snail produces only left-hand (sinistral) progeny on selfing. What are the genotypes of the F_1 snail and its parents?

Answer: If the coil direction were controlled by an extranuclear gene, then the progeny would always exhibit the phenotype of the mother, owing to maternal inheritance. If the trait were determined by a nuclear gene that did not show a maternal affect, then one would expect a "true-breeding" individual to give rise to individuals just like itself. Since neither of these are observed, these data are best explained by considering a nuclear gene that shows a maternal effect. In this case, the shell-coiling phenotype is determined by the genotype of the mother. One can infer the maternal genotype from the progeny phenotype. If dextral is *D*, and sinistral is *d*, the F_1 genotype must have been *dd*, as it only gives sinistral offspring. Since the F_1 itself has a dextral pattern, its maternal parent must have had a *D* allele. To give the F_1 a *d* allele, the maternal parent must have been *Dd.* The paternal parent must have had a *d* allele, and could have been either *dd* or *Dd.*

CHAPTER 21

POPULATION GENETICS

I. CHAPTER OUTLINE

GENETIC STRUCTURE OF POPULATIONS
- Genotypic Frequencies
- Allelic Frequencies

THE HARDY-WEINBERG LAW
- Assumptions of the Hardy-Weinberg Law
- Predictions of the Hardy-Weinberg Law
- Derivation of the Hardy-Weinberg Law
- Extensions of the Hardy-Weinberg Law to Loci with More Than Two Alleles
- Extensions of the Hardy-Weinberg Law to Sex-linked Alleles
- Testing for Hardy-Weinberg Proportions
- Using the Hardy-Weinberg Law to Estimate Allelic Frequencies

GENETIC VARIATION IN SPACE AND TIME

GENETIC VARIATION IN NATURAL POPULATIONS
- Models of Genetic Variation
- Measuring Genetic Variation with Protein Electrophoresis
- Measuring Genetic Variation with RFLPs and DNA Sequencing

CHANGES IN GENETIC STRUCTURE OF POPULATIONS
- Mutation
- Genetic Drift
- Migration
- Natural Selection
- Simultaneous Effects of Mutation and Selection
- Nonrandom Mating

SUMMARY OF THE EFFECTS OF EVOLUTIONARY PROCESSES ON THE GENETIC STRUCTURE OF A POPULATION
- Changes in Allelic Frequency Within a Population
- Genetic Divergence Among Populations
- Increases and Decreases in Genetic Variation Within Populations

SUMMARY OF THE EFFECTS OF EVOLUTIONARY PROCESSES ON THE CONSERVATION OF GENETIC RESOURCES

MOLECULAR GENETIC TECHNIQUES AND EVOLUTION
- DNA Sequence Variation
- DNA Length Polymorphisms
- Evolution of Multigene Families Through Gene Duplication

Evolution in Mitochondrial DNA Sequences
Concerted Evolution
Evolutionary Relationships Revealed by RNA and DNA Sequences

II. REVIEW OF KEY TERMS, SYMBOLS AND CONCEPTS

Without consulting the text and in your own words, write a brief definition of each term in the groups below. Then, either using a short phrase or a simple diagram, identify the relationship(s) between specific pairs of terms within a set. Finally, consult the text (and perhaps a friend who has also done the exercise) to check your answers.

1	2	3	4
quantitative genetics genetic structure genotypic frequency allelic frequency Mendelian population gene pool Hardy-Weinberg law	Hardy-Weinberg equilibrium random mating mutation migration natural selection genetic drift	models of genetic variation classical model balance model neutral model neutral mutation model proportion of polymorphic loci heterozygosity	genetic drift effective population size sampling error founder effect bottleneck effect migration gene flow population viability analysis neutral theory of molecular evolution

5	6
natural selection Darwinian fitness antagonistic pleiotropy selection coefficient protected polymorphism heterosis overdominance heterozygote superiority positive assortative mating negative assortative mating inbreeding outbreeding	molecular basis of evolution synonymous change nonsynonymous change multigene family concerted evolution molecular drive evolutionary tree

III. THINKING ANALYTICALLY

Population genetics shifts away from a consideration of molecular mechanisms underlying biochemical processes in processes to a statistical evaluation of the effects of such processes at the level of the group, population, or species. At the conceptual core of population genetics lies the Hardy-Weinberg law. Therefore, focus your initial efforts on understanding the assumptions and predictions of this law. Practice the examples in the text and problems to become fluent in analyzing whether a population is in Hardy-Weinberg equilibrium. By doing this, you will be better suited to understand extensions and deviations of this law.

Be aware that the allelic symbolism used in population genetics is often different from that used in Mendelian or molecular genetics. Consider the following example. In earlier chapters, the genotypes for normal and sickle-cell hemoglobin were either written as $\beta^A\beta^A$, $\beta^A\beta^S$, $\beta^S\beta^S$ or Hb^AHb^A, Hb^AHb^S, Hb^SHb^S. Here they are written as *Hb-A/Hb-A, Hb-A/Hb-S, Hb-S/Hb-S.* Similarly, care should be taken with the letters p and q, which symbolize the frequency within a population of dominant and recessive members of an allelic pair (but do not symbolize the alleles themselves!).

One of the main concerns of population genetics is how to model changes in allele and genotypic frequencies under a set of specified conditions. The models are often very elegant, and employ equations that relate a set of variables identified in a population. While simple memorization of the equations used to analyze changes in populations takes little time, it is insufficient to understand the models. It will be difficult, if not impossible, to solve even moderately challenging word problems if you can only plug numbers into equations. While you will find it helpful to recognize by sight several of the equations, you need to do more. Take the time to understand the model that leads to the equation, and how the mathematical and statistical analysis in the model is related to a biological question.

A few hints for solving some of the problems are in order. First, you will often find that the frequency of one class of homozygote (usually the recessive homozygotes, given by q^2) is the only piece of hard data available. As you work back from this information, keep track of the assumptions that you are making. Otherwise, you may enter into the realm of circular reasoning. For example, you can legitimately calculate q as the square root of q^2, and if only two alleles exist, calculate p by equating it with $1 - q$. However, if you then determine that heterozygotes exist at a frequency of $2pq$, you are <u>assuming</u> that random mating is occurring and the conditions of Hardy-Weinberg equilibrium are satisfied. Further work with your calculated values of p^2, $2pq$, and q^2 will continue to reflect that assumption. Often, you will need to look for more information, or another approach, to prove that the population in question is in Hardy-Weinberg equilibrium. Second, doing the math requires care and, if you are not exceptionally fluent with it, patience. It will sometimes help to factor out common multipliers [e.g., $2pq^2 + q^2 = q^2(2p + 1)$]. At other times, it will be helpful to recognize members of a binomial expansion [e.g., $p^2 + 2pq + q^2 = (p + q)^2$].

After you have mastered the concepts underlying the Hardy-Weinberg law, focus on understanding how genetic variation can be measured. A number of conceptually important models have been developed that have been supported and challenged by data gathered using a variety of methods, including those of molecular genetics. Models have been proposed to address the substantial amount of genetic variation that exists in a population, and the factors that can lead to changes in a population's genetic structure. Some of these employ quantitative analysis and equations. Approach these models just as you did the Hardy-Weinberg law. First become familiar with the conceptual issues of the model, and then relate the variables in an equation to the key factors considered in the model.

During the past decade or so, the powerful techniques of molecular genetics have led to substantial new insights into some of the evolutionary questions implicit in population genetics. In the final sections of this chapter, you will explore how evolutionary questions can be addressed using relationships derived from changes in nucleic acid sequence. Mastering this material requires you to recall a body of previously covered information, learn new factual material and integrate it into the conceptual framework of population genetics.

IV. QUESTIONS FOR PRACTICE

A. Multiple Choice Questions

1. In a small population, 30 percent of the individuals have blood type M, 40 percent of the individuals have blood type MN and 30 percent of the individuals have blood type N. If p equals the frequency of the L^M allele and q equals the frequency of the L^N allele, what are p and q?
   a. $p = 0.30, q = 0.30$.
   b. $p = 0.50, q = 0.50$.
   c. $p = 0.30, q = 0.70$.
   d. $p = 0.50, q = 0.30$.

2. Is the population described in question 1, above, in Hardy-Weinberg equilibrium?
   a. Yes, because the calculated genotypic frequencies equal the expected genotypic frequencies.
   b. Yes, because at equilibrium you always have equal numbers of recessive and dominant homozygotes.
   c. No, the frequency of heterozygotes is too large and the frequency of homozygotes is too low.
   d. No, the frequency of heterozygotes is too low and the frequency of homozygotes is too large.

3. Which of the following is <u>not</u> an assumption about a population in Hardy-Weinberg equilibrium?
   a. The population is isolated.
   b. Random mating occurs in the population.
   c. The population is free from mutation.
   d. The population is free from natural selection.
   e. The population is free from migration.

4. The frequency of one form of X-linked color-blindness varies among human ethnic groups. What can be said about whether each ethnic group is in Hardy-Weinberg equilibrium? (Let q = the frequency of the normal allele, and p = the frequency of the color-blind allele.)
   a. None of the ethnic populations can be in equilibrium, since all have different values for p and q.
   b. Only the entire human population is in equilibrium.
   c. Some of the ethnic populations may be in equilibrium, providing that the frequency of the trait in males is p and the frequency in females is p^2.
   d. Some of the ethnic populations may be in equilibrium, providing that the frequency of the trait in both sexes is p^2.
   e. All of the ethnic populations will be in equilibrium, since each satisfies the criteria for a population in Hardy-Weinberg equilibrium.

5. In a large, randomly mating population, 80 percent of the individuals have dark hair and twenty percent are blond. Assuming that hair color is controlled by one pair of alleles, is the allele for dark hair <u>necessarily</u> dominant to the one for blond hair?

a. Yes, because otherwise the population would not be dominated by dark-haired individuals.
b. Yes, because more of something (in this case, hair color) is always dominant to less of that thing.
c. No, because relative frequencies of alleles in a randomly mating population is unrelated to issues of dominance and recessiveness.
d. No, because although there is a relationship between dominance and allele frequency, that relationship is not seen in this example.

6. The amounts of genetic variation in a population can be explained in a number of ways. In one model, a large amount of genetic variation is explained by recurrent mutation and random changes in allele frequencies. In this model, natural selection selects against some of the variation affecting fitness but does not select for or against much of the genetic variation. This model is termed
   a. the classical model.
   b. the balance model.
   c. the neutral mutation model.
   d. the random mutation model.

7. Which of the following populations are in Hardy-Weinberg equilibrium?

	Genotypes		
Population	*AA*	*Aa*	*aa*
a	0.72	0.20	0.08
b	0.12	0.80	0.08
c	0.08	0.01	0.91
d	0.25	0.50	0.25

8. How is the rate of forward mutation likely to be related to the rate of back mutation?
   a. The rate of forward mutation is generally lower because there is mutational pull back to a specific form.
   b. The rate for forward mutation is generally higher because once an allele has changed, it is nearly as likely that a subsequent change will be to yet another new form.
   c. The rate of forward and back mutations are co-dependent and are therefore usually equal.
   d. The relative rates of forward and back mutation are so highly variable that one can not formulate an accurate generalization comparing the two.

9. Which of the following can result in genetic drift?
   a. A sampling error.
   b. Random factors producing unexpected mortality.
   c. The establishment of a population by a small number of breeding individuals.
   d. A drastic reduction in the size of a population.
   e. A change in environmental conditions that affects selection.
   f. All of the above.

10. What is the effective population size for a population consisting of 10 breeding males and 2 breeding females?

a. 12
b. 6
c. 7
d. 2

11. Which of the following is <u>not true</u> concerning the effect of migration among populations?
    a. Migration tends to increase the effective size of the populations, leading to a reduction in genetic drift.
    b. Migration is associated with gene flow, which introduces new alleles to the population.
    c. If the allelic frequencies of migrants and the recipient population differ, migration can lead to the further differentiation of two populations.
    d. Migration and genetic drift have opposite effects on size and variability. Migration effectively increases size and variability, whereas drift acts in opposition.

12. What can one generally state about measuring the fitness associated with a specific genotype?
    a. It is relatively easy to assess, and can be based on the number of offspring of an individual.
    b. Since it is based on the reproductive ability of a genotype, it is an absolute term, which requires no assumptions.
    c. It will remain the same from generation to generation.
    d. It is difficult to measure because of antagonistic pleiotropy.

13. Consider a recessive trait that results in a complete lack of reproductive success of homozygotes. How might such a recessive trait be maintained at a high level in a population?
    a. Through new mutation
    b. Through heterosis
    c. Through overdominance
    d. Through heterozygote superiority
    e. All of the above

14. For a particular eukaryotic gene, which rate of DNA change is expected to be the lowest?
    a. The relative rate of evolutionary change in non-functional pseudogenes.
    b. The relative rate of evolutionary change in introns.
    c. The relative rate of evolutionary change for synonymous substitutions in coding sequences.
    d. The relative rate of evolutionary change for nonsynonymous substitutions in coding sequences.
    e. The relative rate of evolutionary change in leaders and trailers.

15. Which of the rates in question 14 is expected to be the highest?

16. Which of the following are true concerning concerted evolution?

a. It allows for rapid differentiation among species.
b. It directs the nature of mutation.
c. It results in homogeneity among multiple copies of a gene.
d. It acts only on coding sequences.
e. All of the above.
f. a and c

Answers: 1b; 2d; 3a; 4c; 5c; 6c; 7d; 8b; 9f; 10c; 11c; 12d; 13e; 14c; 15a; 16f.

B. Thought Questions

1. Consider a population that is not in Hardy-Weinberg equilibrium for a pair of alleles. Show why equilibrium values for genotypic frequencies are reached in one generation after the onset of random mating for autosomal alleles, but more than one generation of random mating is required for sex-linked alleles.
2. Distinguish the concept of heterozygosity with that of the proportion of polymorphic loci.
3. Mutation pressure is rarely the most important determinant of gene frequency. What other factors are important?
4. Can a population be in Hardy-Weinberg equilibrium for one, but not another, pair of alleles?
5. What is the most likely explanation if the gene frequencies agree with those predicted by the Hardy-Weinberg Law, but the genotype frequencies do not? (Hint: Consider mating systems and heterosis.)
6. Why do mutation, migration and drift not necessarily lead to adaptation?
7. What are five different molecular methods that can be used to measure genetic variation? What advantages and disadvantages are there to each?
8. Distinguish between nonrandom mating, positive assortative mating, negative assortative mating, inbreeding and outbreeding.
9. Morphological, behavioral, ultrastructural and biochemical characteristics have been extensively used to construct evolutionary trees. What limitations are there to relying on these traits? How does using DNA and RNA sequences overcome these limitations? What limitations are there to using DNA and RNA sequences?
10. How can analyzing mitochondrial DNA be used to build evolutionary trees? What special features are associated with animal and plant mitochondrial DNA, and how are these useful or problematic?

V. SOLUTIONS TO TEXT PROBLEMS

21.1 In the European land snail, *Cepaea nemoralis*, multiple alleles at a single locus determine shell color. The allele for brown (C^B) is dominant to the allele for pink (C^P) and to the allele for yellow (C^Y). Pink is recessive to brown, but is dominant to yellow, and yellow is recessive to pink and brown. Thus, the dominance hierarchy among these alleles is $C^B > C^P > C^Y$. In one population of *Cepaea*, the following color phenotypes were recorded:

Brown	236
Pink	231
Yellow	33
Total	500

Assuming that this population is in Hardy-Weinberg equilibrium (large, randomly mating, and free from evolutionary processes), calculate the frequencies of the C^B, C^Y, and C^P alleles.

Answer: Equate the frequency of each color with the frequency expected in Hardy-Weinberg equilibrium, letting $p = f(C^B)$, $q = f(C^P)$, and $r = f(C^Y)$.

Brown: $f(C^BC^B) + f(C^BC^P) + f(C^BC^Y) = p^2 + 2pq + 2pr = 236/500 = 0.472$
Pink: $f(C^PC^P) + f(C^PC^Y) = q^2 + 2qr = 231/500 = 0.462$
Yellow: $f(C^YC^Y) = r^2 = 33/500 = 0.066$

Now solve for p, q, and r, knowing that $p + q + r = 1$.

$r^2 = 0.066$, so $r = \sqrt{0.066} = 0.26$

There are two approaches to solve for q. First, since $q^2 + 2qr = 0.462$, one can substitute in $r = 0.26$, giving $q^2 + 2q(0.26) = 0.462$. Recognize this as a quadratic equation and set it equal to 0, and solve for q: That is, solve the equation
$q^2 + 0.52q - 0.462 = 0$
Solving the quadratic equation for q, one has:

$$q = \frac{-0.52 \pm \sqrt{(0.52)^2 - 4(1)(-0.462)}}{2(1)} = 0.467$$

A second approach to solve for q is to realize that

$q^2 + 2qr = 0.462$ and
$r^2 = 0.066$.

Adding left and right sides of the equations together, one has
$q^2 + 2qr + r^2 = 0.066 + 0.462$
$(q + r)^2 = 0.528$
$q + r = 0.726$
$q = 0.726 - r = 0.726 - 0.26 = 0.467$

Since $p + q + r = 1$, $p = 1 - (q + r) = 1 - (0.26 + 0.467) = 0.273$

21.2 Three alleles are found at a locus coding for malate dehydrogenase (MDH) in the spotted chorus frog. Chorus frogs were collected from a breeding pond, and each frog's genotype at the MDH locus was determined with electrophoresis. The following numbers of genotypes were found:

Genotype	Number
M^1M^1	8
M^1M^2	35
M^2M^2	20
M^1M^3	53
M^2M^3	76
M^3M^3	62
Total	254

a. Calculate the frequencies of the M^1, M^2, and M^3 alleles in this population.
b. Using a chi-square test, determine whether the MDH genotypes in this population are in Hardy-Weinberg proportions.

Answer: a. The tally for M^1 alleles is as follows:

GENOTYPE	# INDIVIDUALS	# M^1 ALLELES
M^1M^1	8	16
M^1M^2	35	35
M^1M^3	53	53
	Total	104

The total number of individuals is 254, thus the total number of alleles is 254 x 2, or 508. The frequency of M^1 alleles is 104/508 = 0.205. The frequency of the other alleles is obtained similarly. One has $f(M^1) = 0.20 = p$, $f(M^2) = 0.30 = q$, and $f(M^3) = 0.50 = r$.

b. For three alleles with frequencies p, q, and r, a population in Hardy-Weinberg equilibrium will have $p^2 + 2pq + 2pr + q^2 + 2qr + r^2 = 1$. To test the hypothesis that the population is in Hardy-Weinberg equilibrium, calculate the numbers of individuals expected in each class using this relationship and the values for p, q, and r obtained in part a. Calculate the value of χ^2 as shown in the following table.

Genotype	Observed Value (o)	Expected Frequency	Expected Value(e)	d (o–e)	d^2/E
M^1M^1	8	$p^2 = 0.04$	10	–2	0.40
M^1M^2	35	$2pq = 0.12$	30	5	0.83
M^2M^2	20	$q^2 = 0.09$	23	–3	0.39
M^1M^3	53	$2pr = 0.20$	51	2	0.08
M^2M^3	76	$2qr = 0.30$	75	1	0.01
M^3M^3	62	$r^2 = 0.25$	64	–2	0.06
					χ^2=1.76

Since the six phenotypic classes are completely specified by three allele frequencies, the number of phenotypes (6) minus the number of alleles (3) determines the degrees of freedom (6 – 3 = 3). With 3 degrees of freedom, $0.70 < P < 0.50$. The hypothesis can be accepted as possible. It would appear that the population is in Hardy-Weinberg equilibrium.

21.3 In a large interbreeding population 81 percent of the individuals are homozygous for a recessive character. In the absence of mutation or selection, what percentage of the next generation would be homozygous recessives? Homozygous dominants? Heterozygotes?

Answer: The conditions of this problem meet the requirements for a population in Hardy-Weinberg equilibrium. In such a population, if p equals the frequency of A, and q equals the frequency of a, one expects p^2 AA, $2pq$ Aa, and q^2 aa genotypes after random mating. Here, $q^2 = 0.81$, so $q = 0.9$. Since $p + q = 1$, $p = 1 - 0.9 = 0.1$. In the next generation, one would expect $p^2 = (0.1)^2 = 0.01$ (or 1%) AA

genotypes, $2pq = 2(0.1)(0.9) = 0.18$ (or 18%) *Aa* genotypes and $q^2 = (0.9)^2 = 0.81$ (or 81%) *aa* genotypes.

21.4 Let *A* and *a* represent dominant and recessive alleles whose respective frequencies are *p* and *q* in a given interbreeding population at equilibrium (with $p + q = 1$).

a. If 16 percent of the individuals in the population have recessive phenotypes, what percentage of the total number of recessive genes exist in the heterozygous condition?

b. If 1.0 percent of the individuals were homozygous recessive, what percentage of the recessive genes would occur in heterozygotes?

Answer: a. One has $q^2 = 0.16$, so $q = \sqrt{0.16} = 0.40$. Since $p + q = 1$, $p = 0.60$. The frequency of heterozygotes is $2pq = 2(0.40)(0.60) = 0.48$. Each of the 48% of the heterozygotes have one recessive allele, while each of the 16% of the homozygous recessive individuals have two. Thus, the percentage of the total number of recessive alleles in heterozygotes is $(0.48)/[0.48 + 2(0.16)] = 0.48/0.80 = 0.60$, or 60%.

b. If $q^2 = 0.01$, then $q = 0.1$, and $p = 0.9$. $2pq = 2(0.1)(0.9) = 0.18$. The percentage of the total number of recessive alleles in heterozygotes is $(0.18)/[0.18 + 2(0.1)] = 0.18/0.20 = 0.90$, or 90%.

21.5 A population has eight times as many heterozygotes as homozygous recessives. What is the frequency of the recessive gene?

Answer: The frequency of heterozygotes in a population in equilibrium is $2pq$, and the frequency of homozygous recessives is q^2. Here, there are eight times as many heterozygotes as homozygous recessives, so $2pq = 8q^2$ Since $p + q = 1$, $p = 1 - q$, and one can substitute $1 - q$ for p. This gives $2(1 - q)q = 8q^2$. Dividing both sides by q, and multiplying through, one has $2 - 2q = 8$. Thus, $2 = 10q$, and $q =$ the frequency of the recessive gene $= 0.20$.

21.6 In a large population of range cattle the following ratios are observed: 49 percent red (*RR*), 42 percent roan (*Rr*), and 9 percent white (*rr*).

a. What percentage of the gametes that give rise to the next generation of cattle in this population will contain allele *R*?

b. In another cattle population only 1 percent of the animals are white and 99 percent are either red or roan. What is the percentage of *r* alleles in this case?

Answer: a. In the red, *RR* animals, all of the gametes contain the *R* allele, while in the roan, *Rr* animals, half of the gametes contain the *R* allele. Therefore, [49 + (42/2)] = 70% of the gametes will contain the *R* allele. Another way to look at this problem is to realize that the frequency of gametes bearing a certain allele is the same as the frequency of the allele in the population. Let *p* equal the frequency of *R* in the population. Since there are 49 percent red animals, $p^2 = 0.49$, so $p = 0.70$ (or 70%).

b. If one lets *q* represent the frequency of *r*, since 1 percent of the animals are white, one has $q^2 = 0.01$. Hence, $q = 0.1$.

21.7 In a gene pool the alleles *A* and *a* have initial frequencies of *p* and *q*, respectively. Prove that the allelic frequencies and zygotic frequencies do not change from generation to generation as long as there is no selection, mutation, or migration, the population is large, and the individuals mate at random.

Answer: Let the frequency of allele A equal p and the frequency of allele a equal q, with $p + q = 1$. Then, in the initial generation, the frequency of AA genotypes is p^2, the frequency of aa genotypes is q^2. The frequency of the remaining genotypes (i.e., Aa heterozygotes) must be $1 - (p^2 + q^2)$. Since $p + q = 1$, $(p + q)^2 = 1^2$, and $p^2 + 2pq + q^2 = 1$. Therefore $1 - (p^2 + q^2) = 2pq$ and the frequency of Aa heterozygotes must be $2pq$.

Assume there is no selection, mutation or migration and that the individuals mate at random. Since there are three different genotypes in the population nine crosses are possible. The frequency of each type of cross is determined by the frequency of each parental genotype. The types of crosses, the frequency of each cross, the types of progeny, and the frequency of each progeny class are listed in the table below:

Cross	Cross Frequency	Progeny Ratios AA	Aa	aa	Progeny Frequencies AA	Aa	aa
$AA \times AA$	$p^2 \times p^2 = p^4$	all			p^4		
$AA \times Aa$	$p^2 \times 2pq = 2p^3q$	1/2	1/2		p^3q	p^3q	
$AA \times aa$	$p^2 \times q^2 = p^2q^2$		all			p^2q^2	
$Aa \times AA$	$2pq \times p^2 = 2p^3q$	1/2	1/2		p^3q	p^3q	
$Aa \times Aa$	$2pq \times 2pq = 4p^2q^2$	1/4	1/2	1/4	p^2q^2	$2p^2q^2$	p^2q^2
$Aa \times aa$	$2pq \times q^2 = 2pq^3$		1/2	1/2		pq^3	pq^3
$aa \times AA$	$q^2 \times p^2 = p^2q^2$		all			p^2q^2	
$aa \times Aa$	$q^2 \times 2pq = 2pq^3$		1/2	1/2		pq^3	pq^3
$aa \times aa$	$q^2 \times q^2 = q^4$			all			q^4

To determine the frequency of a particular zygotic class, add up the frequency of the progeny in that class. Then factor out a common multiplier, and note that $p + q = 1$ and that $(p + q)^2 = p^2 + 2pq + q^2$. One has:

$$\begin{aligned}\text{frequency } (AA) &= p^4 + 2p^3q + p^2q^2\\ &= p^2(p^2 + 2pq + q^2)\\ &= p^2(p + q)^2\\ &= p^2(1)^2 = p^2\end{aligned}$$

$$\begin{aligned}\text{frequency } (Aa) &= 2p^3q + 4p^2q^2 + 2pq^3\\ &= 2pq(p^2 + 2pq + q^2)\\ &= 2pq(p + q)^2\\ &= 2pq(1)^2 = 2pq\end{aligned}$$

$$\begin{aligned}\text{frequency } (aa) &= pq + 2pq^3 + q^4\\ &= q^2(p^2 + 2pq + q^2)\\ &= q^2(p + q)^2\\ &= q^2(1)^2 = q^2\end{aligned}$$

Thus, the zygotic frequencies do not change from one generation to the next.

Since all of the gametes of the *AA* parents and half of the gametes of the *Aa* parents will bear the *A* allele, the frequency of *A* in the gene pool of the next generation is $p^2 + pq = p(p + q) = p$. Since all of the gametes of the *aa* parents and half of the gametes of the *Aa* parents will bear the *a* allele, the frequency of *a* in the gene pool of the next generation is $q^2 + pq = q(q + p) = q$. Thus, the gene frequencies do not change from one generation to the next.

21.8 The *S-s* antigen system in humans is controlled by two codominant alleles, *S* and *s*. In a group of 3,146 individuals the following genotypic frequencies were found: 188 *SS*, 717 *Ss* and 2,241 *ss*.

a. Calculate the frequency of the *S* and *s* alleles.

b. Determine whether the genotypic frequencies conform to the Hardy-Weinberg equilibrium by using the chi-square test.

Answer: a. Let p equal the frequency of *S* and q equal the frequency of *s*. Since homozygotes have two identical alleles and heterozygotes have one recessive and one dominant allele, one has:

$$p = \frac{2(188)[SS] + 717[Ss]}{2(3{,}146)} = \frac{1093}{6292} = 0.1737$$

$$q = \frac{717[Ss] + 2(2{,}241)[ss]}{2(3{,}146)} = \frac{5199}{6292} = 0.8263$$

b. Remember that in a χ^2 test, one uses the actual numbers of progeny observed and expected, and not the frequencies. With a hypothesis that the population is in Hardy-Weinberg equilibrium, one has:

Class	Observed (o)	Expected Frequency	Expected (e)	d ($o - e$)	d^2/E
SS	188	$p^2 = 0.0302$	95	93	91.2
Ss	717	$2pq = 0.287$	903	–186	38.3
ss	2,241	$q^2 = 0.683$	2,148	93	4.0
	3,146	1	3,146	0	133.5

There is only one degree of freedom because the three genotypic classes are completely specified by two allele frequencies, namely, p and q. (df. = number of phenotypes – number of alleles = 3 – 2 = 1.) The χ^2 value of 133.5, for one degree of freedom, gives $P < 0.0001$. Therefore, the distribution of genotypes differs significantly from that expected if the population were in Hardy-Weinberg equilibrium.

21.9 Refer to Problem 21.8. A third allele is sometimes found at the *S* locus. This allele S^u is recessive to both the *S* and the *s* alleles and can only be detected in the homozygous state. If the frequencies of the alleles *S*, *s* and S^u are p, q, and r, respectively, what would be the expected frequencies of the phenotypes *S*–, *Ss*, *s*–, and S^uS^u?

Answer: The frequencies are:

$f(S\text{–}) = f(SS) + f(SS^u) = p^2 + 2pr$

$f(Ss) = 2pq$

$$f(s-) = f(ss) + f(sS^u) = q^2 + 2qr$$
$$f(S^uS^u) = r^2$$

21.10 In a large interbreeding human population 60 percent of individuals belong to blood group O (genotype *i/i*). Assuming negligible mutation and no selective advantage of one blood type over another, what percentage of the grandchildren of the present population will be type O?

Answer: The conditions described in the problem indicate that the population is in Hardy-Weinberg equilibrium. Under equilibrium conditions, neither the allele, nor the zygotic frequencies change from one generation to the next. Therefore, in two generations there should still be 60 percent type O individuals.

21.11 A selectively neutral, recessive character appears in 0.40 of the males and in 0.16 of the females in a randomly interbreeding population. What is the gene's frequency? How many females are heterozygous for it? How many males are heterozygous for it?

Answer: There are several possible explanations for the difference in the frequency of the trait in males and females. Two are sex-linkage and autosomal linkage with sex-influenced expression. Sex-linkage can be readily examined. If the population is in Hardy-Weinberg equilibrium (which this one is), the frequency of the recessive allele causing the trait is q, and the gene is X-linked, then the frequency in XY males would be q, while the frequency in XX females would be q^2. Since the frequency in males is 0.4 and the frequency in females is $(0.4)^2 = 0.16$, the data fit a model of sex-linkage with $q = 0.4$. The frequency of heterozygous XX (female) individuals is $2pq = 2(0.6)(0.4) = 0.48$. Since the trait appears to be sex-linked, no heterozygous males exist.

21.12 Suppose you found two distinguishable types of individuals in wild populations of some organism in the following frequencies:

	TYPE 1	TYPE 2
Females	99%	1%
Males	90%	10%

The difference is known to be inherited. What is its genetic basis?

Answer: As in problem 21.11, there are several explanations for differences in the frequency of a trait between males and females. Test the possibility of sex-linkage first. Suppose the trait is sex-linked and recessive and the population is in equilibrium. Let q equal the frequency of the recessive allele, and p equal the frequency of the dominant allele. If the gene is X-linked, one would expect XY males to express the recessive trait at a frequency of q, and XX females to express the recessive trait at a frequency of q^2. By inspecting the data given, one can see that the frequency of type 2 individuals in females is 0.01, which is the square of the frequency of type 2 individuals in males: $(0.10)^2 = 0.01$. Thus, this trait appears to be controlled by a sex-linked pair of alleles, occurring with allele frequencies of $q = 0.1$ recessive and $p = 0.9$ dominant.

21.13 Red-green color blindness is due to a sex-linked recessive gene. About 64 women out of 10,000 are color-blind. What proportion of men would be expected to show the trait if mating is random?

Answer: Let q equal the frequency of the recessive allele, and p equal the frequency of the dominant allele. One expects homozygotes showing the trait to appear at a frequency of q^2 in a population at equilibrium. If $q^2 = 64/10,000 = 0.0064$, $q = 0.08$. Thus, one would expect 8 percent of XY, male individuals to show the trait.

21.14 About 8 percent of the men in a population are red-green color-blind (owing to a sex-linked recessive gene). Answer the following questions, assuming random mating in the population, with respect to color blindness.

a. What percentage of women would be expected to be color-blind?
b. What percentage of women would be expected to be heterozygous?
c. What percentage of men would be expected to have normal vision two generations later?

Answer:

a. Let q equal the frequency of the recessive allele, and p equal the frequency of the dominant allele. The frequency of males with the trait will be q, and the frequency of females will be q^2. Here, $q = 0.08$, so $q^2 = 0.0064$, or 0.64 percent.
b. Since $q = 0.08$, $p = 0.92$. The frequency of heterozygotes is $2pq = 0.1472$, or 14.72 percent. Only women can be heterozygotes, so the frequency of heterozygous women is 14.72 percent.
c. If the population is in Hardy-Weinberg equilibrium, the frequencies of alleles, and the frequency of zygotic phenotypes will not change in two generations. Since $p = 0.92$, 92 percent of the (XY) males will have normal vision.

21.15 List some of the basic differences in the classical, balance, and neutral-mutation models of genetic variation.

Answer: In the classical model, natural populations possess little variation. Within each population, a single allele is strongly favored by natural selection because it "functions the best." This is the wild-type allele, and most individuals are homozygous for this allele. Occasionally, a mutation arises. Usually, this mutation is deleterious, and strong selection occurs against it. In the rare case that a new mutation is advantageous for survival or reproduction, the new allele will increase in frequency and eventually, become the new wild-type.

In the balance model, there is significant genetic variation within a population. Many alleles exist at each locus, and appear in the population in intermediate frequencies. Members of a population are heterozygous at numerous loci. Natural selection <u>actively</u> maintains genetic variation within a population by balancing selection to prevent any single allele from reaching a high frequency.

In a neutral mutation model, there is also significant genetic variation within a population. As in the balance selection model, many alleles exist in intermediate frequencies and members of the population are heterozygous at numerous loci. In the neutral mutation model, much of this genetic variation is explained by recurrent mutation and random changes in allele frequency, and not necessarily by natural selection. Hence, in this model, variation is neutral with regard to selection.

21.16 Two alleles of a locus, *A* and *a*, can be interconverted by mutation:

$$A \underset{v}{\overset{u}{\rightleftarrows}} a$$

u is a mutation rate of 6.0 x 10^{-7}, and *v* is a mutation rate of 6.0 x 10^{-8}. What will be the frequencies of *A* and *a* at mutational equilibrium, assuming no selective difference, no migration, and no random fluctuation caused by genetic drift?

Answer: Let *q* equal the frequency of *a*, and *p* equal the frequency of *A*, with $q + p = 1$. As discussed on text p. 717, when the population is at equilibrium, the frequency of *p* and *q* is given by:

$$q = \frac{u}{u+v} = \frac{6 \text{ x } 10^{-7}}{(6 \text{ x } 10^{-7}) + (6 \text{ x } 10^{-8})} = \frac{6 \text{ x } 10^{-7}}{(6 \text{ x } 10^{-7}) + (0.6 \text{ x } 10^{-7})} = \frac{6}{6.6} = 0.91$$

$$p = 1 - q = 1 - 0.91 = 0.09$$

Thus, the frequencies are (q^2 =) 0.0081 *AA*, 0.1638 *Aa*, and 0.8281 *aa*.

21.17 a. Calculate the effective population size (N_e) for a breeding population of 50 adult males and 50 adult females.
b. Calculate the effective population size (N_e) for a breeding population of 60 adult males and 40 adult females.
c. Calculate the effective population size (N_e) for a breeding population of 10 adult males and 90 adult females.
d. Calculate the effective population size (N_e) for a breeding population of 2 adult males and 98 adult females.

Answer: As discussed on text p. 720, the effective breeding size of a population is given by the equation:

$$N_e = \frac{4 \text{ x } N_f \text{ x } N_m}{N_f + N_m}$$

where N_f equals the number of breeding females and N_m equals the number of breeding males. Apply this equation to each of the situations described as follows:

a. (4 x 50 x 50)/100 = 100
b. (4 x 60 x 40)/100 = 96
c. (4 x 10 x 90)/100 = 36
d. (4 x 2 x 98)/100 = 7.8

21.18 In a population of 40 adult males and 40 adult females, the frequency of allele *A* is 0.6 and the frequency of allele *a* is 0.4.

a. Calculate the 95 percent confidence limits of the allelic frequency for *A*.
b. Another population with the same allelic frequencies consists of only 4 adult males and 4 adult females. Calculate the 95 percent confidence limits of the allelic frequency for *A* in this population.
c. What are the 95 percent confidence limits of *A* if the population consists of 76 females and 4 males?

Answer: a.

$$N_e = \frac{4 \times N_f \times N_m}{N_f + N_m} = \frac{4 \times 40 \times 40}{40 + 40} = 80$$

$$s_p = \sqrt{\frac{pq}{2N_e}} = \sqrt{\frac{0.6 \times 0.4}{2 \times 80}} = 0.039$$

b. The 95% confidence limits are given by $p \pm 2s_p$, or $0.522 \leq p \leq 0.678$.

21.19 The land snail *Cepaea nemoralis* is native to Europe but has been accidentally introduced into North America at several localities. These introductions occurred when a few snails were inadvertently transported on plants, building supplies, soil, or other cargo. The snails subsequently multiplied and established large, viable populations in North America.

Assume that today the average size of *Cepaea* populations found in North America is equal to the average size of *Cepaea* populations in Europe. What predictions can you make about the amounts of genetic variation present in European and North American populations of *Cepaea*? Explain your reasoning.

Answer: Since the gene pool in the present, large North American population is derived from a small number of individuals, a founder effect is likely to have occurred. One would predict that genetic drift (random change in allelic frequency due to chance) will influence the North American populations to a greater degree than the European populations. One would expect to see less variation within and greater genetic differentiation among the North American populations.

21.20 A population of 80 adult squirrels resides on campus, and the frequency of the Est^1 allele among these squirrels is 0.70. Another population of squirrels is found in a nearby woods, and there, the frequency of the Est^1 allele is 0.5. During a severe winter, 20 of the squirrels from the woods population migrate to campus in search of food and join the campus population. What will be the allelic frequency of Est^1 in the campus population after migration?

Answer: Let p_I equal the frequency of *A* in population I, and p_{II} equal the frequency of *A* in population II. If individuals in population I migrate to population II and make up proportion *m* of population II', the new frequency of *A* in population II' (p'_{II}) is given by:

$$p'_{II} = mp_I + (1 - m)p_{II}$$

Here, $p'_{II} = [20/(20 + 80)](0.50) + \{1 - [20/(20 + 80)]\}(0.70) = 0.66$

21.21 Upon sampling three populations and determining genotypes, you find the following three genotype distributions. What would each of these distributions imply with regard to selective advantages of population structure?

POPULATION	*AA*	*Aa*	*aa*
1	0.04	0.32	0.64
2	0.12	0.87	0.01
3	0.45	0.10	0.45

Answer: For each population, determine the frequency of the *A* and *a* alleles, and compare how the population structure compares to that expected if no selection was occurring. To do this calculation, suppose there were 100 individuals in each population containing 200 alleles.

In population 1, there would be (0.04)(100)(2) + (0.32)(100)(1) = 40 *A* alleles, and (0.64)(100)(2) + (0.32)(100)(1) = 160 *a* alleles. Thus, $f(A) = p = 40/200 = 0.20$, and $f(a) = q = 160/200 = 0.80$. If the population was in Hardy-Weinberg equilibrium, one would expect $p^2 = 0.04$ *AA*, $2pq = 0.32$ *Aa*, and 0.64 *aa* individuals, as is observed. *If selection is acting at all,* it may be acting to maintain the observed frequencies of the *A* and *a* alleles in equilibrium.

In population 2, there would be (0.12)(100)(2) + (0.87)(100)(1) = 111 *A* alleles, and (0.01)(100)(2) + (0.87)(100)(1) = 89 *a* alleles. Here, $p = 0.555$ and $q = 0.455$. If the population was in Hardy-Weinberg equilibrium, one would expect $p^2 = 0.308$ *AA*, $2pq = 0.505$ *Aa*, and 0.207 *aa* individuals. The population is not in equilibrium, as there are far more heterozygotes, and far fewer homozygotes that would be expected. This suggests that the heterozygote is being selected for in this population.

In population 3, there would be (0.45)(100)(2) + (0.10)(100)(1) = 100 *A* alleles, and (0.10)(100)(1) + (0.45)(100)(2) = 100 *a* alleles. Here, $p = q = 0.5$. If the population was in Hardy-Weinberg equilibrium, one would expect $f(AA) = f(aa) = p^2 = q^2 = 0.25$ and $f(Aa) = 2pq = 0.50$. The population is not in equilibrium, as there are far more homozygotes (of either type) and far fewer heterozygotes than expected. This suggests that the heterozygote is being selected against in this population.

21.22 The frequency of two adaptively neutral alleles in a large population is 70 percent *A*: 30 percent *a*. The population is wiped out by an epidemic, leaving only four individuals, who produce many offspring. What is the probability that the population several years later will be 100 percent *AA*? (Assume no mutations.)

Answer: There are a variety of situations in which all individuals would become *AA*. One is that the four "founding" individuals are all *AA*. Since the probability of a single individual being *AA* is $(0.7)^2 = 0.49$, the probability of the four founding individuals being *AA* is $(0.49)^4 = 0.0576$, about 1/17. Even if all four founding individuals are not *AA*, there will be some chance that the subsequent population may become all *AA*. For example, there is a low, but distinct probability that only *AA* offspring may be had by *Aa* x *AA* parents. Given that "many" offspring are produced, the likelihood of this will be less than 1/17 [e.g., if ten offspring are produced, the likelihood of all being *AA* in such a cross would be $(1/2)^{10} = 1/1{,}024$]. Since the alleles are adaptively neutral, it would appear that 1/17 is an upper bound for the likelihood of a population being 100 percent *AA*.

21.23 A completely recessive gene, owing to changed environmental circumstances, becomes lethal in a certain population. It was previously neutral, and its frequency was 0.5.

a. What was the genotype distribution when the recessive genotype was not selected against?
b. What will be the allelic frequency after one generation in the altered environment?
c. What will be the allelic frequency after two generations?

Answer: a. Let $p = f(A)$, and $q = f(a)$. Initially, $p = q = 0.5$, so $p^2 = f(AA) = q^2 = f(aa) = (0.5)^2 = 0.25$. $f(Aa) = 2pq = 0.50$.

b. If *aa* individuals are now lethal, they will not contribute to the gene pool in the next generation. Only *AA* and *Aa* individuals will contribute to the gene pool in the next generation. The allele frequency will be

$$f(A) = \frac{(0.25 \text{ x } 2) + (0.50 \text{ x } 1)}{(0.25 \text{ x } 2) + (0.50 \text{ x } 2)} = \frac{1.0}{1.5} = 0.66$$

$$f(a) = \frac{(0.50 \text{ x } 1)}{(0.25 \text{ x } 2) + (0.50 \text{ x } 2)} = \frac{0.5}{1.5} = 0.33$$

c. Given the result of b, one will have $q^2 = (0.66)^2 = 0.436$ *AA*, $2pq = 2(0.66)(0.33) = 0.444$ *Aa*, and $p^2 = (0.33)^2 = 0.109$ *aa* progeny genotypes in the progeny. As in b, only *AA* and *Aa* individuals will contribute to the gene pool in the next generation. The allele frequency will be

$$f(A) = \frac{(0.436 \text{ x } 2) + (0.444 \text{ x } 1)}{(0.436 \text{ x } 2) + (0.444 \text{ x } 2)} = \frac{1.316}{1.76} = 0.75$$

$$f(a) = \frac{(0.444 \text{ x } 1)}{(0.436 \text{ x } 2) + (0.444 \text{ x } 2)} = \frac{0.444}{1.76} = 0.25$$

(Can you show that, under the conditions of this problem, the frequency of *A* after *n* generations will be $n/(n + 1)$ and the frequency of *a* will be $1/(n + 1)$?)

21.24 Human individuals homozygous for a certain recessive autosomal gene die before reaching reproductive age. In spite of this removal of all affected individuals, there is no indication that homozygotes occur less frequently in succeeding generations. To what might you attribute the constant rate of appearance of recessives?

Answer: There are a number of reasons that recessives may appear in a constant frequency. First, (and probably, most importantly) new mutations of *A* to *a* could occur at a low, but constant rate. Second, there could be a selective advantage to heterozygotes (overdominance). Third, there could be non-random mating within the population (e.g., positive assortative mating, inbreeding). Fourth, there could be a low, but steady frequency of migration of heterozygotes into the population. Each of these possibilities violates the assumptions behind a population in Hardy-Weinberg equilibrium, and would allow for maintenance of the recessive allele that was deleterious as a homozygote. (See problem 21.23 to consider the consequences of such an allele if the assumptions were not violated.)

21.25 A completely recessive gene (Q^1) has a frequency of 0.7 in a large population, and the Q^1Q^1 homozygote has a relative fitness of 0.6.

a. What will be the frequency of Q^1 after one generation of selection?

b. If there is no dominance at this locus (the fitness of the heterozygote is intermediate to the fitnesses of the homozygotes), what will the allelic frequency be after one generation of selection?

c. If Q^1 is dominant, what will the allelic frequency be after one generation of selection?

Answer: See Table 21.12. For parts (a) and (b), $p = 0.3$, $q = 0.7$,, and $s = 1 - W = 0.4$.

a. If there is selection against the recessive homozygotes, after one generation, one has:

$$\Delta q = \frac{-spq^2}{1 - sq^2} = \frac{-(0.4)(0.3)(0.7)^2}{1 - (0.4)(0.7)^2} = -0.073$$

$$q^1 = 0.7 - 0.073 = 0.627$$

b. If there is selection with no dominance, so that the fitness of the heterozygote is intermediate between the two homozygotes, one has:

$$\Delta q = \frac{-spq/2}{1 - sq} = \frac{-(0.4)(0.3)(0.7)/2}{1 - (0.4)(0.7)} = -0.0583$$

$$q^1 = 0.7 - 0.0583 = 0.642$$

c. To select against Q^1 as a dominant allele, let p equal the frequency of the dominant allele being selected against: $p = 0.7$, $q = 0.3$ and $s = 0.4$.

$$\Delta \mathrm{p} = \frac{-spq^2}{1 - s + sq^2} = \frac{-(0.4)(0.7)(0.3)^3}{1 - s + sq^2} = \frac{-(0.4)(0.7)(0.3)^2}{1 - 0.4 + (0.4)(0.3)^2} = -0.04$$

$$p^1 = 0.70 - 0.04 = 0.66$$

21.26 As discussed earlier in this chapter, the gene for sickle-cell anemia exhibits overdominance. An individual who is an *Hb-A/Hb-S* heterozygote has increased resistance to malaria and therefore has greater fitness than the *Hb-A/Hb-A* homozygote, who is susceptible to malaria, and the *Hb-S/Hb-S* homozygote, who has sickle-cell anemia. Suppose that the fitness values of the genotypes in Africa are as presented below:

Hb-A/Hb-A = 0.88
Hb-A/Hb-S = 1.00
Hb-S/Hb-S = 0.14

Give the expected equilibrium frequencies of the sickle-cell gene (*Hb-S*).

Answer: From the text discussion on pp. 734-5, one has that at equilibrium, $p = f(Hb\text{-}A) = t/(t + s)$, and $q = f(Hb\text{-}S) = s/(s + t)$, where t equals the selection coefficient of *Hb-S/Hb-S*, and s equals the selection coefficient of *Hb-A/Hb-A*. Since *fitness = 1 – selection coefficient*, one has $f(Hb\text{-}S) = 0.12/(0.12 + 0.86) = 0.122$.

21.27 Achondroplasia, a type of dwarfism in humans, is caused by an autosomal dominant gene. The mutation rate for achondroplasia is about 5×10^{-5} and the fitness of achondroplastic dwarfs has been estimated to be about 0.2, compared with unaffected individuals. What is the equilibrium frequency of the achondroplasia gene based on this mutation rate and fitness value?

Answer: For a dominant allele, the frequency at equilibrium $q = u/s$, where u equals the mutation rate, and s is the selection coefficient (= 1 – fitness). Here, $q = 5 \times 10^{-5}/0.8 = 6.25 \times 10^{-5}$.

21.28 The frequencies of the L^M and L^N blood group alleles are the same in each of the populations I, II, and III, but the genotypes' frequencies are not the same, as shown below. Which of the populations is most likely to show each of the following characteristics: random mating, inbreeding, genetic drift. Explain your answers.

	M–M	*M–N*	*N–N*
I	0.50	0.40	0.10
II	0.49	0.42	0.09
III	0.45	0.50	0.05

Answer: First, calculate the frequencies of each allele. In population I, one has $[(0.5 \times 2) + (0.4 \times 1)]/2 = 0.7$ *M*, and 0.3 *N*. This is the case for all three populations. A population in Hardy-Weinberg equilibrium would have $q^2 = (0.7)^2 = 0.49$ *M–M*, $2pq = 2(0.3)(0.7) = 0.42$ *M–N*, and $q^2 = (0.3)^2 = 0.09$ *N–N*. As population II shows these features, it must be in Hardy-Weinberg equilibrium, and would be expected to exhibit random mating. Inbreeding results in an increase in the frequency of homozygotes, and would be associated with population I (but not III). Genetic drift is random change in gene frequency due to chance, and can be explained by random effects, a small effective population size, founder or bottleneck effects, or sampling error. Genetic drift could be associated with either population I (e.g., given how close the frequencies are to those of an equilibrium population, sampling error in a population in Hardy-Weinberg equilibrium could explain the results) or population III.

21.29 DNA was collected from 100 people randomly sampled from a given human population and was digested with the restriction enzyme *Bam*HI, the fragments were separated by electrophoresis, and then transferred to a membrane filter using the Southern blot technique. The blots were probed with a particular cloned sequence. Three different patterns of hybridization were seen on the blots. Some DNA samples (56 of them) showed a single band of 6.3 kb, others (6) showed a single band at 4.1 kb, and yet others (38) showed both the 6.3 and the 4.1 kb bands.

a. Interpret these results in terms of *Bam*HI sites.
b. What are the frequencies of the restriction site alleles?
c. Does this population appear to be in Hardy-Weinberg equilibrium for the relevant restriction site(s)?

Answer: a. One explanation is that a restriction fragment length polymorphism exists in the population. On some chromosomes, an RFLP having a size of 4.1 kb is found, while on others, the 6.3 kb fragment is found. Individuals having just one size band are homozygotes, while individuals have two different bands are heterozygous. The difference in sizes of the fragments could result from a missing site in the 6.3 kb individuals, or the insertion of a 2.2 kb piece of DNA between two sites that are normally 4.1 kb apart.

b. Homozygotes have two identical alleles, while heterozygotes have two different alleles. There are (56 x 2) + 38 = 150, 6.3 kb alleles, and (6 x 2) + 38 = 50, 4.1 kb alleles. Let q equal the frequency of 6.3 kb alleles, and p equal the frequency of 4.1 kb alleles. Then $q = 0.75$ and $q = 0.25$.

c. For a population in Hardy-Weinberg equilibrium, one would expect $q^2 = 0.5625$ 6.3-kb homozygotes (compared to the 0.56 seen), $2pq = 0.375$ heterozygotes (compared to the 0.38 seen) and $p^2 = 0.0625$ 4.1-kb homozygotes (compared to the 0.06 seen). The population appears to be in Hardy-Weinberg equilibrium.

21.30 DNA was isolated from 10 nine-banded armadillos and cut with the restriction enzyme *Hin*dIII. *Hin*dIII recognizes the six-base sequence: `5'-AAGCTT-3'` / `3'-TTCGAA-5'`. The DNA fragments that resulted from the restriction reaction were separated with agarose electrophoresis and transferred to nitrocellulose using Southern blotting. A labeled probe for the β-hemoglobin gene was added, which resulted in the following set of restriction patterns. Note: +/+ indicates that the restriction site was present on both chromosomes of the individual, +/– indicates that the restriction site was present on one chromosome and absent on one chromosome of the individual, and –/– indicates that the restriction site was absent on both chromosomes of the individual:

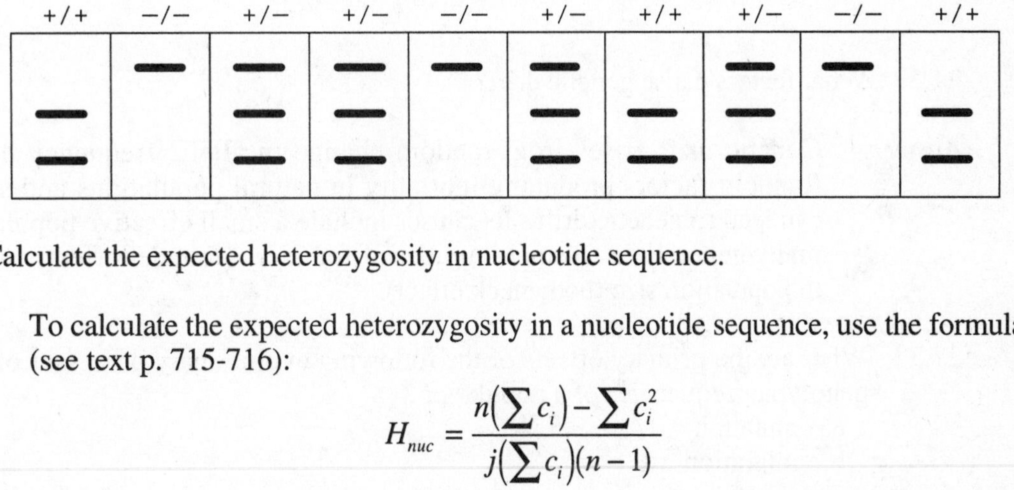

+/+ –/– +/– +/– –/– +/– +/+ +/– –/– +/+

Calculate the expected heterozygosity in nucleotide sequence.

Answer: To calculate the expected heterozygosity in a nucleotide sequence, use the formula (see text p. 715-716):

$$H_{nuc} = \frac{n\left(\sum c_i\right) - \sum c_i^2}{j\left(\sum c_i\right)(n-1)}$$

Here, n equals the number of homologous DNA molecules examined. In this example, ten armadillos were examined, each having two homologous chromosomes, so $n = 20$. The quantity j equals the number of nucleotides in the restriction site. In this example, *Hin*dIII recognizes a six-base-pair sequence, so $j = 6$. For each restriction site, c_i represents the number of molecules in the sample that were cleaved at that restriction site. Here, a single restriction site was cleaved at 10 of the 20 sites, so $\Sigma c_i = 10$. (Individuals with two bands had sites on both chromosomes cleaved, individuals with three bands had a site on one, but not the other chromosome cleaved, individuals with one band had no sites cleaved.) Thus, we obtain

$H_{nuc} = [20(10) - 10^2]/[6(10)(19)] = 100/1{,}140 = 0.088$

21.31 Fifty tiger salamanders from one pond in west Texas were examined for genetic variation by using the technique of protein electrophoresis. The genotype of each salamander was determined for five loci (AmPep, ADH, PGM, MDH, and LDH-1). No variation was found at AmPep, ADH, and LDH-1; in other words, all individuals were homozygous for the same allele at these loci. The following numbers of genotypes were observed at the MDH and PGM loci.

MDH Genotypes	Number of Individuals
AA	11
AB	35
BB	4

PGM Genotypes	Number of Individuals
DD	35
DE	10
EE	5

Calculate the proportion of polymorphic loci and the heterozygosity for this population.

Answer: Since five loci were examined, and only two have more than one allele, (2/5)(100%) = 40% of the loci are polymorphic. Heterozygosity is calculated by averaging the frequency of heterozygotes for each locus. The frequency of heterozygotes for the AmPep, ADH, and LDH-1 loci is zero. At MDH, 35 out of 50 individuals were heterozygous (0.70). At PGM, 10 out of 50 individuals were heterozygous (0.20). Thus, the average heterozygosity is

$$\frac{0 + 0 + 0 + 0.7 + 0.2}{5} = 0.18$$

21.32 What factors cause genetic drift?

Answer: Genetic drift arises from random change in allelic frequency due to chance. Random factors producing mortality in natural populations and sampling error can lead to genetic drift. Its causes include a small effective population size over many generations, a small number of founders (founder effect), and a reduction in population size (bottleneck effect).

21.33 What are the primary effects of the following evolutionary processes on the gene and genotypic frequencies of a population?

a. mutation
b. migration
c. genetic drift
d. inbreeding

Answer:

a. Mutation will lead to change in gene frequencies within a population if no other forces are acting. It will introduce genetic variation. If the effective population size is small, mutation may lead to genetic differentiation among populations.
b. Migration will increase the population size, and has the potential to disrupt a Hardy-Weinberg equilibrium. It can increase genetic variation and may influence the evolution of allelic frequencies within populations. Over many generations, migration will reduce divergence among populations, and equalize gene frequencies among populations.

c. Genetic drift produces changes in allelic frequencies within a population. It can reduce genetic variation and increase the homozygosity within a population. Over time, it leads to genetic change. When several populations are compared, genetic drift can lead to increased genetic differences among populations.
d. Inbreeding will increase the homozygosity within a population and decrease its genetic variation.

21.34 Explain how overdominance leads to an increased frequency of sickle-cell anemia in areas where malaria is widespread.

Answer: Overdominance results when a heterozygote genotype has higher fitness than either of the homozygotes. The two alleles of the heterozygote are maintained in a population because both are favored in the heterozygote genotype. In the case of the sickle cell gene, heterozygotes for *Hb-A*/*Hb-S* are at a selective advantage because the hemoglobin mixture in these individuals provides an unfavorable environment for the growth of malarial parasites. Thus, heterozygotes have higher fitness than *Hb-A*/*Hb-A* homozygotes who are susceptible to malaria. Heterozygotes also have higher fitness than *Hb-S*/*Hb-S* homozygotes who suffer from sickle cell anemia. The favoring of the sickle-cell allele *Hb-S* in the heterozygote results in its relatively high frequency in areas with malaria.

21.35 Suppose we examine the rates of nucleotide substitution in two 300-nucleotide sequences of DNA isolated from humans. In the first sequence (sequence A), we find a nucleotide substitution rate of 4.88×10^{-9} substitutions per site per year. The substitution rate is the same for synonymous and nonsynonymous substitutions. In the second sequence (sequence B), we find a synonymous substitution rate of 4.66×10^{-9} substitutions per site per year and a nonsynonymous substitution rate of 0.70×10^{-9} substitutions per site per year. Referring to Table 21.14 (p. 740), what might you conclude about the possible functions of sequence A and sequence B?

Answer: Nonsynonymous substitutions are those that code for different amino acids, while synonymous substitutions code for the same amino acid. In sequence A, the finding of the same, relatively high rate for synonymous and nonsynonymous substitutions suggests that this sequence may not code for a functional protein, and may derive from a pseudogene. The rate seen for synonymous and nonsynonymous substitutions in sequence A is similar to that seen for pseudogenes shown in Table 21.14 on text p. 740.

The different rates seen for synonymous and nonsynonymous substitutions in sequence B (and in particular, the low rate of nonsynonymous substitutions) suggest that this sequence encodes a protein. A low nonsynonymous substitution rate would be expected if nonsynonymous changes resulted in changes in protein function that were detrimental to fitness. Since most nonsynonymous substitutions would not "improve" a protein's function, such mutations would be eliminated by natural selection. Synonymous substitutions would be tolerated and seen at a higher frequency, as they would not alter protein function.

21.36 What are some of the characteristics of mitochondrial DNA evolution in animals?

Answer: Mitochondrial DNA evolution differs from that of nuclear DNA. Nucleotide sequences in animal mtDNA evolve at a rate 5 to 10 times faster than animal nuclear DNA, while those in plant mtDNA evolve at a slower rate than plant nuclear DNA. The increased rate of change in animal mtDNA may result from a higher mutation rate and/or from relaxed selection pressure.

21.37 What is concerted evolution?

Answer: Concerted evolution is maintenance of sequence uniformity in multiple copies of a gene. It leads to similar sequences in multiple copes of the same gene within a species but different sequences among different species. It can be found in both coding and non-coding sequences (e.g., ribosomal RNA genes). It is intriguing that the process(es) used in concerted evolution to continually enforce uniformity among multiple copies of a sequence must also allow for rapid differentiation among species.

21.38 What are some of the advantages of using DNA sequences to infer evolutionary relationships?

Answer: Using DNA sequence information to infer evolutionary relationships and constructing evolutionary trees has several advantages. DNA sequences provide highly accurate and reliable information, allow direct comparison of the genetic differences among organisms, are easily quantified, and can be used in all organisms.

CHAPTER 22

QUANTITATIVE GENETICS

I. CHAPTER OUTLINE

THE NATURE OF CONTINUOUS TRAITS
- Why Some Traits Have Continuous Phenotypes
- Questions Studied in Quantitative Genetics

STATISTICAL TOOLS
- Samples and Populations
- Distributions
- The Mean
- The Variance and the Standard Deviation
- Correlation
- Regression
- Analysis of Variance

POLYGENIC INHERITANCE
- Inheritance of Ear Length in Corn
- Polygene Hypothesis for Quantitative Inheritance
- Determining the Number of Polygenes for a Quantitative Trait

HERITABILITY
- Components of the Phenotypic Variance
- Broad-Sense and Narrow-Sense Heritability
- Understanding Heritability
- How Heritability is Calculated

RESPONSE TO SELECTION
- Estimating the Response to Selection
- Genetic Correlations

II. REVIEW OF KEY TERMS, SYMBOLS AND CONCEPTS

Without consulting the text and in your own words, write a brief definition of each term in the groups below. Then, either using a short phrase or a simple diagram, identify the relationship(s) between specific pairs of terms within a set. Finally, consult the text (and perhaps a friend who has also done the exercise) to check your answers.

1	2	3	4
quantitative genetics discontinuous trait continuous trait polygenic trait norm of reaction multifactorial trait familial trait penetrance expressivity pleiotropy epistasis	population sample frequency distribution normal distribution binomial distribution binomial expansion	mean variance standard deviation analysis of variance (ANOVA) covariance correlation *vs.* cause and effect correlation *vs.* identity regression regression line, regression co-efficient	polygene contributing alleles polygene or multiple gene hypothesis for quantitative inheritance

5	6	7	8
heritability broad-sense *vs.* narrow-sense heritability	phenotypic variance (V_P) genetic variance (V_G) environmental variance (V_E) genetic-environmental interaction(V_{GxE}) genetic-environmental covariation (COV_{GxE})	genetic variance (V_G) additive genetic variance (V_A) dominance variance (V_D) interaction variance (V_I)	environmental variance (V_E) general environmental effects (V_{Eg}) special environmental effects (V_{Es}) maternal effects (V_{Em})

9	10
evolution artificial selection selection response selection differential narrow-sense heritability	phenotypic correlation genetic correlation pleiotropy positive *vs.* negative correlation

III. THINKING ANALYTICALLY

Quantitative genetics deals with important, but complex issues, and some of the concepts used in the analysis of quantitative traits are subtle. Throughout the chapter, clear and precise thinking is required. In order to sort out the basis for a quantitative trait, it is important to define the contribution of several different factors. These factors can include the environment as well as multiple genes that themselves may or may not interact. Central to the analysis of a quantitative trait is identifying the degree to which variation in a phenotype can be associated with one or more of these factors. Consequently, understanding how variation is measured and used is of the utmost importance in understanding quantitative genetics.

Approach this material by first gaining a solid understanding of the terms used and the important concepts they convey. Geneticists have developed a substantial conceptual framework in which to consider quantitative traits. Some concepts are easy to misunderstand, and to understand what something *is*, it is often important to define what it *is not*. For example, the idea of heritability has important qualifications and limitations. Heritability is a measure of the *proportion of phenotypic variance* that results from genetic differences. It is *not* a measure of the extent to which a trait is genetic, or what proportion of an individual's phenotype is genetic. It is also not fixed for a particular trait, or a measurement of genetic differences between populations. The concepts underlying quantitative genetics have significant utility, but only if they are correctly and clearly applied.

After you develop a strong conceptual foundation, concentrate on understanding the statistical methods that are used to analyze data on quantitative traits. Your understanding should consist of more than just knowledge of how to crunch numbers by plugging them into an equation. You need to get a feel for what a statistic tells you about a data set.

Finally, relate how measurements are made and analyzed to the concepts developed to explain the inheritance of quantitative traits. For example, relate a measurement of heritability to a selection response. Then, explain how you would obtain a particular objective through selection. By doing this, you will achieve a better understanding of the power and utility of quantitative genetics.

IV. QUESTIONS FOR PRACTICE

A. Multiple Choice Questions

1. Which of the following could be used to describe coat colors in mice?
   a. continuous trait
   b. discontinuous trait
   c. polygenic trait
   d. quantitative trait
   e. both b and c

2. What can generally be said about quantitative traits such as crop yield, rate of weight gain, human birth weight, blood pressure and the number of eggs laid by *Drosophila*?
   a. They are intractable to molecular genetics.
   b. They are multifactorial.
   c. They are discontinuous.
   d. They are not heritable because they are polygenic.

3. Which statement below best describes the aims of a quantitative genetic analysis of a trait.
   a. To determine whether genes or the environment control a trait.
   b. To determine the best breeding strategy to retain a trait.
   c. To determine how much of the phenotypic variation associated with a trait in a population is due to genetic variation and how much is due to environmental variation.
   d. To determine whether a trait is controlled by nature versus nurture.

4. The description of a population in terms of the number of individuals that display varying degrees of expression of a character or range of phenotypes is a
   a. polygraph.
   b. polynomial.
   c. normal distribution.
   d. frequency distribution.

5. The general expression for the binomial expansion is
   a. $(p + q)^n$.
   b. $(a^2 + b^2)$.
   c. $(a + b)^2$.
   d. $a^2 + 2ab + b^2$.

6. Which two of the following accurately describe the term *variance?*
   a. It is a reflection of the accuracy of an estimated measurement.
   b. It is a measure of how much a set of individual measurements is spread out around the mean.
   c. It is the average value of a set of measurements.
   d. It is equal to $(\Sigma x_i/n)$.
   e. It is equal to $\dfrac{\sum(x_i - \bar{x})^2}{n-1}$.

7. What is heritability?
   a. The proportion of a population's phenotype that is attributable to genetic factors.
   b. The proportion of a population's phenotypic variation that is attributable to genetic factors.
   c. The degree to which family members resemble one another.
   d. The degree to which a continuous trait is controlled by genetic factors.

8. The proportion of the phenotypic variance that consists of genetic variance, additive or otherwise, is called
   a. heritability.
   b. broad-sense heritability.
   c. narrow-sense heritability.
   d. phenotypic variance derived from genetic-environmental interactions.

9. The narrow-sense heritability of a trait is determined to be very close to 1.0. Which of the following inferences can be made?
   a. The trait will be difficult to select for.
   b. The trait can be readily selected for.
   c. Most of the phenotypic variance results from additive genetic variance.
   d. Most of the phenotypic variance results from environmental variance.
   e. Both a and c.
   f. Both b and c.

10. In a population of chickens raised under controlled conditions, body weight is negatively correlated with egg production, but positively correlated with egg weight. Because of market conditions, a farmer is interested in producing lots of small eggs. What selection strategy might be beneficial?
   a. select for smaller hens.
   b. select for larger hens.

Answers: 1e; 2b; 3c; 4d; 5a; 6b & e; 7b; 8b; 9f; 10a

B. Thought Questions

1. Assume that mature fruit weight in pumpkins is a quantitative trait. In the following experiment, environmental factors (weather, soil, etc.) are uniform. Two pumpkin varieties, both of which produce fruit with a mean weight of 20 lb., are crossed. The F_1 produces 20 lb. pumpkins. The F_2 plants, however, give the following results:

Mean fruit wt. (lb.)	5	12.5	20	27.5	35
Number of plants	19	82	119	79	21

Explain these results: postulate how many genes are involved and how much each contributes to fruit weight.

2. Distinguish between heritability, broad-sense heritability and narrow-sense heritability. In what ways can the latter two quantities be measured, and how can they be put to use by plant and animal breeders?
3. What statistics from a regression analysis are used to estimate narrow-sense heritability?
4. What sources contribute to the phenotypic variance associated with a quantitative trait?
5. Distinguish between a genetic correlation and a phenotypic correlation. In particular, address how a phenotypic correlation might exist when the trait is, or is not, influenced by a common set of genes.
6. As discussed in the text, even though a positive correlation exists between alcohol consumption and the number of Baptist ministers, there is thought to be no causal relationship between these two phenomena. Can you think of other correlations that are likely to be true, but which clearly demonstrate that correlations (whether positive or negative) do not imply a cause and effect, and that a correlation is not the same thing as an identity?
7. Defend each of the following statements:
    a. Broad-sense heritability does not indicate the extent to which a trait is genetic.
    b. Heritability does not indicate what proportion of an individual's phenotype is genetic.
    c. Heritability is not fixed for a trait.
    d. If heritability is high in two populations, and the populations differ markedly in a particular trait, one cannot assume that the populations are genetically different.
    e. Familial traits do not necessarily have high heritability.
8. If a population is phenotypically homogeneous for a particular trait, can you predict whether it has a high narrow-sense heritability? If not, what additional information would you need?
9. Why is calculating narrow-sense heritability values for some quantitative human traits especially difficult?
10. Distinguish between a selection differential and a selection response.

V. SOLUTIONS TO TEXT PROBLEMS

22.1 The following measurements of head width and wing length were made on a series of steamer-ducks:

SPECIMEN	HEAD WIDTH (cm)	WING LENGTH (cm)
1	2.75	30.3
2	3.20	36.2
3	2.86	31.4
4	3.24	35.7
5	3.16	33.4
6	3.32	34.8
7	2.52	27.2
8	4.16	52.7

a. Calculate the mean and the standard deviation of head width and of wing length for these eight birds.
b. Calculate the correlation coefficient for the relationship between head width and wing length in this series of ducks.
c. What conclusions can you make about the association between head width and wing length in steamer-ducks?

Answer: a. The mean is obtained by summing the individual values and dividing by the total number of values. The mean head width is 25.21/8 = 3.15 cm, and the mean wing length is 281.7/8 = 35.21 cm.

The standard deviation equals the square root of the variance (the square root of s^2). The variance is computed by squaring the sum of the difference between each measurement and the mean value, and dividing this sum by the number of measurements minus one. One has:

$$s_{headwidth} = \sqrt{s^2} = \sqrt{\frac{\sum(x_i - \bar{x})^2}{n-1}} = \sqrt{\frac{1.68}{7}} = \sqrt{0.24} = 0.49$$

$$s_{winglength} = \sqrt{s^2} = \sqrt{\frac{\sum(x_i - \bar{x})^2}{n-1}} = \sqrt{\frac{413.3}{7}} = \sqrt{59.04} = 7.68$$

b. The correlation coefficient, r, is calculated from the covariance, *cov*, of two quantities. Let head width be represented by x, and wing length be represented by y. r is defined as:

$$r = \frac{\text{cov}_{xy}}{s_x s_y} = \frac{\dfrac{\sum x_i y_i - \frac{1}{n}\left(\sum x_i \sum y_i\right)}{n-1}}{s_x s_y}.$$

The first factor $(\Sigma x_i y_i)$ is obtained by taking the sum of the products of the individual measurements of head width and the corresponding measurements for wing length. The next factor $(\Sigma x_i \Sigma y_i)/n$ is the product of the sums of the two sets of measurements divided by the number of pairs of measurements. The difference between these values is then divided by (n–1), and then by the products of the standard deviations of each measurement. One has:

$$r = \frac{\frac{913 - (1/8)(25.21 \times 281.7)}{7}}{0.49 \times 7.68} = \frac{3.61}{3.76} = 0.96$$

c. Head width and wing length show a strong positive correlation, nearly 1.0. This means that ducks with larger heads will almost always have longer wings, and ducks with smaller heads will almost always have smaller wings.

22.2 Answer the following questions:

a. In a family of 6 children, what is the probability that three will be girls and three will be boys?

b. In a family of five children, what is the probability that one will be a boy and four will be girls?

c. What is the probability that in a family of six children, all will be boys?

Answer: a. Three boys and three girls can be obtained in a variety of different orders, and each of these must be considered to determine the probability of obtaining this result. One could determine the number of different orders that are possible, and then use the sum rule and add together the probability of obtaining three boys and three girls in each possible order. It is easier to use the binomial expansion than to work out all of the ways in which this result can be obtained. The binomial expansion is given by

$$p(s,t) = \frac{n!}{s!t!} a^s b^t$$

Here, s equals the number of the first class (here, say the number of girls, so $s = 3$), t equals the number of the other class (here, the number of boys, so $t =$ 3), n equals the total number in both classes (here $n = 6$), a is the probability of the first event (*p(girl)* = 1/2) and b is the probability of the second event (*p(boy)* = 1/2). Substituting these values into the equation, one has:

$$p(3 \text{ girls, } 3 \text{ boys}) = [6!/(3!3!)] \times (1/2)^3 \times (1/2)^3 = 20/64 = 0.3125$$

b. Again, use the binomial expansion, with $n = 5$, $s = 4$, and $t = 1$. One has p(4 girls, 1 boy) = $[5!/(4!1!)] \times (1/2)^1 \times (1/2)^4 = 5/32 = 0.15625$.

c. One could again use the binomial expansion ($n = 6$, $s = 0$, $t = 6$, $a = b = 1/2$, note that $0! = 1$), but since there is only one way of obtaining six boys, it is straightforward to do this calculation without the binomial expansion. The chance of obtaining one boy is 1/2. The chance of obtaining six boys (use the product rule) is $(1/2)^6 = 1/64 = 0.015625$.

22.3 In flipping a coin, there is a 50 percent chance of obtaining heads and a 50 percent chance of obtaining tails on each flip. If you flip a coin 10 times, what is the probability of obtaining exactly 5 heads and 5 tails?

Answer: As in part a of Problem 22.2, there are a variety of different ways to obtain this result. Use the binomial expansion, with $n = 10$, $a = b = 1/2$, $s = t = 5$. $p =$ $[10!/(5! \times 5!)] \times (1/2)^5 \times (1/2)^5 = 252/1{,}024 = 0.246$

22.4 The F_1 generation from a cross of two pure-breeding parents that differ in a size character is usually no more variable than the parents. Explain.

Answer: The degree of phenotypic variability is related to the degree of genetic variability. Since each pure-breeding parent is homozygous for the genes (however many there are) controlling the size character, each parent is homogeneous in type. A cross of two pure-breeding strains will generate an F_1 heterozygous for those loci controlling the size trait. Since the F_1 consists of all heterozygotes, it is genetically as homogeneous as the parents. Therefore, it shows no greater variability than the parents.

22.5 If two pure-breeding strains, differing in a size trait, are crossed, is it possible for F_2 individuals to have phenotypes that are more extreme than either grandparent (i.e., be larger than the largest or smaller than the smallest in the parental generation)? Explain.

Answer: If the two grandparents of the F_2 individuals (i.e., the two parents in the parental generation) represent the two extremes for the size trait that is found in the population, the answer is no. However, if neither of the grandparental types is at an extreme end of the size range, and if each strain differs in the appropriate genes, then it is possible to obtain an F_2 that is more extreme than their grandparents. As an example, consider a quantitative trait that is specified by three allelic pairs: *A/a, B/b,* and *C/c*. Suppose each capital allele contributes a certain amount to the quantitative character. Then, individuals in the population with six capital alleles (*AA BB CC*) show one extreme, while individuals with six lowercase alleles (*aa bb cc*) show the other extreme. If the parental cross were *AA BB cc* x *aa bb CC*, the F_1 would be *Aa Bb Cc*, and some of the F_2 would be *AA BB CC* and *aa bb cc*. In this scenario, some of the F_2 would have more capital alleles than either grandparent, and so exhibit a more extreme phenotype. Similarly, some of the F_2 will have fewer capital alleles than either grandparent, and so exhibit a less extreme phenotype than their grandparents.

22.6 Two pairs of genes with two alleles each, *A/a* and *B/b* determine plant height additively in a population. The homozygote *AA BB* is 50 cm tall, the homozygote *aa bb* is 30 cm tall.

a. What is the F_1 height in a cross between the two homozygous stocks?
b. What genotypes in the F_2 will show a height of 40 cm after an $F_1 \times F_1$ cross?
c. What will be the F_2 frequency of the 40-cm plants?

Answer:

a. Since the cross is *AA BB* x *aa bb*, the F_1 genotype will be *Aa Bb*. Since capital alleles determine height additively, and individuals with four capital alleles have a height of 50 cm, while individuals with no capital alleles have a height of 30 cm, each capital allele appears to confer (50 – 30)/4 = 5 cm of height over the 30 cm base. *Aa Bb* individuals with two capital alleles should have an intermediate height of 40 cm.
b. Any individuals with two capital alleles will show a height of 40 cm. Thus, *Aa Bb, AA bb,* and *aa BB* individuals will be 40 cm high.
c. In the F_2, 1/16 of the progeny are *AA bb*, 4/16 are *Aa Bb* and 1/16 are *aa BB*. Thus, 6/16 = 3/8 of the progeny will be 40 cm high.

22.7 Three independently segregating genes (*A, B, C*), each with two alleles, determine height in a plant. Each capital-letter allele adds 2 cm to a base height of 2 cm.

a. What are the heights expected in the F_1 progeny of a cross between homozygous strains *AA BB CC* (14 cm) x *aa bb cc* (2 cm)?
b. What is the distribution of heights (frequency and phenotype) expected in an F_1 x F_1 cross?
c. What proportion of F_2 plants will have heights equal to the heights of the original two parental strains?
d. What proportion of the F_2 will breed true for the height shown by the F_1?

Answer: a. The cross *AA BB CC* x *aa bb cc* will produce all *Aa Bb Cc* progeny. The three capital letter alleles will add 3 x 2 = 6 cm to the 2 cm base height, giving a total of 8 cm.

b. There are a variety of ways to approach this problem. One way is to use a modified Punnett Square and another is to use the binomial theorem [use the coefficients of $(a + b)^6$]. Shown here is still another approach, a branch diagram:

Aa x *Aa*	*Bb* x *Bb*	*Cc* x *Cc*	Proportion	# Capital Alleles	Ht. (cm)
1/4 *AA*	1/4 *BB*	1/4 *CC*	1/64	6	14 cm
		1/2 *Cc*	2/64	5	12 cm
		1/4 *cc*	1/64	4	10 cm
	1/2 *Bb*	1/4 *CC*	2/64	5	12 cm
		1/2 *Cc*	4/64	4	10 cm
		1/4 *cc*	2/64	3	8 cm
	1/4 *bb*	1/4 *CC*	1/64	4	10 cm
		1/2 *Cc*	2/64	3	8 cm
		1/4 *cc*	1/64	2	6 cm
1/2 *Aa*	1/4 *BB*	1/4 *CC*	2/64	5	12 cm
		1/2 *Cc*	4/64	4	10 cm
		1/4 *cc*	2/64	3	8 cm
	1/2 *Bb*	1/4 *CC*	4/64	4	10 cm
		1/2 *Cc*	8/64	3	8 cm
		1/4 *cc*	4/64	2	6 cm
	1/4 *bb*	1/4 *CC*	2/64	3	8 cm
		1/2 *Cc*	4/64	2	6 cm
		1/4 *cc*	2/64	1	4 cm
1/4 *aa*	1/4 *BB*	1/4 *CC*	1/64	4	10 cm
		1/2 *Cc*	2/64	3	8 cm
		1/4 *cc*	1/64	2	6 cm
	1/2 *Bb*	1/4 *CC*	2/64	3	8 cm
		1/2 *Cc*	4/64	2	6 cm
		1/4 *cc*	2/64	1	4 cm
	1/4 *bb*	1/4 *CC*	1/64	2	6 cm
		1/2 *Cc*	2/64	1	4 cm
		1/4 *cc*	1/64	0	2 cm

Add up the proportion of individuals with like heights to obtain:

1/64 → 14 cm

6/64 → 12 cm

15/64 → 10 cm

20/64 → 8 cm

15/64 → 6 cm

6/64 → 4 cm

1/64 → 2 cm.

c. 1/64 are 2 cm and 1/64 are 14 cm, so the answer is 2/64.

d. None of the F_2 with the height of 8 cm can breed true, because this height depends upon a genotype having three capital-letter alleles, which in turn requires heterozygosity for at least one gene locus.

22.8 Repeat Problem 22.7, but assume that each capital-letter allele acts to double the existing height; for example, *Aa bb cc* = 4 cm, *AA bb cc* = 8 cm, *AA Bb cc* = 16 cm, and so on.

Answer: a. 16 cm

b. 1/64 → 128 cm

6/64 → 64 cm

15/64 → 32 cm

20/64 → 16 cm

15/64 → 8 cm

6/64 → 4 cm

1/64 → 2 cm.

c. 2/64

d. The 16 cm F_1 height is determined by the presence of three dominant alleles. This can only occur if one gene pair is heterozygous. Consequently, the 16 cm strain could not breed true.

22.9 Assume three equally and additively contributing pairs of alleles control flower length in nasturtiums. A completely homozygous plant with 10-mm flowers is crossed to a completely homozygous plant with 30-mm flowers. F_1 plants all have flowers about 20-mm long. F_2 plants show a range of lengths from 10 to 30 mm, with about 1/64 of the F_2 having 10-mm flowers and 1/64 having 30-mm flowers. What distribution of flower length would you expect to see in the offspring of a cross between an F_1 plant and the 30-mm parent?

Answer: Use the clue that 1/64 of the F_2 progeny show either extreme trait. Note that (as in problems 22.8 and 22.9, and in text Table 22.5, p. 768) in an F_2 resulting from a trihybrid cross, the proportion of one kind of homozygote is 1/64. This suggests that three allelic pairs control this quantitative trait. The completely homozygous 10 cm plant would have six lower case alleles at three loci, *aa bb cc*, while the completely homozygous 30-cm plant would have six capital alleles at three loci, *AA BB CC*. If each capital allele contributes 1/6 of the 20 cm difference between the two extremes, or 20/6 = 3.33 cm, the F_1 trihybrid individual would be 20 cm (=10 cm base + 3 x (3.33 cm/capital allele) = 20 cm). Thus, the trait fits a model in which three capital alleles contribute additively to this quantitative trait.

If an F_1 plant is crossed to the 30-cm parent, one would have *Aa Bb Cc* x *AA BB CC*. This cross would produce 1/8 each of *Aa Bb Cc, Aa Bb CC, AA Bb Cc, Aa BB Cc, AA BB Cc, AA Bb CC, Aa BB CC,* and *AA BB CC*. Thus, there will be 1/8 that are 30 cm high, 1/8 that are 20 cm high, 3/8 that are 23.33 cm high and 3/8 that are 26.67 cm high.

22.10 In a particular experiment the mean internode length in spikes of the barley variety *asplund* was found to be 2.12 mm. In the variety *abed binder* the mean internode length was found to be 3.17 mm. The mean of the F_1 of a cross between the two varieties was approximately 2.7 mm. The F_2 gave a continuous range of variation from one parental extreme to the other. Analysis of the F_3 generation showed that in the F_2 8 out of the total 125 individuals were of the *asplund* type, giving a mean of 2.19 mm. Eight other individuals were similar to the parent *abed binder,* giving a mean internode length of 3.24 mm. Is the internode length in spikes of barley a discontinuous or a quantitative trait? Why?

Answer: Internode length shows the characteristics of a quantitative trait. These characteristics include F_1 progeny that show a phenotype intermediate between the two parental phenotypes, and an F_2 showing a range of phenotypes with extremes in the range of the two parents.

22.11 From the information given in Problem 22.10, determine how many gene pairs involved in the determination of internode length are segregating in the F_2.

Answer: As shown in Table 22.5, p. 768, when two allelic pairs control a trait, one expects 1/16 of the F_2 progeny to exhibit the parental phenotypes. Given that 8/125, or about 1/16, of the F_2 show a phenotype similar to each parent, one could try to model internode length as a quantitative character controlled by two allelic pairs. *AA BB* individuals would represent one extreme, at about 3.2 mm, while *aa bb* individuals would represent the other extreme, at about 2.1 mm.

If each capital letter allele contributes equally and cumulatively to the trait, each capital letter allele would add (3.2 – 2.1)/4 = 0.275 mm (about 0.3 mm) of internode length to a base length of about 2.1 cm. The relative proportions of F_2 phenotypes would be given by the coefficients of the binomial expansion of $(a + b)^4$: 1:4:6:4:1. One would therefore expect an F_2 of the following type:

1/16 $\rightarrow$ 2.1 mm internodes

4/16 $\rightarrow$ 2.3 - 2.4 mm internodes

6/16 $\rightarrow$ 2.6 - 2.7 mm internodes

4/16 $\rightarrow$ 2.9 - 3.0 mm internodes

1/16 $\rightarrow$ 3.2 - 3.3 mm internodes.

One could measure the F_2 and assess whether these proportions were found.

22.12 Assume that the difference between a type of oats yielding about 4 g per plant and a type yielding 10 g is the result of three equal and cumulative multiple-gene pairs *AA BB CC*. If you cross the type yielding 4 g with the type yielding 10 g, what will be the phenotypes of the F_1 and the F_2? What will be their distribution?

Answer: If six equal and cumulatively acting capital alleles result in the addition of 6 g to the base weight of 4 g, each allele contributes 1 g. The F_1 will have the genotype *Aa Bb Cc*, and so have a yield of 4 + 3 = 7 g. The F_2 will appear in a distribution with relative proportions according to the seven coefficients of the binomial expansion $(a + b)^6$ (see Table 22.5). One expects 1/64 = 4 g, 6/64 = 5 g, 15/64 = 6 g, 20/64 = 7 g, 15/64 = 8 g, 6/64 = 9 g, 1/64 = 10 g.

22.13 Assume that in squashes the difference in fruit weight between a 3-lb type and a 6-lb type is due to three allelic pairs, *A/a*, *B/b*, and *C/c*. Each capital-letter allele contributes a half pound to the weight of the squash. From a cross of a 3-lb plant (*aa bb cc*) with a 6-lb plant (*AA BB CC*), what will be the phenotypes of the F_1 and the F_2? What will be their distribution?

Answer: The F_1 will have the genotype *Aa Bb Cc*, and so have a weight of 4.5 lb. The F_2 is expected to show a distribution predicted by the seven coefficients of the binomial expansion $(a + b)^6$ (see Table 22.5). One should see 1/64 = 3 lb, 6/64 = 3.5 lb, 15/64 = 4 lb, 20/64 = 4.5 lb, 15/64 = 5 lb, 6/64 = 5.5 lb, and 1/64 = 6 lb.

22.14 Refer to the assumptions stated in Problem 22.13. Determine the range in fruit weight of the offspring in the following squash crosses: (a) *AA Bb CC* x *aa Bb Cc;* (b) *AA bb Cc* x *Aa BB cc;* (c) *aa BB cc* x *AA BB cc.*

Answer: In each case, one needs to consider the maximum and minimum number of capital letter alleles that can be contributed to the progeny, noting that each contributes 0.5 lb above the 3 lb base weight.

a. *AA Bb CC* x *aa Bb Cc.* At least two, and at most five capital alleles can be contributed to the progeny. The progeny weights will range from 4.0 to 5.5 lbs.

b. *AA bb Cc* x *Aa BB cc.* At least two, and at most four capital alleles can be contributed to the progeny. The progeny weights will range from 4.0 to 5.0 lbs.

c. *aa BB cc* x *AA BB cc.* Exactly three capital alleles will be contributed to the progeny. The progeny weight will be 4.5 lbs.

22.15 Assume that the difference between a corn plant 10 dm (decimeters) high and one 26 dm high is due to four pairs of equal and cumulative multiple alleles, with the 26-dm plants being *AA BB CC DD* and the 10-dm plants being *aa bb cc dd.*

a. What will be the size and genotype of an F_1 from a cross between these two true-breeding types?

b. Determine the limits of height variation in the offspring from the following crosses:

i. *Aa BB cc dd* x *Aa bb Cc dd;*

ii. *aa BB cc dd* x *Aa Bb Cc dd;*

iii. *AA BB Cc DD* x *aa BB cc Dd;*

iv. *Aa Bb Cc Dd* x *Aa bb Cc Dd.*

Answer: Since four pairs of alleles are equally responsible for the 16 dm height increase, each allele contributes 2 dm.

a. A cross of *AA BB CC DD* x *aa bb cc dd* will produce *Aa Bb Cc Dd* progeny. With four capital letter alleles, the F_1 height will be 18 dm.

b. i. *Aa BB cc dd* x *Aa bb Cc dd.* The minimum number of capital alleles contributed to the progeny would be one, and the maximum number of capital alleles contributed to the progeny would be four. The height range would be 12 to 18 dm.

ii. *aa BB cc dd* x *Aa Bb Cc dd.* At least one and at most four capital alleles would be contributed to the progeny. The height range would be 12 to 18 dm.

iii. *AA BB Cc DD* x *aa BB cc Dd.* At least four and at most six capital alleles would be contributed to the progeny. The height range would be 18 to 22 dm.

iv. *Aa Bb Cc Dd* x *Aa bb Cc Dd.* A minimum of none and a maximum of seven capital alleles would be contributed to the progeny. The height range would be 10 to 24 dm.

22.16 Refer to the assumptions given in Problem 22.15. But for this problem two 14-dm corn plants, when crossed, give nothing but 14-dm offspring (case A). Two other 14-dm plants give one 18-dm, four 16-dm, six 14-dm, four 12-dm, and one 10-dm offspring (case B). Two other 14-dm plants, when crossed, give one 16-dm, two 14-dm, and one 12-dm offspring (case C). What genotypes for each of these 14-dm parents (cases A, B, and C) would explain these results? Would it be possible to get a plant taller than 48 dm by selection in any of these families?

Answer: In each of cases A, B, and C, both 14-dm parents must have two capital letter alleles. Consideration of the possible arrangement of these alleles leads to the solution of cases A, B and C.

Case A: For two 14-dm plants to only give rise to more 14-dm plants, each of the plants must be homozygous for one of the capital alleles. Both plants could be homozygous for the same capital allele (e.g., *AA bb cc dd* x *AA bb cc dd*), or each could be homozygous for a different capital allele (e.g., *AA bb cc dd* x *aa bb CC dd*). The progeny plants would have two capital alleles, and be 14-dm high.

Case B: The 1 four capital alleles:4 three capital alleles:6 two capital alleles:4 one capital allele:1 no capital allele proportions are the coefficients in the binomial expansion of $(a + b)^4$. This results suggests that the parents were heterozygous at the two (identical or different) loci, e.g., *Aa Bb cc dd* x *Aa Bb cc dd*, or *Aa Bb cc dd* x *aa bb Cc Dd.*

Case C: The 1 one capital allele:2 two capital alleles:1 three capital allele proportions are the coefficients in the binomial expansion of $(a + b)^2$. One would expect this ratio if one of the parents was homozygous at one locus, and the other heterozygous at two loci. For example, the parents could be *aa BB cc dd* x *Aa Bb cc dd*, or *aa BB cc dd* x *Aa bb Cc dd.*

22.17 A quantitative geneticist determines the following variance components for leaf width in a population of wild flowers growing along a roadside in Kentucky:

Additive genetic variance (V_A)	=	4.2
Dominance genetic variance (V_D)	=	1.6
Interaction genetic variance (V_I)	=	0.3
Environmental variance (V_E)	=	2.7
Genetic-environmental variance (V_{GE})	=	0.0

a. Calculate the broad-sense heritability and the narrow-sense heritability for leaf width in this population of wildflowers.

b. What do the heritabilities obtained in part a indicate about the genetic nature of leaf width variation in this plant?

Answer: a. The broad-sense heritability of a trait represents the proportion of the phenotypic variance in a particular population that results from genetic differences among individuals, while the narrow-sense heritability measures only the proportion of the phenotypic variance in that population that results from additive genetic variance.

Broad-sense heritability
= genetic variance/phenotypic variance = V_G/V_P
= (4.2 + 1.6 + 0.3)/(4.2 + 1.6 + 0.3 + 2.7 + 0.0)
= 6.1/8.8 = 0.69

Narrow-sense heritability
= additive genetic variance/phenotypic variance
= V_A/V_P
= 4.2/8.8 = 0.48

b. About 69 percent of the phenotypic variation in leaf width observed in this population is due to genetic differences among individuals. About 48 percent of the phenotypic variation is due to additive genetic variation. The phenotypic variation due to additive genetic variation represents the part of the phenotypic variance that responds to natural selection in a predictable manner.

22.18 Assume all genetic variance affecting seed weight in beans is genetically determined and is additive. From a population where the mean seed weight was 0.88 g, a farmer selected two seeds, each weighing 1.02 g. He planted these and crossed the resulting plants to each other, then collected and weighed their seeds. The mean weight of their seeds was 0.96 g. What is the narrow-sense heritability of seed weight?

Answer: In this example, there was 0.96 – 0.88 = 0.08 gm of change after one generation of selection. Thus, the selection response was 0.08 gm. This selection response is dependent on both the narrow-sense heritability and the selection differential. The selection differential is the difference between the mean phenotype of the selected parents and the mean phenotype of the unselected population. In this case the selection differential is 1.02 – 0.88 = 0.14 gm. The narrow sense heritability is given by:

selection response = narrow-sense heritability x selection differential.

$0.08 = h^2 \times 0.14$

$h^2 = 0.08/0.14 = 0.57$

22.19 Members of the inbred rat strain SHR are salt sensitive: they respond to a high salt environment by developing hypertension. Members of a different inbred rat strain, TIS, are not salt sensitive. Imagine you placed a population consisting only of SHR rats in an environment that was variable in regard to distribution of salt, so that some rats would be exposed to more salt than others. What would be the heritability of blood pressure in this population?

Answer: SHR rats will continue to respond to salt by developing hypertension. Since the strain is inbred, any variation in blood pressure will result from the amount of exposure to salt, and not from genetic variation. Therefore heritability for this population will be zero. Similarly, the inbred TIS rats would also have a heritability of zero (and retain a low blood pressure).

22.20 In Kansas a farmer is growing a variety of wheat called TKI38. He calculates the narrow-sense heritability for yield (the amount of wheat produced per acre) and finds that the heritability of yield for TK138 is 0.95. The next year he visits a farm in Poland and observes that a Russian variety of wheat, UG334, growing there has only about 40 percent as much yield as TK138 grown on his farm in Kansas. Since he found the heritability of yield in his wheat to be very high, he concludes that the American variety of wheat (TK138) is genetically superior to the Russian variety (UG334), and he tells the Polish farmers that they can increase their yield by using TK138. What is wrong with his conclusion?

Answer: Heritability is a measurement of the genetic variance of a *particular* population in a *specific* environment. Heritability is not fixed for a specific trait. It depends on genetic makeup *as well as* the specific environment of the population in which it is measured. Consequently, heritability cannot be used to make inferences about the basis for differences between two distinct populations. As the environmental conditions on the two farms differ, the heritability calculated for a population in Kansas cannot be used to infer the future performance of the same strain in another environment. The yield of TK138 grown in Poland would most likely be different than when grown in Kansas, perhaps even lower than the yield of the UG334 variety.

22.21 Dermatoglyphics are the patterns of the ridged skin found on the fingertips, toes, palms, and soles. (Fingerprints are dermatoglyphics.) Classification of dermatoglyphics is frequently based on the number of triradii; a triradius is a point from which three ridge systems separate at angles of 120°. The number of triradii on all ten fingers was counted for each member of several families, and the results are tabulated below.

FAMILY	MEAN NUMBER OF TRIRADII IN THE PARENTS	MEAN NUMBER OF TRIRADII IN THE OFFSPRING
I	14.5	12.5
II	8.5	10.0
III	13.5	12.5
IV	9.0	7.0
V	10.0	9.0
VI	9.5	9.5
VII	11.5	11.0
VIII	9.5	9.5
IX	15.0	17.5
X	10.0	10.0

a. Calculate the narrow-sense heritability for the number of triradii by the regression of the mean phenotype of the parents against the mean phenotype of the offspring.

b. What does your calculated heritability value indicate about the relative contributions of genetic variation and environmental variation to the differences observed in number of triradii?

Answer: a. The narrow-sense heritability of the number of triradii will equal the slope, b, of the regression line of the mean offspring phenotype on the mean parental phenotype.

$$b = \frac{\text{cov}_{xy}}{s_x^2} = \frac{\dfrac{\sum x_i y_i - \frac{1}{n}\left(\sum x_i \sum y_i\right)}{n-1}}{\dfrac{\sum\left(x_i - \bar{x}\right)^2}{n-1}} = \frac{\sum x_i y_i - \frac{1}{n}\left(\sum x_i \sum y_i\right)}{\sum\left(x_i - \bar{x}\right)^2}$$

For this data set, x is the mean number of triradii in the parents and y is the mean number of triradii in the offspring. One has:

$\bar{x} = 11.1$ $\qquad \sum x_i = 111$

$\sum (x_i - \bar{x})^2 = 51.4$ $\qquad \sum y_i = 108.5$

$\sum x_i y_i = 1275.5$

$$b = \frac{1257.5 - \dfrac{111 \times 108.5}{10}}{51.4} = 1.03$$

b. A narrow-sense heritability value of 1.03 indicates that all of the observed variation in phenotype can be attributed to additive genetic variation among the individuals.

22.22 A scientist wishes to determine the narrow-sense heritability of tail length in mice. He measures tail length among the mice of a population and finds a mean tail length of 9.7 cm. He then selects the ten mice in the population with the longest tails; mean tail length in these selected mice is 14.3 cm. He interbreeds the mice with the long tails and examines tail length in their progeny. The mean tail length in the F_1 progeny of the selected mice is 13 cm.

Calculate the selection differential, the response to selection, and the narrow-sense heritability for tail length in these mice.

Answer: The selection differential equals 14.3 – 9.7 = 4.6 cm. The response to selection equals 13 – 9.7 = 3.3 cm. The narrow-sense heritability equals 3.3/4.6 = 0.72.

22.23 Suppose that the narrow-sense heritability of wool length in a breed of sheep is 0.92, and the narrow-sense heritability of body size is 0.87. The genetic correlation between wool length and body size is –0.84. If a breeder selects for sheep with longer wool, what will be the most likely effects on wool length and on body size?

Answer: The strong narrow-sense heritability of both wool length and body size indicates that these traits will respond to selection. The negative correlation coefficient between wool length and body size indicates that if longer wool is selected for, smaller body size will also be obtained.

22.24 The heights of nine college-age males and the heights of their fathers are presented below.

HEIGHT OF SON (INCHES)	HEIGHT OF FATHER (INCHES)
70	70
72	76
71	72
64	70
66	70
70	68
74	78
70	74
73	69

a. Calculate the mean and the variance of height for the sons and do the same for the fathers.
b. Calculate the correlation coefficient for the relationship between the height of father and height of son.
c. Determine the narrow-sense heritability of height in this group by regression of the son's height on the height of father.

Answer: a. Sons: mean = 70.0, variance = 10.25
Fathers: mean = 71.9, variance = 11.6

b and c.

Let y equal the height of sons, and x equal the height of the fathers. One can calculate $\Sigma x_i y_i = 45{,}333$, $\Sigma x_i = 647$, $\Sigma y_i = 630$, $n = 9$ so that

$$r = \frac{\text{cov}_{xy}}{s_x s_y} = \frac{\dfrac{\sum x_i y_i - \frac{1}{n}\left(\sum x_i \sum y_i\right)}{n-1}}{\sqrt{s_x^2 s_y^2}}$$

$$r = \frac{\dfrac{45333 - [(630)(647)/9]}{8}}{\sqrt{11.6 \times 10.25}} = \frac{5.375}{10.904} = 0.493$$

$$b = \frac{\text{cov}_{xy}}{s_x^2} = \frac{5.375}{11.6} = 0.463$$

When, as in this case, the mean phenotype of the offspring is regressed against the phenotype of only one parent, the narrow-sense heritability is $2b = 2(0.46) = 0.92$. Since h^2 is very close to 1, most of the phenotypic variation is determined by genes with additive affects. Non-additive factors (genes with dominance, gene with epistasis, environmental factors) contribute little to the phenotypic variation.

22.25 The narrow-sense heritability of egg weight in a particular flock of chickens is 0.60. A farmer selects for increased egg weight in this flock. The difference in the mean egg weight of the unselected chickens and the selected chickens is 10 g. How much should egg weight increase in the offspring of the selected chickens?

Answer: Selection response = narrow-sense heritability x selection differential
Selection response = 0.60 x 10 g = 6 g.

NOTES

NOTES

NOTES

Notes

NOTES

NOTES

NOTES

NOTES

NOTES

NOTES